ENVIRONMENTAL RISK: IDENTIFICATION AND MANAGEMENT

ALBERT R. WILSON

LEWIS PUBLISHERS INC.
121 S. MAIN STREET, P.O. DRAWER 519, CHELSEA, MI 48118

Library of Congress Cataloging-in-Publication Data

Cataloging information is on file with the Library of Congress.

ISBN 0-87371-388-5

Direct all inquiries to CRC Press, Inc., 2000 Corporate Blvd., N.W., Boca Raton, Florida 33431.

PRINTED IN THE UNITED STATES OF AMERICA
2 3 4 5 6 7 8 9 0
Printed on acid-free paper

DEDICATION

This book is dedicated to my wife, Marilyn, and to our children, Laura and Ian, and to their contemporaries and successors who will inherit the fruits of our decisions.

Albert R. Wilson

Mr. Wilson is President of Environmental Analysis and Valuation, Inc. (EAV) and the Hazardous Materials Institute, Inc. (HMI). Both organizations are involved with the impacts of environmental risks on real property and businesses. HMI specializes in research and education associated with the identification, management and financial impacts, while EAV specializes in the development of valuation opinions for the assignment of damages to real property and businesses. Mr. Wilson developed the concept and associated techniques for an auditable and trackable inventory of environmental risks, and the selection of an optimum feasible strategy for dealing with an inventory of such risks. Mr. Wilson's particular area of expertise is in the development of systems, methods and programs for the achievement of the strategic objective of "maximum risk reduction at minimum feasible cost" utilizing classic inventory management techniques and mathematical tools such as multi-objective linear programming.

FOREWORD

This book is the product of research, analysis, and frustration. For over twenty years editorials have been written; speeches given; books produced; and laws, rules and regulations promulgated on the subject of environmental risks. At the present time, we have over 9,000 pages of federal rules and regulations dealing with various aspects of environmental risk. We have an accrued liability for environmental risks against real property in at least the amount of $2 trillion against a total estimated value of $10 to $12 trillion. Industry, government, and the individual citizen are being asked to shoulder an ever-increasing burden to both clean up past problems and prevent or reduce the impact of future problems.

From a personal point of view, I am deeply concerned about the environmental future for my children and grandchildren. I love the unspoiled sky and sea and mountains and want them to know what I have known. As an engineer and a business manager, as a father, and as a concerned individual, I am deeply frustrated with both sides of the environmental debate. This book is partially a product of that frustration.

The real source of the frustration is, however, a surprising one. This is a debate conducted without observance of the first rule of debate, the terms have not been defined. For example define, in precise terms an ''environmental risk,'' or a ''Phase I Environmental Audit,'' or ''due diligence.'' Organizations are being asked to sign the ''Valdez Principles,'' a very worthy statement of responsible environmental behavior, but are refusing to do so, and for very good reason. One of the principles requires an ''environmental audit'' of the organization, but what is an ''environmental audit?'' Although many will at first disagree, when pressed you will find that there is no operable definition of an environmental audit with even remotely the precision of its presumed analog—a financial audit.

The problem with a lack of definitions is fundamental. If the problem cannot be measured with a reasonable degree of accuracy, it cannot be brought under control and managed. To date the way out has been a call for the total elimination of environmental risks, something that is, in general, absolutely a physical impossibility. Even if the total elimination of environmental risks were possible, would it be desirable? Cleaning an operating room without hazardous chemicals is difficult to impossible, removing ticks from the dog to prevent Lyme disease would not be possible, elimination of asbestos from certain products will increase risk to human health. Our civilization and standard of living depend on the existence and intelligent use of environmentally risky substances or processes.

This book is designed to be a beginning for a responsible debate on the definition of terms and an application of existing knowledge to develop a basis for the long-term control of environmental risks. It is to be hoped that you will find many areas of agreement with the ideas, definitions and concepts outlined here, but not too much agreement. The real purpose is to begin a responsible debate, a debate based on defined terms and concepts that may need to be modified, but are at least a starting point.

Albert R. Wilson
Columbus, Ohio
May, 1990

INTRODUCTION

There will be four products of this book, albeit in a somewhat rudimentary form.

1. A uniform methodology for the identification and measurement of *environmental risk.* To accomplish this, an explicit definition of *environmental risk* will be developed and used throughout.
2. A uniform methodology for the reporting of environmental risks—an Environmental Risk Inventory System (ERIS).
3. A management program for the long-term control of environmental risks—the Environmental Risk Management Program (ERMP).
4. A set of definitions of Phase I, II and III Environmental Risk Assessments, Compliance Audits, and "due diligence" based on the data requirements of the ERIS and ERMP. Definitions derived from the end use requirements of the ERMP.

This book is explicitly written from the point of view of the manager; a manager required to control environmental risks under the strategic objective of "maximum feasible reduction of environmental risk." There is no intention to avoid any responsibility for environmental risk, but instead an explicit recognition of the risks to be dealt with, a means of measuring the risks and the performance of the risk management program, and a recognition that the requirement for the feasibility of the solution cannot simply be ignored. The legal basis for the approach taken here comes from two primary sets of environmental rules and regulations; (1) the USEPA Hazard Ranking System (HRS) as set forth in *29 CFR 300* Appendix A; and (2) the *Asbestos Hazard Emergency Response Act of 1986* (AHERA) as set forth in *40 CFR 763 et seq*. The choice of these two sets of rules and regulations was deliberate. The HRS deals with the analysis of the consequences of an uncontrolled release of a hazardous substance into the environment, a condition that management control will seek to prevent. While the HRS has been the subject of considerable criticism and can be improved, it is nevertheless a part of the current body of environmental law and does establish a uniform framework for analysis of environmental risk consequences. The HRS is used here in a highly modified format intended to allow the development of a set of uniformly arrived at risk scores, risk scores that do not have any intrinsic meaning, but that will have strong comparative value. Environmental risks can only rarely, if ever, be reduced to an absolute zero value. This fact implies that a relative risk value will be as meaningful for management control purposes as an absolute value provided that the risk values are arrived at consistently.

The choice of AHERA is based on the fact that it is the best description of a long-term environmental risk management control program existent in the law today. While AHERA also has its legitimate critics, the basic fact of the outline of a risk management control program of a high degree of sophistication cannot be disputed.

To bring the three concepts together (an inventory of risks, measured in relative risk terms, to develop a long-term management control program) required one additional element—a definition of environmental risk. It somehow seems strange that the concept of environmental risk has not been explicitly defined, but if it has been, the author has not seen it. Environmental risk must be defined as a system of components, and once viewed in this light it can be seen that it is not the substance that will be of primary management concern, but whether or not the substance is under control.

This book uses asbestos as an example material because it is convenient to do so, but it explicitly is not a book about asbestos. As far as the author has been able to test the concepts, ideas, and approaches outlined here, it will work for any type of environmental risk from underground storage tanks, to threats to wetlands, to lead in paint or car exhausts.

The importance of environmental risk assessment and long-term management can be seen from a single statistic: the liability for environmental risks associated with real property is greater than $2 trillion against a total real property value estimated to be in the range of $10 to $12 trillion; 17% to 20% of total value. This fact clearly points out that the nation as a whole, let alone a specific individual property owner, will not have sufficient resources to both clean up existing environmental risk liabilities and to prevent future liabilities in a short period of time. Given a severe constraint on resources, the issue then becomes how to allocate those resources most efficiently to reduce risk to the maximum feasible extent. This is what is meant by the statement ''maximum feasible reduction of environmental risks'' and a rational approach toward achieving this objective is the underlying theme of this book.

There are many people whose effort, both direct and indirect, have gone into this book and many will unfortunately remain nameless. Those who can and should be acknowledged, in alphabetical order, include Walter Bovaird, Joe Davis, Roger Ferland, James Harrington, W. Lee Minnerly, Brian Muehling, Robert Neil, and R. Eric Zimmerman. Each have contributed research, or a major concept or idea, although the responsibility for the use of those remains solely with the author.

TABLE OF CONTENTS

8. Analysis of Potential Targets and Risk Sources

9. Environmental Risk — The Inventory Risk "Value"

10. Cost Models and Cost Estimation—Benefit Models and Benefit Estimation

11. Environmental Risk Management Decision Making

PART II — ENVIRONMENTAL RISK INVESTIGATION
Data Sources and Investigation Procedures for Phase I, II and III Environmental Risk Assessments

12. Phase I Environmental Risk Assessments
PART I — GENERAL CONTEXT OF PHASE I ERAs

13. Phase I Environmental Risk Audits PART II — EXAMINATION OF THE LAND

14. Substance Identification: Sampling Theory and Techniques

PART I

THE CONTEXT OF ENVIRONMENTAL RISK MANAGEMENT

THE DEVELOPMENT OF THE ENVIRONMENTAL RISK INVENTORY

1 ENVIRONMENTAL RISK MANAGEMENT

BASIC CONCEPTS AND TERMINOLOGY

The language of environmental risk management is far from completely developed and the concepts behind it are equally incomplete. This is a natural result of the *ad hoc* nature of the evolution of environmental regulation, regulation that has developed more as a response to crisis than in response to well-formulated underlying policies. To some extent, this book will be an exploration of the common themes of existing regulations and environmental protection goals in an attempt to develop a basis for regularized management of environmental risk. The purpose of this chapter is to introduce some of the basic concepts and ideas upon which the rest of the text will rely and to begin the exploration of the underlying themes of environmental regulation and management.

Fundamental to the environmental risk management process is the idea that management must know what is to be managed and to understand the parameters within which the management process must take place. The basic process is one of discovery (often referred to as inspection), planning, execution, and auditing. Fundamental to the management control of environmental risk is the development of a inventory of those risks in a manner that will allow for the measurement of progress in controlling and managing that inventory, an Environmental Risk Inventory System (ERIS). The development of an auditable ERIS is the goal of this part of this book.

Negative Default Concept

Virtually all of the environmental laws, rules, regulations and cases may be said to rest upon a single, generally unstated concept: *negative default*.

By this it is meant that for any given environmental risk, we must assume the worst and then attempt to prove that a better situation exists.

For example, under the AHERA rules and regulations, it is required that if there is a reasonable suspicion that a material might contain asbestos, then it must be assumed to contain asbestos and be treated accordingly until and unless it can be proven to not contain asbestos. Under OSHA, the worker protection rules require that a worker be protected from the worst conditions unless and until it can be shown that conditions are better and therefore less protection is necessary.

Whenever dealing with environmental risks, the negative default principle must be kept in mind and actions adjusted accordingly.

Proof of Zero Existence

One of the most pervasive of hazardous materials myths, and the one probably most inimical to rational decision making concerning hazardous materials, is the idea that any property can be given an absolutely clean bill of health. That is, that a statement may be made that no hazardous materials exist in a given property. Most people are familiar with the difficulty associated with the proverbial task of "finding a needle in a haystack." The task of proving that hazardous materials do not exist in a given property may be framed in terms of the even more difficult task of proving that "the needle does not exist in the haystack."

A simple analogy will serve to demonstrate the point. Suppose that prior to purchasing a given 100 ft by 100 ft parcel of land we require that it be proven that not one single 55-gallon drum of hazardous substance exists on or in that parcel. Assuming that we would be willing to accept some reasonable level of doubt, a search must be conducted. Let us suppose that we are willing to accept the idea that the drum cannot have been buried more than 10 feet under the surface of the ground as our definition of a reasonable doubt. Given this definition, and assuming that a 55-gallon drum is 30 inches in its smallest cross-section, it will be necessary to drill more than 1,600 holes to a depth of 10 feet on the parcel in order to show that no such drum exists, and then only to the level that we were willing to accept the concept that the drum could not have been buried deeper than 10 feet. In other words, even after having drilled 1,600 holes, we still have not absolutely proven that the drum—let alone a paint can or materials leaking from a drum on a neighboring site—does not exist. (See Figure 1.1)

Clearly, some criteria less than 100% certainty must be accepted in virtually all environmental risk situations. Those who require or claim to provide absolute certainty do not understand the issues.

The criteria of certainty is formally known as the *confidence interval*

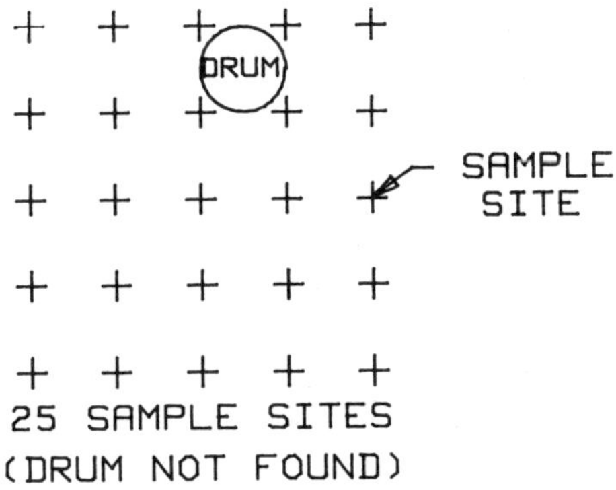

Figure 1.1 Search for a Drum on a 10 ft × 10 ft Plot.

for a given statistical sampling strategy. In general, a confidence interval is a measure of the range within which we are likely to find the correct answer. Associated with the confidence interval is the *confidence coefficient,* which measures the likelihood that a given answer lies within the confidence interval. In sampling statistics the objective is formally known as the hypothesis, and ours will be that "No hazardous materials are present in the plot." To prove the hypothesis we must first specify the confidence coefficient, that is, to state to what level of certainty we will require that a result of the testing showing no hazardous materials to be present is the correct answer. Without this information, we cannot design a sampling strategy or evaluate the results of any sampling done. To specify a given confidence coefficient, say .95, means that we are willing to accept a risk of approximately 5% that a finding of "no hazardous substances present" will be in error.

The design of an appropriate sampling strategy for a given situation is a highly technical and complex subject and it is probably sufficient to say here that a confidence coefficient of .7 to .8 is relatively inexpensive to obtain, but that increasing the confidence coefficient above this level will require exponentially increasing costs for the sampling necessary. It should also be noted that a number of "environmental audits" currently in use for commercial property transfers probably do not have a confidence coefficient exceeding .3, that is there is a greater than 70% chance that existing risk sources have **not** been found. This is not to say, however, that sampling is the only method of assuring the parties involved in real property of a reasonable certainty of the non-existence of risk sources, nor is it the most cost efficient method by itself.

In summary, it must be clearly kept in mind when dealing with risk sources in property that it is relatively easy to prove that a risk source exists, but it is virtually impossible to prove that one does not exist.

Zero Tolerance

The myth of *zero tolerance* is much like the myth of *zero existence* in that it is virtually impossible to state that no hazardous materials will be permitted on or in a property. This is especially the case when we consider that such common items as floor cleaning compounds, copy machine toners, drain cleaners, solvents, and many other items commonly used to simply maintain a given structure or operation. All contain hazardous materials (and may come under the OSHA Hazard Communications Rule). Hazardous materials are a fact of modern technological existence—a fact that will not change in the foreseeable future—and therefore one that must be understood, accepted, and allowance/adjustments made in order to proceed in an orderly and business-like fashion.

The mere existence of a risk source does not imply a problem. It is an uncontrolled risk source that represents the primary problem. Among other things, control of a risk source implies:

1. Knowledge of the source's existence and its current condition,
2. The existence of a plan to bring the source under control and keep it under control,
3. The open acknowledgement of the existence of the source to all affected or potentially affected parties, and
4. The existence of trained and equipped personnel to deal with the risk.

All of these elements are a part of our statutory and common law.

Zero Responsibility

As with zero existence and zero tolerance, *zero responsibility* is generally an impossible condition. Any party associated in a meaningful way with a property containing a risk source has some responsibility for the possible consequences of the existence of that source. This is not to say that the responsibility of all parties is equally shared, or that a given party's responsibility cannot be limited, but rather the fact that some responsibility exists must be a recognized part of any involvement. Debt and equity participants and tenants have obvious stakes in the existence of risk sources on or in a property. The involvement of agents, appraisers, employees, inspectors, and guests are less obvious, but nevertheless real, and may be of substantial extent.

End to Responsibility

The myth that there is an end to responsibility for an owner upon completion of a removal project is commonly held and, with the exception of removals whereby the materials are reduced to non-hazardous components or permanently rendered harmless, universally untrue. The "Superfund" laws, CERCLA[1] and SARA[2], clearly make the owner of the property at the time of removal responsible for the material removed **forever**. Under these laws, the owner at the time of removal may be held jointly, strictly and severally liable for the disposal of a risk source even if, at a later date, that substance is only a part of another problem at the disposal site.

This set of laws, and similar ones at the state and local levels, may result in a removal operation increasing (as opposed to decreasing) the owner's risks. This will be particularly the case if the disposal of the materials is improper or is included in a disposal site where other risk sources exist. At the very least, the removal of a substance that is not permanently rendered harmless will result in a contingent liability risk for the owner.

Negative Leverage

The ideas of negative default, zero existence, zero responsibility and the lack of a defined end to responsibility give rise to the idea of negative leverage associated with a hazardous substance. *Negative leverage* is the concept that there is no basis for assuming a relationship between the potential cost of an environmental risk liability and the value of a property or operation that may give rise to the liability.

In our economy, it is commonplace to think that assets and liabilities bear some direct relationship to each other in a mathematical sense. For example, a mortgage on a property is normally set up in terms of some fraction of the total asset value of the property, or that the liabilities on a corporate balance sheet are some fraction or relatively low multiple of the related assets. With respect to environmental risks, no such relationship exists. An asset worth a few thousand dollars may give rise to a multi-million dollar environmental liability.

Suppose that an underground storage tank leaks on a small piece of property. The material lost through the leak is only worth a few thousand dollars and the property containing the tank only worth a few thousand more. If the leak results in the semi-permanent contamination of an underground aquifer, the owner of the originating property will generally be held responsible for all of the consequences of that contamination no matter how far away from the property the consequences may spread even to the total cost of replacing a municipal water supply costing hundreds of thousands or millions of dollars. This is a very realistic example of negative leverage.

Control Mechanisms

It may be argued that a risk source does not present an actual (as opposed to a potential) threat to human health and/or the environment so long as the risk source is under proper control. One of the primary functions of an environmental risk inspection is to determine if the risk source is under proper control and to determine how that control might fail. The inspector must identify the control mechanism(s), evaluate their efficiency, and the possible failure mechanisms.

Material Condition

The phrases *material condition* or *condition of the materials* are often used in environmental rules and regulations and can lead to substantial confusion because they do not refer to the condition of the subject hazardous substance, but rather to the condition of the physical component of the primary control mechanisms that govern the risks associated with the hazardous substance.

Material condition for our purposes, refers to the physical condition of the primary control mechanism for the material relative to the risk of injury to human health and/or the environment. For example, the matrix material of an asbestos-containing building product, such as the calcium or magnesium carbonate that traps the asbestos fibers and holds them in place, is the physical component of the primary asbestos control mechanism. So long as the matrix is intact—that is so long as the matrix is capable of reliably performing the holding function with respect to the asbestos fibers—then we would consider the material to be in good condition. When the matrix becomes damaged in a manner that will destroy the holding property, the material would be considered in relatively poor condition. The basic criteria of concern is whether or not the matrix is capable of holding the fibers in place, thereby protecting human health and/or the environment. This ''holding ability'' is a specific example of a general class of primary control mechanisms.

Condition Assessment

Based on the foregoing discussion, it should be obvious that the inspector will need to evaluate the environmental risk with respect to the current condition of the control mechanisms and the probable future condition of those mechanisms.

Again, using an asbestos-containing building material (ACBM) as the example, suppose that the ACBM is a spray-on fire-proofing material in the ceiling of a gymnasium of a elementary school. The material may be currently in near-perfect condition, meaning that the matrix is intact and successfully holding the asbestos fibers in place, and may therefore be judged to be in good condition. However, if the roof was to leak, or the use of the gym change from elementary to high school, the failure mechanisms of water or physical damage (resulting from the ability of the high school students to reach the ceiling where the elementary students could not) the probable future condition of the material might have to be judged uncertain or likely to sustain damage that would impair the functioning of the control mechanism.

Transport Mechanisms

Transport mechanisms are the means by which the risk source can move from its present location to a location where human health and/or environmental damage might result. Common transport mechanisms are the movement of air or water acting as the carriers of a substance, but others in the form of the actual migration of a hazardous substance, direct contact through the skin or ingestion, or in the case of nuclear substances, the emission of radioactive particles, are possible.

One of the inspector's most often overlooked functions is to identify the availability and accessibility of transport mechanisms to the risk source. Without this information the assessment of the risk to human health and/or the environment is impossible.

Occupancy Characteristics (Structures)

The underlying objective of all inspections is to develop the information necessary to evaluate the risk (potential or actual) to a target. For this reason, particularly with substances having a direct effect on human health (as opposed to those that effect the general environment such as the action of CFCs[3] on the ozone layer) the inspector must develop data on the occupancy characteristics of the structures within which the risk source exists (exterior risk sources will be discussed later).

Of particular concern in the development of this data will be the classification of the individuals into groups with respect to the influence of their activities on the control and transport mechanisms. For example, maintenance personnel who must work directly with the risk source or the control mechanisms should be clearly differentiated from occupants who are passive in the presence of the material. The differentiation should be along the lines of the probability of the presence of the individual resulting in the failure of a control mechanism.

Impact (Risk) Assessment

The data developed by the inspector in the form of condition assessments and occupancy characteristics are utilized by an experienced individual, called a *management planner* under AHERA or the *environmental risk manager* (ERM) throughout this manual, to develop a risk or hazard assessment. The specific purpose of the risk assessment is to determine which risks present greater potential for harm to human health and/or the environment than others as a basis for action planning.

Risk assessment under AHERA is approached through the use of a decision tree that allows the subjective judgment of the inspector and management planner to be regularized. The decision tree concept will be explored in some detail later. The specific purpose of the AHERA decision tree is to analyze the likely impact of a potential release of a specific risk source within the confines of a structure; an application that will be generalized from asbestos to any risk source.

Although not usually shown in the form of a decision tree, the Hazard Ranking System (HRS) is also a decision-making process that will be utilized extensively throughout to regularize the judgment of the expert in the evaluation of the impact or likely impact of a risk source released into the environment. The HRS was originally intended for the evaluation of a release for purposes of ranking Superfund sites, but it will also be generalized here for use as an evaluation tool with respect to the likely impact of a possible release.

Similar to decision trees, but with significant differences, are *fault trees*. The fault tree concept was originally developed by A. D. Little & Company to analyze the engineering design of nuclear reactors. Fault trees will be combined with decision tree and HRS analyses to quantify the possible impacts of environmental problems as a basis for the allocation of resources. Each of these techniques are used to perform a risk assessment; the basis for risk management.

Risk Measurement

Fundamental to risk management is a method for the measurement of risk in quantitative terms for use in decision support systems. Basically, *risk measurement* has at least two components from the point of view of environmental risk management; exposure or opportunity, and quantitative impact. Exposure relates to the length of time over which the target is available for contact with the environmental risk. Obviously if there is no opportunity to be exposed, no risk exists. The quantitative impact relates to the kind, type, and severity of the likely results of exposure.

With some exceptions, exposure will be measured in terms of the number of thousands of person-hours of time per year spent in the presence of an environmental risk, possibly weighted by a factor indicating the type of activity the average individual is engaged in while in the presence of the substance. An individual who is simply present in the immediate vicinity of a risk but is passive with respect to the control mechanism for the risk would reasonably be at less of a risk than an individual actively engaged in maintenance of the control mechanism.

With respect to the quantitative impact, more complex issues arise dealing with the persistence, toxic/carcinogenic properties, and latency periods associated with the specific human health effects, or the physical/chemical/biological mechanisms through which a substance impacts the global or regional environment. These will be dealt with as specialized risk measurement topics.

Response Actions

In general there are six generic response actions available for dealing with an environmental risk:

1. Operations and Maintenance (O&M) programs designed for the long-term maintenance of the control mechanisms.
2. *Repair,* or the restoration of the control mechanism to a functional status.
3. *Encapsulation,* or the application of an impermeable membrane to prevent the source from gaining access to a transport mechanism.
4. *Enclosure,* or the construction of an impermeable physical barrier to prevent the source from gaining access to a transport mechanism.
5. *Isolation,* or to simply seal off large areas of risk sources from access to human beings or the environment, but possibly permitting access to the source containing area to trained, equipped and qualified individuals.
6. *Removal and disposal,* or the physical relocation of the source from a high-risk location to a relatively lower risk location or possibly reduction to harmless components or condition.

Management Decision-Making

Management decision-making relies on six major types of data:

1. The condition assessment including the inspector's evaluation of the control and transport mechanisms,
2. The occupancy characteristics,
3. The likely health/environmental impact associated with the risk, (1, 2 and 3 being combined into the risk assessment),

4. The evaluation of the costs of the available response actions,
5. The risk reduction impact of each available response action, and
6. The resources available to implement the response actions.

The problem of management decision-making is not that of selecting the course of action that will meet the objective of reducing the risk to human health and/or the environment to the lowest level, but the balancing of that objective with the available resources required to accomplish the task. If it were possible to eliminate all risk by some immediately available means and the resources required were available, the management decision would be simple. In the real world, this set of circumstances rarely occurs.

For example, in the removal of ACBMs the current situation is that the number of competent contractors available to perform the work is very small compared to the demand, resulting in a very high percentage of the work performed (estimates range from 60% to 80% of all work in 1988[4]) contaminating the building to a greater extent than leaving the material in place. In addition to this problem, the costs of removal are such that the funds are simply not available, and even if they were, we do not have the disposal sites necessary for the safe burial of the estimated 700 million tons of ACBMs.

Management decision-making then requires the balancing of all of these factors to develop a feasible plan of action; that is, a plan of action that can actually be accomplished given the objectives and available resources for dealing with risk sources.

General Area Concerns

Several major pieces of Federal legislation, notably the *Clean Air, Clean Water* and *Safe Drinking Water Acts,* may result in major influences on property or operations from widely distributed sources of risk; sources not on or adjacent to the subject property. For example, in the Los Angeles Basin, the entire area is under a "new source ban" with respect to the construction of new sources of ozone production such as dry-cleaning establishments. This ban is a result of the failure to reduce ozone production within the basin, primarily from cars and trucks. The USEPA uses a "bubble" concept illustrated below in Figure 1.2, that recognizes the idea that all sources within the "bubble" defined by the air over the basin must be generally controlled in order to reduce the production of ozone.

The Environmental Risk Management Program

This book develops the theory and implementation requirements of a system of positive control for environmental risks. The *Asbestos Hazard*

Emergency Response Act of 1986 (AHERA) rules and regulations set out just such a system for the long term management of a specific environmental risk, asbestos in a school, and the basic ideas behind AHERA have been generalized into the Environmental Risk Management Program (ERMP).

The ERMP can be defined as follows:

- *Operative Basis:* An uncontrolled environmental risk represents unacceptable human health and financial risks.
- *Strategic Objective:* Maximization of risk reduction at minimum feasible cost[5].
- *Tactical Methodology:*
  1. To control an environmental risk:
     a. The risk must be identified, located and described, and the current and likely future condition of the control mechanisms for the risk must be evaluated.
     b. A feasible plan of action must be developed that will place the risk into a benign state and assure that it will remain in that state until it can be permanently reduced to a zero risk level, if possible.
  2. That a qualified individual must be made responsible for the execution of the plan of action.
  3. That all individuals who may be affected by the risk have the right to know that the risk exists, that it is being dealt with, and to be assured

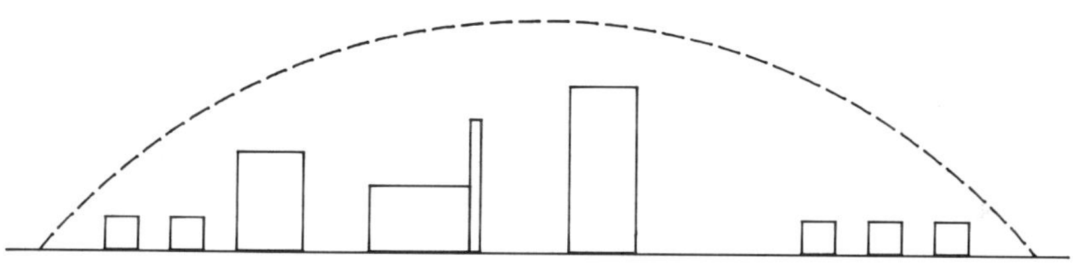

IF AIR or WATER STANDARDS EXCEEDED
WITHIN THE "bubble" THEN:

1. NO NEW SOURCES

2. POSSIBLY NO EXPANSION OF SOURCES

Figure 1.2 EPA "bubble" Concept.

that the methods and procedures for dealing with the risk are proper and effective.

4. That only properly trained individuals should be permitted to come into direct contact with the risk, and particularly the control mechanisms for the risk.
5. That regular audits of the performance of the risk control program are an integral part of the positive control of the risk.

If this program sounds like Management 101, it is. With several technical exceptions, the control of environmental risks is little different from the management of an inventory in the classic sense. From this point the environmental risks will be dealt with as an inventory to be identified, quantified, protected and managed, and the measurement of management performance will be evaluated from a similar point of view.

Definition of Environmental Risk — The Risk System

Usually, the phrase *environmental risk* is taken to mean a specific substance, such as asbestos or PCBs or CFCs or lead, that is the source of the risk. In reality, the issue is far more complex and until the full complexity is understood, the impact of an environmental risk cannot be evaluated from either a management or a property value impact point of view.

Environmental risks are not simply the presence or absence of a single substance. They are in fact the source of the risk (asbestos, PCBs, CFCs, etc.) and the conditions which surround that risk source; how it is controlled, how a target may become endangered, and the extent of the danger. This book treats the risk source and the surrounding context of the existence of that risk source as a whole (the risk system), discusses how this approach defines and determines the identification and management of the risk system, and how the management and control of the risk system may impact the value of stakeholders'[6] interests.

The Environmental Risk System

An environmental risk must be viewed not as a substance, but as a system of components that govern the risk. Figure 1.3 sets out the four major components of the environmental risk system and their relationship to each other.

The Risk Source The risk source is what we most commonly refer to when we think about environmental risks; the asbestos, PCBs, CFCs, or whatever.

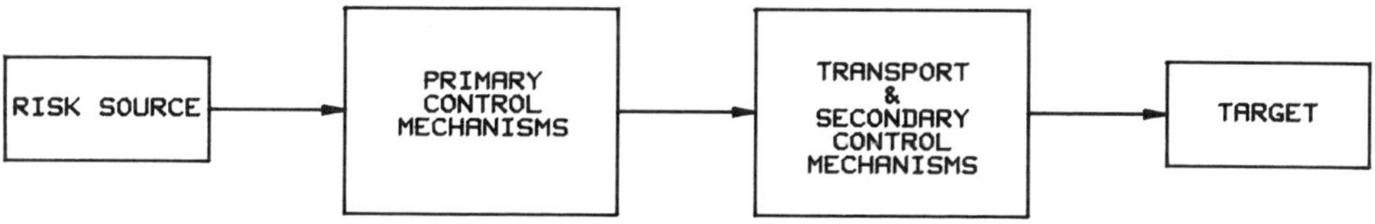

Figure 1.3 The Environmental Risk System.

These are the sources of the risk and without them there simply would be no environmental risk. Some risk sources can conceivably be eliminated, such as the removal and proper destruction of PCBs, while others cannot, such as radon produced by the natural atomic decomposition of the base rocks of the earth. Most risk sources have both positive and negative aspects. Asbestos, for example, is both an excellent fire retardant and a carcinogen. The question is, do the benefits outweigh the risks?

The Primary Control Mechanism The primary control mechanism is the means by which a risk source is controlled to render it both useful and relatively benign. Useful with respect to the function it performs and benign with respect to damage to human health and/or the environment. The basic problem of environmental risk management is not whether or not the risk source exists, but whether or not it is under control. The costs to establish and maintain control, or to remove the risk source if adequate control cannot be established and maintained, determine the value impact of the presence of a risk source.

The primary control mechanism almost always consists of two major components: a physical component in the form of some device that directly maintains the risk source in a benign state, and a human component that interacts to insure that the physical component functions as intended. Both components must function for the risk source to be both useful and benign.

The Transport and Secondary Control Mechanisms If the primary control mechanism(s) fail, the question will become one of whether or not a transport mechanism is available to move the risk source into the immediate presence of a target, and how efficiently that transport mechanism is likely to function. The USEPA has identified in its Hazard Ranking System[7], five transport modes: surface water, ground water, air movement, direct contact, and fire/explosion. The point is that without a transport mechanism the risk source cannot be brought into immediate proximity to a target and cannot therefore do harm to the target.

Inextricably associated with the transport mechanism are a number of possible secondary control mechanisms that may be either natural or artificial in origin. Like the primary control mechanisms, the secondary control mechanisms must be evaluated in order to determine the efficiency of the transport mechanism in moving the risk source to the target in sufficient concentration to result in damage. In many cases, the analysis of secondary control mechanisms can lead to the development of significant risk reduction strategies that improve the risk/benefit relationship.

An example of the transport/secondary control mechanism can be found in the situation surrounding the failure of a fuel oil storage tank in Pittsburgh. In that situation the primary control mechanism, the steel storage tank, failed, and several hundred thousands of gallons of fuel oil were released and flowed into the Ohio River system. The transport mechanism involved was the surface water drainage pattern of the Ohio River. The secondary control mechanisms were the diking around the fuel oil tank and the slope of the ground toward the river system.

The Target(s) Targets are basically of two forms: human and sensitive environments. Of the two, the human target is of the greatest concern and considerable resources must be allocated to the analysis of the relationship between the components of the risk system to determine the likely impacts on targets.

The overall analysis of a target is governed by the negative default principle. This implies that the number of possible targets, the probable impacts on targets, and the failure of primary control mechanisms must be assumed in a "worst case" scenario. The mitigating factors are the efficiency and availability of the transport/secondary control mechanisms, the probability of use of the transport mechanism, and the probability of failure of the primary control mechanism(s).

Target examples are obviously human beings, wetlands, endangered species, and less obviously the impact on designated monuments and environmental areas such as air pollution control "bubbles," national parks, primitive areas, and the like.

The Evaluation of the Risk System

Figure 1.4 provides the primary topics to be considered with respect to the risk system beginning with those associated with the risk source, such as its quantity, toxicity and persistence, and ending with considerations of the potential targets. The evaluation process involves the explicit consideration of at least the factors shown in Figure 1.4 and is generally labeled a Phase III Environmental Risk Analysis (ERA).

Because of the lack of standards for the conduct of environmental risk

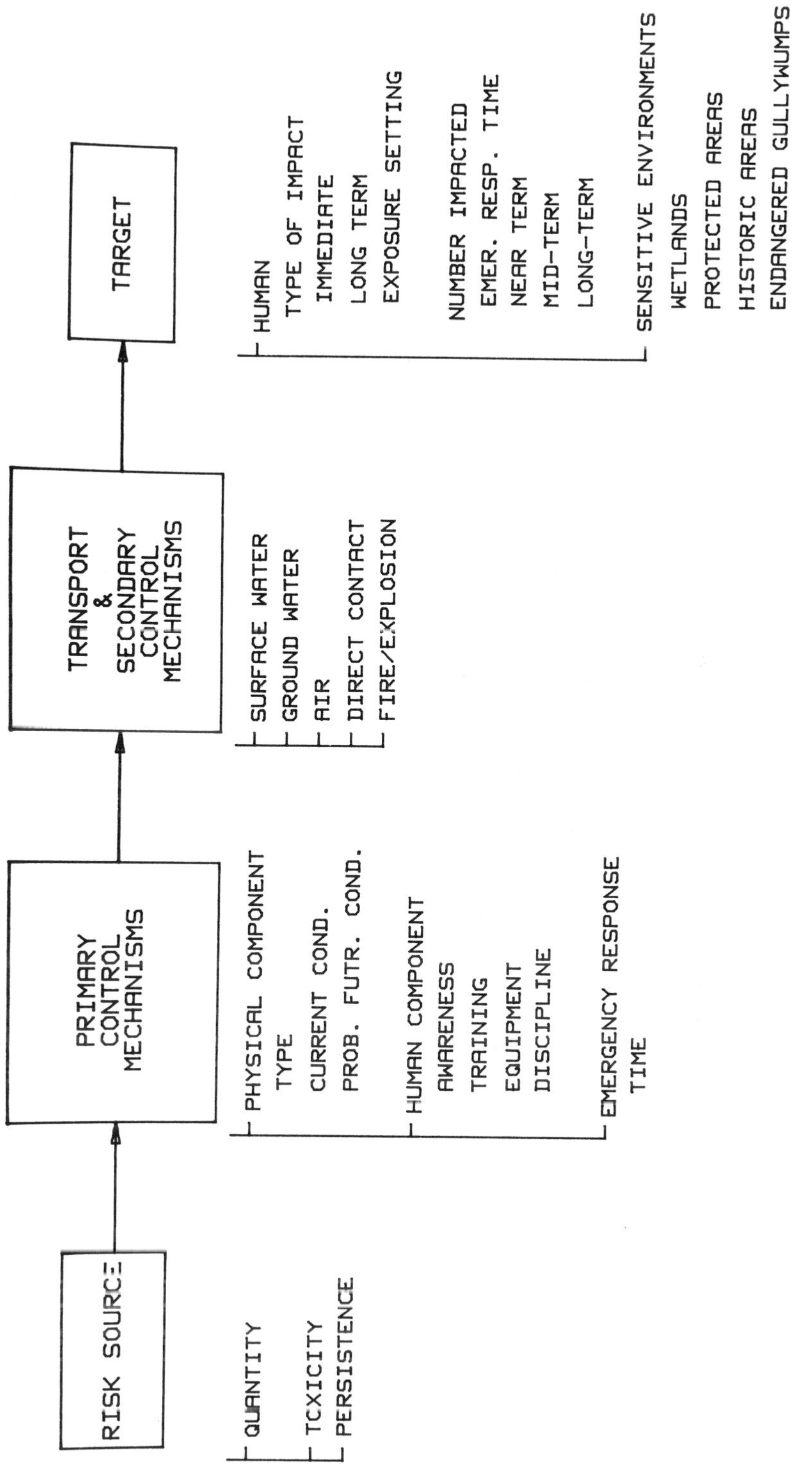

Figure 1.4 Risk System Primary Analysis Factors.

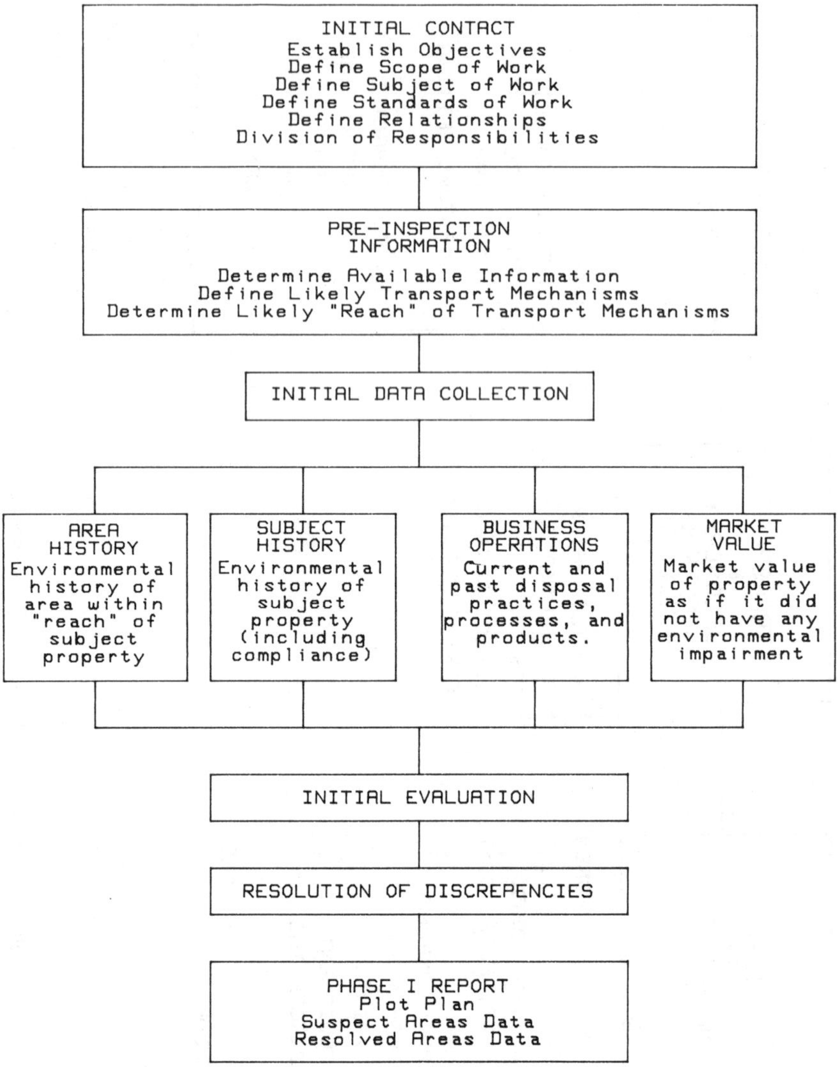

Figure 1.5 Risk Evaluation Process.

evaluations, it is appropriate at this point to provide at least a working definition of each of the major evaluation phases. The total evaluation/valuation and environmental assessment process is shown in Figure 1.5.

Phase I ERAs For purposes of this discussion, a Phase I ERA is defined as determining the answer to the following question: Is there a reasonable basis to suspect the presence of an environmental risk? To answer this question a complete check of a property's history, the history of surrounding properties, and the current condition of the subject and immediately adjacent properties will be required. Note that a Phase I ERA cannot and will not provide the quantitative information required for the valuation or management of a property containing an environmental risk.

The Phase I ERA is the first and most important step in the "due diligence" effort required to protect the "innocent purchaser" defense available under the Superfund laws (CERCLA, SARA). Although much has been written concerning "due diligence" there are only a few concrete guidelines available to define what specifically constitutes "due diligence." The USEPA provided some indication of the requirements of "due diligence" in its publication of *Guidance on Landowner Liability under Section 107(a)(1) of CERCLA, de minimus Settlements under Section 122(g)(1)(8) of CERCLA, and Settlements with Prospective Purchasers of Contaminated Property: MEMORANDUM* (hereafter referred to as *de minimus*) issued June 6, 1989. This document implies that, among other things, due diligence will constitute proof that the purchaser of a property found to be contaminated performed a Phase I ERA that included at least:

1. A search of the subject property's history back to at least virgin land status. This search must include all reasonably available public records such as the CERCLA lists, state environmental protection agency records, underground storage tank records, aerial photographs, ownership records, operator records, and almost any other publicly available sources of information.
2. A search of the surrounding properties' histories to a least-virgin land status. Again the search must include reasonably available public records. The key here is the definition of surrounding property and, although this is not precisely defined, the implication is that a surrounding property is one within reasonable reach of the available transport mechanisms relative to the subject property.
3. A physical examination of the subject property, and to the extent reasonably possible the immediately adjacent properties, to determine if there is a visible indication of the presence of an environmental impairment.

Beyond the *de minimus* concerns are considerations primarily associated with ongoing business operations. In this case a "compliance audit" must be included as a part of a Phase I ERA to determine if all of the applicable rules and regulations have been observed by the business enterprise. In addition, the possibility of the acquisition of environmental liabilities through the purchase of the operating assets of a product line must be considered.

Without the completion of a comprehensive Phase I ERA, the buyer cannot be reasonably assured of compliance with the requirements of due diligence. A balance between the costs of a Phase I analysis and the amount of risk associated with a failure to perform an adequate due diligence investigation will be required. Without the existence of a Community Environmental Risk Inventory (CERI's are covered in a later chapter), a comprehensive Phase I ERA may cost tens of thousands of dollars. The tradeoff must recognize that the concept of negative leverage applies to environmental risks, that is that there is absolutely no correlation between the magnitude of an environmental liability and the value of a property. Environmental liabilities of hundreds of thousands or millions of dollars may attach to properties having unencumbered values of only a few thousands.

Phase II ERAs A Phase II ERA may be defined as follows: Given a reasonable basis to suspect the presence of an environmental risk, confirm or deny the presence of the risk in a scientifically acceptable manner. Note that again the negative default principle must govern the Phase II ERA process. Specifically, the existence of a risk may be assumed or it may be demonstrated by a single analytical finding. It may be denied only through the completion of a properly designed and executed analytical sampling protocol. Phase II ERAs are often combined into Phase III ERAs.

Phase III ERAs The Phase III ERA is conducted on the basis that an environmental risk exists, with the explicit purpose of providing the engineering information required to evaluate the magnitude of the risk and the applicable remediation/risk reduction tactics to be employed. It is only a Phase III ERA that can provide the data necessary to value or manage a property or operation when an environmental risk is present.

A comprehensive Phase III ERA will result in the engineering evaluation of the entire risk system and should result in an action plan designed to reduce the possible impact on the targets from the risk sources. The Phase III process is not trivial, often taking many months to complete and requiring the expenditure of hundreds of thousands of dollars. Unfortunately, there is simply no other reliable way to obtain the information required for the evaluation, valuation, or management of environmental risks. It is the cost of implementing the action plan and the expected costs of the failure of the primary control mechanisms that will govern the value of a property or business operation.

The concept and implications of the risk system will be developed throughout this book to provide a uniform theoretical basis for risk analysis. The ultimate objective is to develop a risk inventory that can be audited and used both as a basis for the measurement of the performance of risk management, and as a basis for the allocation of resources to reduce risk as called for in the ERMP.

The Environmental Risk Inventory

One of the primary objectives of this book is to develop the basis for an auditable inventory of environmental risks. With such an inventory, the regular management of environmental risks as a continuing part of business operations becomes possible, along with the ability to measure the performance of such risk management, and the environmental asset and liability position of the organization. Although it is not explicitly stated in the existing environmental rules and regulations, the development of an inventory of environmental risks is clearly contemplated in such diverse requirements as the Security and Exchange Commission's (SEC) requirements for 10-K disclosure of environmental liabilities, the AHERA rules, and the OSHA Hazard Communications Rules.

There are several areas of concern in utilizing the inventory approach. In particular it is difficult to exactly quantify the amount of a hazardous substance that may exist in a number of situations. In addition, the condition of the control mechanisms that govern the level of risk, particularly with respect to the future condition of those control mechanisms, can be difficult or impossible to evaluate in precise terms. Also, the type and likelihood of a specific environmental impact may not be precisely quantifiable. These issues are related to our traditional concepts of an inventory that can be converted into a financial measurement, but are not a problem if we are prepared to recognize that some aspects of business operations simply cannot be quantified in classic financial terms.

The quantification of the size of the inventory of environmental risks will be approached on the basis of establishing a beginning inventory with the recognition that the application of proper inventory management techniques will disclose changes and refinements in the inventory over time. The objective will be to establish the best possible starting point for the inventory given the circumstances, and then to modify the initial inventory in the light of additional information developed later. In some instances it will be possible to establish a very accurate beginning inventory such as for asbestos, PCBs, or lead in a building. In others the initial inventory will be less accurate simply because feasible measurement techniques do not permit a higher level of accuracy.

The inventory of environmental risks will be fundamentally different in practice from the classic form of inventories because of two characteristics not generally associated with the classic inventory. The first is the importance of location with respect to the risks the inventory represents. An inventory of pure asbestos buried a thousand feet underground and sealed off in a manner that will prevent it being reached under any normal circumstances presents a very different level of risk from the same quantity of materials spread throughout a building. The second characteristic is that the risk-inducing substance is not the real issue; the mechanism through which the risk is controlled is the point of actual concern.

Environmental risk inventories then have two additional properties that will modify the record keeping and management approach; the physical location of the inventory, and the control mechanism associated with the inventory component. Both of these properties must be explicitly dealt with in the ERMP.

There is a need to have a uniform means of expressing the quantity of the risk source for the development of a decision-making system. In the context of an ordinary business inventory the uniform measurement unit is financial in nature, the dollar value of the inventory, but for risk management purposes the dollar value is meaningless. Instead, we will express quantities in terms of a dimensionless unit utilizing a conversion table developed for the USEPA Hazard Ranking System. Basically, the risk source will be identified and initially quantified in terms of barrels, gallons, tons of material, but this measurement will then be converted to a dimensionless unit of measure for comparison. This should not trouble anyone any more than converting the number of bolts and the number of gallons of gasoline in an inventory into the number of dollars required to purchase those items in order to develop a "value" for financial reporting purposes.

Characteristics of Environmental Risks

Environmental risks have two characteristics that appear to distinguish them from other types of risks. First is the probabilistic nature of environmental risks. It is rare when a given substance can be said to have a 100% certainty of causing damage. Generally, environmental risks have an associated probability of causing damage and that probability is often given in terms of a complex set of conditional probabilities over an extended span of time. In reality, this issue is not significantly different from the conditional probabilities of more commonly recognized risks such as the risk of human or environmental damage associated with a fire, which may also be considered an environmental risk.

The second characteristic can be substantially different; that of the "reach" associated with a risk. A given environmental risk may need to be judged differently depending on how long in time the risk may exist and the physical distance over which it may be effective. The first part of the problem of "reach" will be the subject of a specialized topic known generally as persistence, and the second will be the subject of transport mechanisms.

A given environmental risk may be measured in terms of a number of different parameters; toxicity, persistence, transportability, and as yet unlabeled concepts dealing with global environmental effects such as acid rain, global warming or ozone depletion. In addition, risks must be judged in terms of our existing knowledge of the probable effects of a given condition on human health and/or the environment. For example, a carcinogenic substance must be evaluated differently from a toxic substance because we perceive a different human health mechanism at work for carcinogens than at work for toxins.

One of the major problems in measuring environmental risks is that the relationship between multiple risk sources may not be linear, or even known. The relationship between smoking and asbestos exposure appears to be synergistic, that is if asbestos exposure may result in *n* incremental deaths per 100,000, then smoking and asbestos combined appears to result in 50*n* incremental deaths per 100,000. Because these relationships are just now being explored and are generally unquantified at this time, this book will treat each risk source as independent of all other risk sources for analysis purposes. This is an unfortunate but necessary simplification given our current state of knowledge.

To measure the extent of a risk source's possible impact on a target, we will again turn to the Hazard Ranking System and use its toxicity/persistence measurement tools and its means of measuring the impact on targets based on the proximity and available transport mechanisms between the target and the risk source.

Feasibility Measurement

Feasibility measurement represents a major problem if it is approached as a purely financial issue. In fact, attempting to measure feasibility in financial terms is a strategic mistake of the first order and is unnecessary in any event.

Feasibility cannot be measured by strictly financial means without opening up the very controversial issue of the value of human health or life, or the equally controversial issues of the value of damage to the general environment. However, with the advent of newer mathematical tools for management such as multi-objective linear programming, it is no longer necessary to measure feasibility in commensurable units in order to develop optimum feasible solutions.

The approach taken here will be to measure program costs in classic financial terms and to measure program benefits in terms of units of risk exposure; units that may change from situation to situation but which will remain consistent within a given context. The optimum feasible solution will be defined as the maximum risk reduction at minimum feasible cost given current technology. It is to be emphasized that this approach does not assign a value to human health or life or environmental damage except in terms of the implications of the overall amount of resources an organization or society is prepared to commit to risk reduction.

Responsibility

One of the basics of environmental risk management is the assignment of responsibility. The laws and courts will assign responsibility for environmental damage wherever possible. Good management requires that responsibility, and the concomitant authority to deal with the risks, be assigned. Without responsibility and authority, environmental risk management is simply not possible.

In the case of the AHERA rules and regulations that we are using as a model, responsibility for dealing with the asbestos inventory in a school is assigned explicitly to the so-called "designated person," an individual charged with the detailed implementation of the asbestos inventory management control program. AHERA requires that this individual be explicitly identified, trained, and given a recognized grant of authority over the execution of the program. Considering the stakes associated with environmental risks, the appointment of an explicitly identified and competent individual cannot be less important in the general case than it is with respect to asbestos in schools, a relatively low-level risk compared to others.

Notification

A careful reading of existing environmental laws and cases will demonstrate that often the most severe penalties are reserved for failure to notify potentially affected parties of the existence of a risk. Criminal penalties are applicable for a deliberate failure to notify, particularly if an injury results. These are obvious negative consequences, but a substantial portion of risk management may be associated with a deliberate policy of prompt and proper notification, an issue dealt with in terms of the public relations aspects of risk management.

Notification can be an opportunity to establish the creditability of an ERMP and to enlist the aid and active support of the potentially affected community in the risk reduction and control effort. The most effective environmental risk management programs will be implemented with the active aid and support of the very lowest levels of the organization, and since these individuals probably must be notified anyway, proper notification can be used to enhance the operating characteristics of the program at minimal cost.

Legal Basis for the Environmental Risk Management Program

Already mentioned as the model for an ERMP are the AHERA rules and regulations, but a number of other instances can be cited. For example, the Occupational Health and Safety Administration's (OSHA) Hazard Communications Rules call for a specific program of identification, notification and training. The Superfund Amendment and Reauthorization Act (SARA) calls for community notification under the Title III Community Right-to-Know provisions. The *Toxic Substances Control Act* (TSCA), *Clean Water Act, Clean Air Act, Resource Conservation and Recovery Act* (RCRA), and others explicitly or implicitly call for the implementation of various forms of control over environmental risks.

Risk Justification for Management Control of Environmental Risks

From the point of view of risk to human health and/or the environment there are several excellent reasons for the detailed definition and implementation of an ERMP. In particular:

1. The current status of removal and disposal practices for most hazardous substances is in its infancy, and in many cases the demand for such services has far exceeded the capacity of competent service organizations. This has resulted in what appears to be a majority of removal/disposal practices yielding worse contamination of the facility after removal, and consequently higher liability for the disposal of the removed materials.
2. A number of environmental risks simply cannot be eliminated from the system. They are necessary to the continued proper functioning of our technological society. In addition, new environmental risks are coming into existence or are being recognized on a daily basis. The net result is that a given organization must develop the ability to maintain control over environmental risks for the foreseeable future, it being generally impossible to eliminate all such risks, giving rise to the need for an ERMP.

Economic/Financial Justification for Management Control of Environmental Risks

The negative leverage that realized environmental risks can exert against the value of an enterprise can be orders of magnitude of the unencumbered value of the entity's assets. Damages may be assessed against the entity for the total cost of cleaning up any of the problems caused, plus up to fifty million dollars in fines, plus liability for real or perceived injury to human health or the environment.

Under the Superfund laws, liability for damages may be assessed against "owners or operators" and such liability is strict, joint and several. The implications are that the owner or operator may be held responsible for the entire cost of remediation even if that owner or operator contributed only a small portion to the total environmental problem and even if that portion represented only a potential problem as opposed to a realized problem.

Further, the liability may result in a manner that will deprive the entity of outside financial support such as insurance funding, even if the entity is "innocent" of creating the environmental risk. In a court case recently decided in the New York State Court of Appeals, in which a landowner purchased a property and five years after the purchase discovered contamination left by the previous owner, it was determined that the owner's comprehensive general liability policy excluded intentional pollution. The court held that there was nothing in the policy language to indicate that the insurer intended to pay for intentional pollution caused by a third party.[8]

Liability may also result from history. In the case of the *Safety Light Corp,*[9] Safety Light was held responsible for the pollution of a property by the United States Radium Corporation in 1917-1926, because United States Radium had been acquired through a chain of corporate acquisitions by Safety Light.

The point is simple, failure to control environmental risks can substantially affect the balance sheet to the point of insolvency. A single 55-gallon drum of a particularly hazardous substance that escapes control can reduce the asset value of an enterprise by hundreds of thousands of dollars.

REFERENCES

1. CERCLA, Comprehensive Environmental Responsibility, Cleanup and Liability Act, 42 USC 9601, *et seq.*
2. SARA, Superfund Amendment and Reauthorization Act, Pub. L. 99-499, 100 Stat. 1613 (Oct. 17, 1986).
3. Chlorofluorocarbons, such as freon.
4. NOTE: These figures originate with an unpublished study conducted for the EPA. Further investigation indicates that, of the number of properties left more contaminated, the following situation appears to exist:
    1. The firms doing the work had less than two years experience.
    2. In over 80% of the occasions, the work was not supervised by an experienced abatement supervisor or organization specializing in the supervision of such work.
    3. Of the jobs performed in such a manner that the work left the property with less contamination, more than half were supervised by an experienced third party and designed by an experienced professional abatement design firm.

5. NOTE: Cost is taken in a very broad context including technological, economic, and health/environmental impact considerations.
6. *Stakeholders* are very broadly defined, ranging from the direct owners of a property, to the financing sources, to the occupants, to neighbors, to the regulators and others as well.
7. *29 CFR 300* Appendix A; "Uncontrolled Hazardous Waste Site Ranking System: A User's Manual."
8. As reported in *The Wall Street Journal;* November 22, 1989; "Legal Beat;" page B3.
9. *T&E Industries v. Safety Light Corp.*, 227 NJ Supra 228, 3 TXLR 424 (App. Dv. 1988) and a related 27 E.R.C. 1953 (D. NJ 1988).

2 THE ENVIRONMENTAL LAWS, RULES, REGULATIONS AND COMMON LAW[10]

THE RELATIONSHIP BETWEEN FEDERAL AND STATE LAWS

In all that follows, the reader must keep in mind that there is a specific relationship between federal and state environmental laws. In particular, almost all of the federal laws contain the provision that an individual state, county, city or town may, at its option, elect to implement laws, rules and regulations that are *more stringent than* the federal laws, rules and regulations. In general, the federal laws form the basis for environmental risk regulation, but superimposed on the federal laws may be a layer of state laws and court decisions that considerably tighten the requirements.

This chapter is not intended to be legal advice in any sense, but rather a discussion of the laws, rules and regulations from the point of view of the informed layperson. In all cases, competent environmental counsel should be sought whenever an issue involving environmental risk arises. The emphasis here is on competent *environmental* legal counsel; attorneys in general practice may not meet this criteria of specialized knowledge.

THE LEGAL TOOLS

There are currently a number of pieces of federal legislation that form the basis of environmental risk regulation and the major components will be briefly discussed in this chapter. In general, there is a structure to these laws that should be understood on a gross level first.

There are three laws that may be said to span the total scope of environmental risk regulation, the *National Environmental Policy Act* (NEPA), the *Resource Conservation and Recovery Act* (RCRA), and the *Toxic Substances Control Act* (TSCA). The NEPA applies to the federal government and its activities in almost all forms including the use of federal monies to fund local projects. It fundamentally requires that the federal government must "look before it leaps" in the sense that environmental impact assessments are required prior to the implementation of a federal project or the use of federal funds (with some exceptions for national defense projects). Many state and local governments have adopted the NEPA requirements in some form and there are indications that the private sector will be held to a similar standard.

The RCRA and TSCA will be discussed in some detail later, but fundamentally they define what shall be considered to be a hazardous substance[11,12] (TSCA) and how that substance shall be controlled from "cradle to grave" (RCRA). These three laws underlie the balance of the laws which tend to be very much situation specific.

For example, the *Clean Water Act* and the *Safe Drinking Water Act* govern surface and ground water contamination along with the older *Rivers and Harbors Act of 1899* and several other more minor pieces of legislation. The *Clean Air Act* governs air pollution, including the provisions of the *National Emissions Standards for Hazardous Air Pollutants* (NESHAPs) amendments to the *Clean Air Act.* As this book is being written, the *Clean Air Act* is undergoing major revision, but it fundamentally will still govern air pollutants of a local, regional and national character with only minor consideration for global pollutants. The Superfund laws, CERCLA and SARA, govern uncontrolled releases or threatened releases of hazardous substances, and so on. The point is that except for NEPA, RCRA, and TSCA, the other environmental laws are specific to the situation to be governed and contain many minute but extremely important provisions applicable to a particular part of the environment.

There are over 9,000 pages of federal rules and regulations that govern environmental risks of one form or another and no attempt can be made in this book to deal with all of them in detail. Instead, this chapter will concentrate on providing an overview of the different approaches and provisions of both state and federal governments and the courts to environmental risk regulation.

One of the consequences of so many essentially uncoordinated federal and state laws is that the definition of a hazardous substance under one law may not be the same as a hazardous substance defined under another law. For example, hazardous substances under CERCLA/SARA are defined to include not only those explicitly defined by CERCLA/SARA, but also those covered under the *Clean Air Act,* the *Clean Water Act,* and TSCA[13].

Comprehensive Environmental Response, Compensation and Liability Act (CERCLA) as Amended by the *Superfund Amendment and Reauthorization Act* (SARA). 42 USC 9601 *et seq.*

While there are many specific provisions in the Superfund laws, one of the key provisions that will govern environmental risks is that concerning who will be held responsible for the uncontrolled release of a hazardous substance into the environment and the extent of that liability. In general, liability for a responsible party is strict, joint and several, and further is perpetually retroactive.

As a highly simplified example, suppose that company A dumped substance X into a site in the early part of this century. The site also contains many other hazardous substances contributed by many other parties both before and since company A was involved. In the year 2010, the substances somehow escape control and present a threat to human health and life and upon investigation, company A is the only contributor to the site that is still around and can be identified. Under CERCLA/SARA, company A may have to pay for the total cost of remediating the site and all surrounding areas affected by the substances from the site without regard to substance X's proportional contribution to the problem. This is at least one interpretation of the meaning of "strict, joint and several" and perpetually retroactive liability.

The costs to be borne by company A[14] are the total costs for response to the situation at the site and its affected surroundings for the remedial actions judged necessary to protect human health and/or the environment and any fines or other costs levied against the site. In addition, company A may be held responsible for damages to natural resources resulting from the site, such as the contamination of water supplies.

The "net" thrown out by CERCLA/SARA[15] to determine who is responsible is extremely large and loosely defined. Specifically, the net may include the current and past owners and/or operators who generated the materials, the transporters who brought the materials to the site, the current and past owners and/or operators of the site, and virtually anyone who may be interpreted to be an owner, operator, or transporter involved in the site.

> Under CERCLA, the United States government through the USEPA has broad powers to seek or impose remedial and response action on property owners, past and present, and on those who generated the hazardous substance. The government may issue an administrative order pursuant to Section 106[16] or may itself clean up the property and seek to recover the costs.[17] Failure to abide by administrative order would subject the recipient thereof to treble damages in addition to the actual costs of removal.[18]

There are also private causes of action which may be brought by any other party whose actions are "consistent with the National Contingency Plan." Recent court decisions indicate that the plaintiff need only have incurred response costs and the cleanup need not have been ordered or approved by state or federal government subject to certain exceptions.[19] The SARA amendments made the right to seek contribution express in the *Act* and further clarified that sites need not be listed on the National Priorities List[20] in order to be the subject of a private cost recovery action.[21]

While there was initially a great deal of argument, it is now fairly clear that CERCLA is a strict, joint and several liability statute.[22] That is, a plaintiff, whether governmental or private, may sue any party who is liable under the action for the entire response costs. USEPA in fact generally sues only major contributors leaving them to seek contribution from other smaller contributors. Joint and several liability is, of course, particularly important when many of those liable are no longer available or unable to respond to costs.

Thus, anyone acquiring a piece of property or acquiring a security interest in property must be concerned with the diminution of value that the presence of hazardous substances will impose as well as the potential personal liability which will be incurred. Even though CERCLA specifically excludes from direct liability those who, without participating in the management of a facility, hold *indicia* of ownership (such as title) primarily to protect their security interest,[23] the inability to close or to sell the subject property greatly diminishes its value. Moreover, lenders must be concerned that their involvement in the property does not arise to the level of making them an "operator." Even though they are not an owner, they may be deemed liable as an operator of the property if they are involved in the management of the property. In one case, a bank whose vice president became directly involved in operating a site was held liable.[24] In another case, a parent corporation that directed the affairs of its subsidiary with respect to environmental compliance was deemed an operator even without trying to "pierce the corporate veil."[25,26]

As indicated, there are some limitations to who can be considered to be responsible,[27] but the limitations are somewhat vague and subject to considerable judicial interpretation. For example, a transporter may not necessarily be held responsible if the transporter did not choose the site. A lender may not be responsible if the lender cannot be interpreted to be an owner or operator, but the means by which a lender may become an owner or operator under CERCLA/SARA are sometimes very subtle. A current owner of a property may not be held responsible if it can show that an environmental investigation was performed with "due diligence" and that as a result the current owner had no basis for knowing that a hazardous substance existed and did not further contribute to the problem. In addition to the defenses against CERCLA/SARA liability outlined above, there are several other provisions in the laws that may be applicable.

The first defense may be whether or not substance X is a hazardous substance. For example, some petroleum products are not considered hazardous as are some construction materials. In addition, some releases of substances may be allowed by permit, such as those governed under proper National Pollution Discharge Elimination System (NPDES) permits, or pesticides applied according to applicable regulations.

Secured lenders may be exempted if the lender does not participate in management decisions (a very tricky area) and simply holds the title to a property as *indicia* of ownership. This area of exemption is constantly being tested in court with some very interesting results. For example, in a suit against a major bank, the allegation is that the bank, in attempting to collect on a debt, became an operator by virtue of the idea that the attempt to collect strongly influenced the disposal of hazardous substances (a concept not yet decided by the courts). In the now infamous Maryland National Bank case, the holding of title after foreclosure exposed the bank to remediation costs (established by court decision). In still a third area, having an officer of the lender participate in management decision-making by being on the board of directors has established liability to the lender and this concept may be extended to include the informal giving of operational advice by another, as yet undecided, court case.

If it can be established that the contamination was placed by an unrelated third party with which the entity has no contractual relationship and the entity took due care after notice of contamination, liability may not apply to the entity. Again, this is a very tricky area of the law with many issues yet to be decided.

Related to the third-party concept is the innocent landowner defense provided in CERCLA/SARA.[28,29] The basic idea of the innocent landowner defense is that if the entity did not know of the contamination at the time of acquisition of the property and would not have reason to know of the contamination after conducting a reasonable inquiry into the possibility of contamination as judged by a number of factors and did not contribute, the entity should not be held responsible. The problem is that the meaning of several of the above phrases are only *very* loosely defined.

Resource Conservation and Recovery Act (RCRA) 42 USC 6901, *et seq.*

Although there are many provisions in RCRA, some of them are more appropriate to this book than others. For example, Subpart C governs the ongoing generation, transportation, treatment, storage and disposal of hazardous wastes. The provisions of Subpart C form a basis for helping to determine the environmental risk exposure of the entity.

RCRA defines hazardous wastes by listing specific substances considered to be hazardous *and* by defining the characteristics of a hazardous waste. The definitional characteristics are: (1) toxicity, (2) reactivity, (3) ignitability, and (4) corrosivity (all to be discussed extensively in a later chapter).

RCRA requires that treatment, storage and disposal (TSD) facilities have permits under Part B. A TSD facility may be subject to corrective actions designed to reduce environmental risks that can substantially effect the operations and risk exposure of the facility. In many ways, the requirements for TSD permits are partial embodiments of the ERMP concepts in that they require groundwater monitoring, closure plans and financial assurance that any environmental risk reduction actions will be within the financial resources of the facility. It should be noted that TSD permits may be required for many different types of facilities; the requirements being determined by the amount(s) and type(s) of substances, the length of storage time, and the distance to appropriately permitted disposal facilities.

Enforcement under RCRA can compel corrective action for contamination due to past TSD activities given the potential for imminent and substantial threat to human health and/or the environment. A Phase III ERA will usually be necessary to determine if ongoing or past activities need corrective action or TSD permits.

Underground Storage Tanks (UST)

On September 23, 1988, the USEPA promulgated its long-awaited underground storage tank regulations.[30] These rules apply to all types of underground storage tanks including petroleum, chemical, and hazardous waste. USEPA estimates that there are approximately 2 million underground storage tank systems in the United States and of these, over 75 percent are constructed of unprotected steel. USEPA further estimates that between 10 and 20 percent of existing underground storage tank facilities have already had a leak. Particularly suspect are those tanks over ten years old.

Because of the magnitude of the problems, USEPA's newly issued rules require strict monitoring of tank content inventories and periodic pressure testing for petroleum underground storage tanks in commercial, non-farm use. Where the monitoring indicates leakage, potentially expensive site investigation and cleanup is required. Therefore, appraisal of any property containing underground storage tanks should include a review of recent leakage tests and of such factors as tank construction and age.

In addition, if closure of the tank is required, USEPA suggests that certain procedures and rules be followed. If the tank was closed prior to the December 22 effective date, the owner/operator is required to reassess his closure and close the underground storage tank in accordance with the new rules if the implementing agency believes that prior releases from the tank present a risk or potential risk to human health or the environment.[31]

The underground storage tank regulations cover both hazardous substances as defined by the various rules and regulations and petroleum products that are often not defined as hazardous substances. These regulations require the retrofitting of USTs to detect leaks, the testing for leaks, and financial assurances that the owner/operator of a leaking underground storage tank (LUST) will be able to pay for any remediation required. The corrective actions to be taken in the event of a leak are contained within the regulations.

Clean Water Act (CWA) 33 USC 1251, *et seq.*

The *Clean Water Act* regulates the quality of surface water primarily by providing a permit system that will govern the amount of contaminants that may be discharged into surface waters. The National Pollutant Discharge Elimination System (NPDES) permits for a facility will specify the kinds and amounts of pollutants that may be discharged by the facility over a given period of time. The existence of an NPDES permit for a facility implies that periodic testing of effluent streams must occur and the results reported to the issuing authority. If the discharge is not directly into a surface water system, but is instead into a municipal sewage treatment system, the treatment system may place pretreatment requirements on the effluent stream to ensure that the municipal system will be able to successfully treat the balance of the effluent stream to its own permitted discharge requirements.

Section 404 of the CWA provides for "dredge and fill" permits whenever a wetland is involved. These dredge and fill permit requirements can strictly regulate or stop activities in wetlands.

Wetlands[32]

One of the most significant restrictions which may impact the value of property is on wetland development. While there has been some long-term basis for these restrictions, Section 404 of the *Clean Water Act*[33] and regulations adopted thereunder[34] make it clear that almost any land in the United States which will support aquatic vegetation or other conditions indicative of long-term periodic inundation would be considered a wetland. As a matter of law, these properties cannot be filled or modified without a permit issued by the USEPA (previously issued by the Corps of Engineers). In issuing such permits, the USEPA must take into account the *National Environmental Policy Act*,[35] *Endangered Species Act*[36] and other applicable requirements. Thus, property which might appear ideally suited for a particular purpose may be subject to restrictions which prevent any development at all or impose significant mitigating requirements. Perhaps even equally significant, such a process of undergoing a wetland review may result in a delay in proceeding with development. Recent amendments to the *Clean Water Act* have established administrative proceedings for enforcing these requirements.[37] The USEPA can issue administrative orders and impose administrative fines for construction and violation of the wetlands regulations. In addition, willful violations of the *Act* are felonies and negligent violations are misdemeanors.[38] Therefore significant criminal penalties are also available.

Safe Drinking Water Act (SDWA) 42 USC 300(f), *et seq.*

The SDWA protects the drinking water system by setting Maximum Contaminant Levels (MCLs) for public drinking water systems. Because the majority of drinking water in the United States comes from underground aquifers (something over 80%) the MCLs are used as the basis for groundwater cleanup criteria for CERCLA and generally for all environmental cleanup that involves potable groundwater.

Clean Air Act (CAA) 42 USC 7401, *et seq.*

The CAA regulates both stationary and mobile air pollution sources and is currently undergoing extensive legislative revision. In general, pollution sources will be regulated by the states which may join together to form regional groups to deal with such issues as acid rain emission regulations or divide into subregions within a state such as the South Coast Air Quality Management District in Southern California to govern local emission situations. Mobile pollution sources are generally governed by the USEPA (tailpipe emission standards), but again the states or subregions may place more stringent restrictions on the emissions.

As contemplated by Congress at the time of this book, the revised *Clean Air Act* will govern stationary sources by imposing limitations on emissions. However, the regulatory approach taken in the revised CAA will apparently differ from older legislation in several important respects. First, the revised CAA will apparently encourage the achievement of emission reductions from stationary sources through the concept of Maximum Achievable Control Technology (MACT); that is, the implementation of the best technological means of reduction as opposed to a specifically required method. Second, the revised CAA will apparently encourage alternative means of achieving emission standards for both stationary and mobile sources through the use of alternative fuels. Third, the revised CAA will apparently encourage the use of so-called "pollution credits" that would be generated by a stationary facility overachieving emission reductions. The "pollution credits" would become a saleable[39] or "bankable" commodity that can be used to offset the costs of achieving the reduction or used to offset the non-achievement of emission reductions elsewhere. Fourth, the revised CAA will also apparently reach much further down the economic scale to the small point sources of emissions, such as the local dry cleaner or paint shop.

In general, new facilities are regulated by stricter new source performance standards and, depending upon location, Prevention of Significant Deterioration (PSD) or new source review criteria. In addition, emission of toxic air pollutants either by stationary sources or by activities (e.g., asbestos removal) are regulated by the National Emission Standards for Hazardous Air Pollutants (NESHAPs) under the CAA.

Toxic Substances Control ACT (TSCA) 15 USC 2601, *et seq.*

TSCA regulates the manufacture and commercial distribution of hazardous chemical products within the United States. It does not govern the export of such products. The major provision of concern for our purposes here are the regulations concerning PCBs (polychlorinated biphenyls).

Briefly, the provisions of TSCA require that substances with more than 500 parts per million (ppm) of PCBs must be incinerated. Substances with 50 to 500 ppm of PCBs must be incinerated or landfilled in a RCRA-permitted facility. Electrical equipment containing PCBs must be labeled and regularly inspected (assuming it contains less than 50 ppm of PCBs). It should be noted that many properties contain PCBs in sometimes unexpected locations such as fluorescent light fixtures, hydraulic equipment, and in waste dumping areas.

Asbestos in schools is also regulated under TSCA through the *Asbestos Hazard Emergency Response Act of 1986* (AHERA).

Definitions of Hazardous Materials

The various environmental rules and regulations provide different definitions of a hazardous material. Asbestos, for example, has three different definitions in current federal usage.

- *First (Oldest) Definition of an Asbestos-Containing Material (ACM):* The 1979 USEPA definition of an ACM used as a building material defined such a material to be one containing more than 1% asbestos and that was friable (i.e. could be crushed or crumbled by hand pressure).
- *Second (AHERA) Definition:* AHERA defined an asbestos-containing building material to be one containing more than 1% asbestos. The friability requirement was dropped, expanding the definition to cover some 3,500 building products.
- *Third (OSHA Hazard Communications Rule) Definition:* The OSHA Hazard

Communications Rule defines a material to be hazardous if it contains, for example, more that 0.1% of a carcinogen. Since asbestos is a carcinogen, this definition considerably expands the definition of an ACM to include tens of thousands of both natural and man-made substances.

The states have further added to this complicated situation by providing their own, more stringent, definitions. The state of New York, for example, defines an asbestos-containing material to be a material with "any detectable amount" of asbestos.

The point is simply that the investigator must be aware of the different definitions for a hazardous substance, the source and application of the definition, and the applicability of a given definition to the investigation at hand, *prior to beginning the investigation.*

Reporting And Disclosure Requirements

The issue of reporting and disclosure has many facets—some of which are not germane to this book—such as the relationship between a seller and a buyer in common law. The concentration here will be on the specific federal and state statutory requirements for disclosure and the potential applications of required disclosures to the evaluation of risk, risk management, and risk control/reduction.

CERCLA

The basic disclosure requirements under CERCLA are contained in Section 103(a)[40] which imposes a reporting obligation upon the release of a hazardous substance into the environment as follows:

> "Any person in charge of a . . . facility shall, as soon as he has knowledge of any release (other than a federally permitted release) of a hazardous substance from such . . . facility in quantities equal to or greater than those determined pursuant to Sections 102 of this title, immediately notify the National Response Center . . .".[41]

The phrases used in the above passage are defined and limited generally as follows:

- *Federally permitted* releases are those allowable under defined federal and state statutes (e.g., NPDES-permitted releases under the *Clean Water Act;* releases in compliance with a Subtitle C hazardous waste management permit under RCRA).[42]

- *Notification of continuous releases* need only be made annually to the appropriate On-scene Coordinator.[43]
- CERCLA's *petroleum exclusion* excuses the reporting under CERCLA of releases of oil or any fraction of oil, such as gasoline.[44]
- Information provided pursuant to CERCLA's notification requirement cannot be used against any person providing the information in a criminal prosecution under CERCLA (except for perjury or giving a false statement for example).[45]
- Failure to make the required report or notification or the making of a false or misleading report or notification may result in both civil and criminal penalties and up to three years imprisonment for a first offense.[46]
- The *reportable quantity* (RQ) of any hazardous substance is set by statute at one pound until determined differently by EPA.[47] Reportable quantities are calculated over a 24-hour period.[48]

  Note: CERCLA does not assign notification responsibilities to specific individuals, it merely states that "any person in charge of a facility" has responsibility to report releases, but does not define "any person in charge."

CERCLA imposed a one-time reporting obligation on owners or operators to report the existence of facilities that may have contained hazardous waste within 180 days after CERCLA's enactment (December 11, 1980).[49] This obligation is not imposed on owners or operators who have discovered the existence of such facilities after that date.

The third party defense to CERCLA liability under Section 107(b)(3)[50] is not available to a seller who "obtained actual knowledge of the release or threatened release of a hazardous substance (during ownership). . . and then subsequently transferred ownership of the property to another person without disclosing such knowledge. . .".[51]

Under Section 120(h) of CERCLA,[52] whenever the United States enters into any contract for the sale or transfer of real property which is owned by the United States and at which hazardous substances have been stored for over one year, known to be released, or disposed of, the contract must include type and quantity of substance and the time at which such storage, release or disposal took place.[53]

The Federal Emergency Planning and Community Right-to-Know Act of 1986 (EPCRA) (Title III of SARA). 42 USC 11001, et seq.

SARA imposes emergency release and other notification requirements in addition to those of CERCLA.[54] The emergency release notification requirements[55] are similar to those of CERCLA, except that: (1) The substances that must be reported include those on a list of "extremely hazardous substances" as well as those on the CERCLA hazardous substances list;[56]

and (2) The notification is given to local and state emergency planning and response personnel rather than the National Response Center.[57]

The facilities that are required to report are only those that produce, use or store hazardous chemicals. Continuous releases and releases that do not have the potential for off-site exposure are exempt from the emergency release reporting requirements.[58] As under CERCLA, "federally permitted" releases need not be reported under Title III.[59] The "emergency" release notification provisions of SARA will usually not apply to the later discovery of a predecessor's release.

Other provisions of SARA Title III, such as the submission of material safety data sheets (MSDS),[60] hazardous chemical inventory forms,[61] and annual reporting of toxic chemical releases[62] will have substantial impact on the continuing operation of certain industries, but these provisions are not designed to discover "historical" releases.[63]

RCRA

A Notification of Hazardous Waste Activity must be submitted to EPA (or authorized state) within 90 days after issuance of a regulation identifying a substance of concern as a hazardous waste.[64] This notification requirement applies to: (1) generators, (2) transporters, and (3) owners or operators of facilities that treat, store or dispose (TSD) of the hazardous waste. The notification must state the location and general description of the hazardous waste activities.

Various notifications and reports are required to be made once a facility is identified as involving hazardous waste activities. For example, the EPA may impose reporting obligations by order if it believes the release of a hazardous waste may present a substantial hazard to human health or the environment.[65] Notifications must be made relating to changes in hazardous wastes managed, increases in design capacity, or changes in TSD processes,[66] or on partial or final closure,[67] or on transfer of permit or changes in ownership.[68]

Underground Storage Tank (UST)

The underground storage tank reporting requirements are based on RCRA Subtitle I and specific state laws designed to implement these reporting requirements. The UST requirements apply to a different range of substances than do the other portions of RCRA, specifically "regulated substances," i.e., CERCLA hazardous substances and petroleum.[69] There are separate requirements for hazardous waste tanks whether above or below ground under RCRA Subtitle C.

The federal rules and regulations define an underground storage tank as:

". . . any one or combination of tanks, including underground pipes connected thereto, which is used to contain an accumulation of regulated substances, and the volume of which, including the volume of underground pipes, is 10 percent or more beneath the surface of the ground."

The following notifications are required by the EPA rule and many state laws regarding compliance with UST technical requirements:

1. Owners of UST systems that were in the ground as of May 8, 1986 were required to notify a designated state or local agency unless the UST system was taken out of operation on or before January 1, 1974.[70]
2. Owners of UST systems brought into use after May 8, 1986, must notify the responsible state agency within 30 days of bringing the UST system into use.[71]
3. Owners of new UST systems must certify compliance with requirements for tank and piping installation, cathodic protection (when required), financial responsibility, and release detection.[72]
4. Owners and operators must report the following suspected releases within 24 hours: (a) discovery of releases of regulated substances; (b) unusual operating conditions that indicate a release; and (c) release detection monitoring results indicating a release.[73]
5. Owners and operators must report confirmed releases within 24 hours.[74]

After a release, the following reporting requirements apply:

1. Owners and operators must summarize initial abatement steps in a report filed within 20 days of release confirmation;[75]
2. Owners and operators must submit site characterization information within 45 days of release confirmation;[76]
3. A report must be submitted on removal of any free product within 45 days after confirmation;[77]
4. Information resulting from a soil and groundwater investigation (when investigation is required) must be submitted as soon as practicable or on a schedule set by the agency;[78]
5. A corrective action plan must be submitted when required by the agency;[79] and
6. The results of implementing any required corrective action plan must be submitted in a format and on a schedule set by the agency.[80]

Owners and operators must report the following spills or overfills within 24 hours:

1. Petroleum exceeding 25 gallons or that causes a sheen on nearby surface water; and

2. A hazardous substance release that equals or exceeds its CERCLA reportable quantity. If the spill or overfill is less than these amounts but cannot be cleaned up within 24 hours, the situation must be reported immediately.[81]

On or after October 24, 1988, persons who sell a tank intended to be used as an UST must notify the purchaser of the obligation, as tank owner, to notify the agency of installation of the tank.[82]

Owners and operators must provide at least 30-day advance notice of permanent tank closure or change in service (storage of non-regulated substance).[83]

The rule setting financial responsibility requirements for owners or operators of petroleum UST systems requires that evidence of financial responsibility be submitted to the agency in certain circumstances described in the rule. Generally, these requirements are triggered by identification of certain releases, by events affecting the financial assurance mechanism, or by an agency request.[84] In an exception to the usual relationship between state and federal laws, the state financial responsibility requirements may be no more stringent than the federal requirements.

The agency to which reports are made may be either an EPA regional office in states without UST authority or the agency appointed by the state to receive such notices when the state has such authority (often the state department of environmental quality or the state fire marshal's office).

Leaks from underground storage tanks are commonly encountered in real estate transactions. Discovery of the existence of underground storage tanks should immediately trigger concern regarding the possibility of an environmental problem. Reporting requirements provide a ready source of government information concerning the existence of underground storage tanks. However, since nonreporting is prevalent, caution should be exercised to assure that the required information is reported accurately and incorrect conclusions are not drawn.

There are a number of classifications of USTs that are not covered by the reporting requirements. For example:[85]

1. Farm or residential tanks with a capacity of 1,100 gallons or less storing motor fuel for noncommercial uses;
2. Tanks storing heating oil for consumption only on the premises where stored;
3. Septic tanks and storm water or wastewater collection systems;
4. Pipeline facilities (including gathering lines) regulated under the *Natural Gas Pipeline Safety Act of 1968*, the *Hazardous Liquid Pipeline Act of 1979*, or state laws comparable to these *Acts;*
5. Surface impoundments, pits, ponds, and lagoons;
6. Flow-through process tanks;
7. Liquid traps or associated gathering lines directly related to oil or gas production processes and gathering operations; and

8. Storage tanks situated on or above the floor of underground areas, such as basements and cellars.

In addition to this list of specifically excluded USTs, consider the number of UST whose present location and condition is unknown. A number of past gas stations, for example, have been converted to other uses without the proper removal or closure of the associated tanks.

The Clean Water Act

The CWA imposes a reporting obligation as follows:

> "Any person in charge of . . . an on-shore facility . . . shall, as soon as he has knowledge of any discharge of oil or hazardous substance from such . . . facility in violation of . . . this subsection, immediately notify the United States Government of such discharge."[86]

Failure to provide the required notification can result in a penalty of up to $10,000 and one year imprisonment.[87]

Toxic Substances Control Act

The owner of a facility must report any release of polychlorinated biphenyls (PCBs) in excess of 50 parts per million or untested mineral oil.[88] Any such report must be made within 24 hours to an EPA regional office. PCBs are most commonly found in transformers and capacitors where they were used as cooling fluids. Although transformers and capacitors containing PCBs must be labeled in accordance with TSCA, they frequently are not.

TSCA imposes a variety of other reporting requirements (e.g., Chemical Substances Inventory, Preliminary Assessment Information, Premanufacture Notifications, etc.[89]).

Examples of State Reporting Requirements

New Jersey Reporting Requirements

Environmental Cleanup Responsibility Act (ECRA)[90] ECRA is a fairly unique and far-reaching approach to environmental cleanup. Under ECRA, disclosure has been elevated to an obligation to investigate for the presence of contamination prior to closing certain transactions. ECRA involves the state government to an unprecedented extent in property transactions and imposes "cleanup" preconditions on the sale, transfer, or closure of "industrial establishments" that handle or dispose of hazardous substances or hazardous wastes.

The seller must secure from the state a "negative declaration" (i.e., a statement that the facility is clean) or a state-approved cleanup plan by the date of closing. Until the State of New Jersey has approved the investigation and made provisions for remedial activities, the transfer or transaction cannot take place and, in fact, can be "unwound."

New Jersey's Hazardous Substances Discharge — Reports and Notices Act[91] This *Act* requires investigation and reporting of any spill of hazardous substances by an owner or operator within 10 days, followed by sampling results within 30 days. This *Act* is limited to industries that fall within certain Standard Industrial Classifications (SICs) promulgated by the Federal Office of Management and Budget.

New Jersey's Spill, Compensation and Control Act[92] This statute requires the report of spills of petroleum and petroleum products, hazardous substances, and certain pesticides that enter or threaten to enter state waters (including groundwater) in harmful quantities.

Connecticut Reporting Requirements

Connecticut Transfer of Establishments Act[93] Unlike New Jersey's ECRA, the *Transfer of Establishments Act* does not place primary reliance on state enforcement but provides private cause of action in strict liability against noncomplying transfers. To establish transferor's liability for cleanup costs and consequential damages, the buyer will need to prove four things:

1. The property is an "establishment."
2. The property was transferred.
3. No "declaration" regarding the presence of hazardous waste was provided by seller.
4. There is hazardous substance contamination on the property.

The seller must give one of three forms of declaration to buyers (or other transferees):

1. A "negative declaration" where there has been no release on the property;
2. That a "discharge, spillage" or other release has occurred but a cleanup has been completed and approved by appropriate state officials or the officials have determined that no cleanup is required; or
3. Where one of the two above declarations cannot be made, a statement that hazardous substances are present on the property.

California Reporting Requirements

Section 25359.7 of the California Health and Safety Code requires any owner of nonresidential real property who knows or "who has reasonable cause to believe" that there has been a hazardous substance release on his property is required, prior to sale, to give written notice of the condition to a buyer. Failure to provide the required notice subjects the seller to a five thousand dollar civil penalty. The same obligation of notice is imposed on a renter or lessee.

The California Health and Safety Code forbids the unlawful disposal of hazardous waste.[94] The Code effectively imposes a reporting obligation as follows:

> "Each day on which the deposit remains and the person had knowledge thereof is a separate additional violation, unless the person immediately files a report of the deposit [with the Department of Health Services] and is complying with any order concerning the deposit. . ."

This requirement applies to a person who "disposes or causes the disposal" of a hazardous waste, and thus should not impact subsequent purchasers of a property that is already contaminated.

California requires inventories of hazardous substances and requires "handlers of hazardous materials" to, immediately upon discovery, report "any release or threatened release" of such material.[95]

Leaks from underground storage tanks are also addressed in California. By statute, leaks from underground storage tanks must be reported by the operator to a local agency within 24 hours after the release has been detected or should have been detected. A full written report must be transmitted by the owner and the operator of the underground storage tanks within five working days of the occurrence of the release.[96] No minimum quantity of material must have been released before this requirement is triggered. This statute applies to innocent, subsequent owners or lessees of the property, except for lessees who do not qualify as "operators" of the tanks.

Local or municipal ordinances should also be reviewed or consulted in specific cases because local California jurisdictions impose registration and reporting requirements.

Other State Disclosure Requirements

Pennsylvania Solid Waste Management Act, 35 Pa. Comm. State Ann. Sec. 6018.45.
West Virginia Hazardous Waste Management Act, W.Va. Code Section 20-5E-20.
Illinois Environmental Protection Act, Ill. Ann. Stat. Ch. 111-1/2 Sec. 1021(n).
Minnesota Environmental Response and Liability Act, Minn. Stat. Ann. Sec. 115B.16.

SECURITIES AND EXCHANGE COMMISSION (SEC) DISCLOSURE REQUIREMENTS

Securities Act of 1933 The *Securities Act of 1933* (SA),[97] imposes civil liability for registration statements[98] and prospectuses[99] that:

> ". . . [contain] an untrue statement of material fact or [omit] a material fact required to be stated therein or necessary to make the statements therein not misleading. . .".[100]

The term "material" has been defined as a matter to which there is a substantial likelihood that an investor would attach importance in determining whether to purchase the security.[101] Furthermore, such information must be added to the registration statement and prospectus to make them not misleading under the circumstances.[102]

Securities Exchange Act of 1934 The *Securities Exchange Act of 1934* (SEA),[103] imposes criminal sanctions for the use of manipulative or deceptive devices in the sale of any security and imposes civil liability for the use of any false or misleading statement.[104] Additionally, Rule 10b-5 makes it unlawful to make any untrue statement of material fact or to omit to state a material fact necessary to make the statement not misleading.[105]

Sections 13(a) and (b) of the SEA[106] and SEC regulations require the disclosure of Management's Discussion and Analysis of Financial Condition and Results of Operations (MD&A) in registration statements and several specified periodic reports, e.g., 10-Ks. The MD&A requirements in Item 303 of Regulation S-K[107] ask not only for a registrant's financial statements but for whatever additional information is necessary to allow an investor to accurately gauge a registrant's current financial status and future potential status. Specifically, an annual 10-K report must describe any "known trends or any known demands, commitments, events or uncertainties that will result in the registrant's liquidity increasing or decreasing in any material way."[108]

SEC Securities Act Release No. 6835[109] SEC Release No. 6835 clarifies the obligations of a company that has registered securities to disclose the impact of being named a potentially responsible party (PRP) by the EPA under CERCLA. The MD&A requirements are applied to a registrant that is a named PRP, has no statutory defenses, and is attempting to determine its remedial action costs. Among other MD&A requirements, the probability of contributions from other PRPs for such costs and the extent to which the costs are covered by insurance must be revealed.

The SEC requires disclosure of PRP status and quantification of the financial impact of such status to the extent quantification is practicable.[110] The only means of escaping disclosure is a determination by management "that a material effect on the registrant's financial condition or results of operations is not reasonably likely to occur." Joint and several liability, uncertainties of contribution and insurance coverage, make it likely that the registrant's 10-K for at least a year will have an MD&A discussion of PRP status.

Since PRP status does not automatically lead to an EPA enforcement order under Section 106 of CERCLA[111] and remedial action costs are not considered sanctions,[112] the mere designation of a registrant as a PRP need not be disclosed as a "Legal Proceeding."[113] However, in a footnote to the "Legal Proceedings" release, the SEC requires disclosure under "Legal Proceedings" if there are "particular circumstances" that lead registrant to reasonably conclude that it will be the target of EPA enforcement action, e.g., it is a non-participating PRP in a Consent Decree proceeding. With EPA's increased issuance of Section 106 orders as prerequisites to Consent Decree negotiations, the "particular circumstances" situation may become more prevalent.

Other State Equivalents[114]

Many states have equivalents to CERCLA and SARA. Many of these have important differences. For example, under the *Illinois Environmental Protection Act,*[115] provisions for liability to the state are virtually identical to liability of the government under CERCLA. However, the state act does not expressly provide for liability to third parties, and has not been amended to specifically include such things as the "good faith purchaser" provisions of the SARA amendments.[116] Under the New York law,[117] the state may require a notice of transfer of property if it is listed pursuant to the *Act* or may even in fact block the transfer.[118]

Perhaps most importantly, under various state Superfund equivalents, the definition of hazardous substance or hazardous waste may differ and things which would not subject property to liability under the federal law may in fact subject it to liability under a state law. For example, Illinois ties the definition of hazardous substance to hazardous waste, which has a broader definition in Illinois in some circumstances than under USEPA regulations.[119]

State Lien Laws Many states have imposed liens related to hazardous waste cleanup. The broadest lien law, that of New Jersey,[120] provides for a ''Superlien'' on contaminated property which takes precedence over all other liens and a lien on all other property in the state owned by the same owner if properly filed. Illinois has just adopted a lien law which started out as a ''Superlien'' in the legislature, but was amended to subordinate it to primary lenders.[121] Numerous other states have such lien laws at this time.

Radon As recent news articles have indicated, major areas across the United States may have radon contamination of property, particularly residential real estate. Radon, a colorless, odorless gas produced by the decomposition of uranium, appears to occur throughout the United States with certain well-known significant pockets in the Pennsylvania and New Jersey area called the ''Reading Prong.'' It is noted, however, that recent articles indicate it is much more widespread. Fortunately, testing for radon is not that complex. If it is discovered, there are fairly simple engineering solutions, such as sealing basements, providing ventilation to prevent the infiltration of radon and to prevent accumulation in the household.

Illinois Responsible Property Transfer Act In contrast to the New Jersey ECRA law, Illinois recently adopted its *Responsible Property Transfer Act* (RPTA).[122] This law does not impose specific cleanup obligations on the seller of the property. Instead, it provides an elaborate system for disclosure of any activity which might give rise to the hazardous liability. The principal behind this law is that there are now adequate remedies to require cleanup of property. Full disclosure to purchasers and sellers and to the state will provide the mechanisms for cleanup. The private sector was the leading proponent of the alternative arguing principally that the New Jersey approach had resulted in such significant delays in transactions that it was having an adverse impact on the economy.

Restrictions on Development

State and Local Restrictions

State and local communities are also imposing restrictions on development of wetland areas. For example, the Village of Mundelein in Illinois has adopted a new ordinance relating to preservation of wetlands. Some such ordinances rely on wetland maps prepared by the Corps of Engineers and prevent any new development or building on wetlands pursuant to such maps. Thus, in looking at development property, not only the federal wetland restrictions, but also local restrictions should be considered.

Other local restrictions on development may be important. In Illinois, the Pollution Control Board may impose hook-up restrictions preventing hook-up of any additional residences or properties to the local sewer system. There may be restrictions on the withdrawal of water or the availability of water related either to its quantity or to its quality. These must also be looked into.

Finally, other restrictions on industrial development may impact the value of property for a particular purpose. Under the *Clean Air Act,* areas may be subject to a New Source Ban to prevent construction or modification of major sources of air pollution because of a failure to adopt federally required regulations.[123] For example in the State of Illinois, there is a New Source Ban for failure to have an improved New Source Rules and there is also a New Source Ban imposed for failure to have regulations protecting the ozone standards.[124]

In addition to outright New Source Bans, significant restrictions on industrial development of property may occur under the *Clean Air Act* prevention of significant deterioration program[125] in clean air areas and non-attainment area provisions[126] in areas of the county not achieving the ambient air quality standards. If they are not under a New Source Ban, development of new facilities must undergo new source review at the state or federal level. In non-attainment review, this will require determining applicable control standards, availability of offsets and the compliance status of any other facilities owned by the developer. Similarly, even in clean air or attainment areas, Prevention of Significant Deterioration regulations[127] must be complied with. They may involve ascertainment of the availability of ''air quality increments.'' When contemplating major industrial development, this is an area which must be looked into by competent professionals to determine the possibility of new source permitting.

Environmental Requirements on Property Transference

If the property to be acquired is an on-going industrial or commercial operation, it may well be subject to detailed environmental requirements. While this discussion is not intended to be comprehensive, at least the general nature of the requirement should be kept in mind when attempting to determine the value of either real estate or on-going operations. Any facility which discharges effluents into the waters of the United States must have a federal permit called the National Pollution Discharge Elimination System (NPDES) Permit,[128] issued pursuant to the *Clean Water Act* which imposes significant detailed limitations on these discharges. If the discharges are to a public sewer system, the discharges must comply with the Federal Pretreatment Program, issued pursuant to Section 307(b) of the *Clean Water Act.*[129] One acquiring an on-going operation must determine whether the facility is in compliance, whether it has had past violations which a new owner may be responsible for, and whether it will be able to comply with future requirements that are being imposed.

Similarly, almost all sources of air pollution and air pollution control equipment require state permits and are subject to detailed and comprehensive regulations. When acquiring an on-going operation, it must be determined whether such permits have been issued, whether they are valid, whether their terms are being met, and whether there are other applicable requirements presently in effect or scheduled to come into effect which will impact the value of the property.

In addition to requirements discussed earlier under CERCLA and state equivalents for hazardous substance releases, there are elaborate provisions regulating hazardous and solid wastes at on-going operation facilities under the *Resource Conservation Recovery Act* (RCRA)[130] and the state hazardous waste programs. Under RCRA, a generator of hazardous waste must have a generator number, and a storer, treater or disposer must have a permit or be under "interim status."[131] The detailed and complex requirements of these regulations may have to be assessed on a site-specific basis by an environmental professional to determine whether the facility is in compliance or may have incurred significant past liabilities. In many states, including Illinois, even non-hazardous waste disposal requires special permitting and manifests.[132]

It is important to remember that the operating requirements and environmental control on air, water, solid waste, and hazardous waste are a regulatory program rather than a remedial program such as CERCLA. This means there are significant penalties, civil and criminal, for any violation. Someone acquiring a company that has 40 violations of an NPDES permit over the last couple of years might want to consider the potential $1 million civil penalty liability. In considering the impacts of the environmental regulatory programs, it is not sufficient to determine whether the state or federal government may have an interest in past or on-going violations. All the federal and environmental laws have *citizen suit*[133] provisions which allow citizens to bring suit even if the government chooses not to. These must all be taken into account.

Investigation for Property Transfers

Given the significant environmental liabilities which may be attached to real estate, as well as the restrictions that environmental requirements may impose on its use and development, an environmental audit or investigation becomes an essential part of most environmental real estate or corporate transactions today. Such investigation is both factual and legal in nature. One of the simplest starting points of such an investigation is the *Freedom of Information Act* (FOIA) request concerning the property and neighboring properties which can be made to the United States Environmental Protection Agency and usually to state and local agencies as well.[134] In a FOIA request to the USEPA, one can inquire as to whether the property has ever been the subject of a USEPA investigation, whether it is on the National Priorities List,[135] whether it is on the CERCLA's list of sites under investigation,[136] whether a notice pursuant to Section 103 (a) of CERCLA hazardous substance release or a notice pursuant to Section 103(c) of hazardous substance disposal has been filed with respect to the property[137] or whether government action has been taken. Such a FOIA request is probably the minimum required on property that has the potential to have been contaminated.

An additional source of information on any property which should not be overlooked is the availability of aerial photographs showing the history over a period of time. Depending upon the location, private companies or government entities may routinely aerially survey the vicinity and make photographs available. One consultant has taken a dramatic series of aerial photographs which show that an apparently clean warehouse area had once been used for creosote treatment of railroad ties and telephone poles and for storage of chemicals. The non-environmental professional may also obtain these records from sources which are familiar to them.

The first and most important step in any environmental audit of real estate or corporate property is a site inspection, preferably by someone knowledgeable in the environmental field. A non-environmental specialist, however, can learn a great deal from a site inspection. An appraiser or other non-environmental specialist should review the property for any signs of past industrial activity, any signs of contamination such as discolored soil or buildings, filled material, any signs of neighboring industrial activity, contamination or disposal which might affect the property, and anything else out of the ordinary. If the non-environmental specialist sees any of these, he is on notice to go further with an inspection by a professional prior to acquisition.

If there is evidence of contamination on the property, it is often necessary to do a more detailed environmental investigation using appropriate environmental consultants. Depending on the nature of the property, the possible contamination, and the need for certainty of result, these inspections and investigations may vary greatly in scope, intensity, and cost. The most complex investigations, which will be similar to a remedial investigation under the CERCLA laws discussed earlier, will include groundwater sampling on the property; upgradient and downgradient sampling to characterize groundwater movement, bore holes at appropriate locations and depth to determine soil contamination, and an intensive chemical analysis of all samples. Costs of several hundreds of thousands of dollars have been routinely quoted for such investigations on industrial property. On the other hand, much simpler surveys may be appropriate depending on the nature of the site.

One word of caution. It is important to determine that the investigation undertaken is appropriate to one's needs. An investigation that is insufficient to reveal potential problems may be a trap for the seller, buyer, appraiser, and the lender. On the other hand, millions can be spent on investigations to no purpose. Confidence in the environmental professionals involved is critical.

REFERENCES

10. Much of the information contained in this chapter and the following chapter, originated with the writings or lecture notes of the following two attorneys specializing in environmental law: Mr. James T. Harrington, Ross & Hardies, Attorneys at Law, Chicago, IL; Mr. Roger Ferland of the law firm of Streich, Land, Weeks and Cardon, Phoenix, AZ. The narrative use of Mr. Ferland's and Mr. Harrington's notes, unless they are directly quoted and so indicated, remain strictly the responsibility of the author.
11. For definition of "hazardous substance," see 42 USC 9601(14). Moreover, EPA must designate additional substances as hazardous which, when released into the environment may present substantial danger to the public health or welfare or the environment. 42 USC 9602(a). A list of all such "hazardous substances" can be found in *40 CFR 302* (1987).
12. The most complete list of hazardous substances of which the author is aware is contained in Corson, Lynn A.; *Master List of Hazardous Substances, Incorporating All Chemicals Included in the Four Lists Cited Lists Cited in the Emergency Planning and Community Right-to-Know Act of 1986;* Environmental Management and Education Program; Purdue University; West Lafayette, IN 49907.
13. 42 USC 9607(a). See *Clean Air Act,* 42 USC 7412(1); *Clean Water Act,* 33 USC 1321(b)(2)(A); and *Toxic Substances Control Act,* 15 USC 2606(f).

14. Subject to certain defenses, all such persons are liable for all costs of removal or remedial action incurred by the United States or a State "not consistent with a national contingency plan," any other costs of response incurred by any other person consistent with a "national contingency plan" and damages for injury to or loss of natural resources. The National Contingency Plan (NCP) details the procedures and standard for responding to release of hazardous substances, pollutants and contaminants. See 42 USC 9605 and *40 CFR 300 et seq.* This quotation is drawn from: Harrington, James T.; *The Legal Implications;* copyright 1988, Ross & Hardies, Attorneys at Law; Chicago, IL.
15. Section 107, 42 USC 9607(a).
16. 42 USC 9606(a).
17. 42 USC 9604.
18. 42 USC 9606(b) provides that "[any person who, without sufficient cause, willfully violates, or fails or refuses to comply with, any administrative] order . . . may . . . be fined not more than $25,000 for each day in which such violation occurs or such failure to comply continues."
19. The courts have almost unanimously concluded that governmental approval is not necessary to maintain a private cost recovery action. See e.g., *Artesian Water Co. v. New Castle County,* 659 F. Suppl. 1269 (D. Del. 1987), aff'd, Nos. 87-3622, 87-3623 (3d Cir. June 29, 1988); *Cadillac Fairview, California, Inc. v. Dow Chemical Co.,* 840 F.2d 691 (9th Cir. 1988); *Pinole v. Point Properties, Inc.,* 596 F. Supp. 283 (N.D. Cal. 1984). However, there are certain circumstances in which private cleanup requires prior governmental approval. For example, when private parties seek reimbursement from the Superfund, the costs must be approved under the NCP and certified by the responsible official. 42 USC 9611(a)(2). Also, EPA approval of a potentially responsible party's remedial action is a prerequisite to 107 recovery when EPA has initiated investigation of the site. 42 USC 9622(e)(6). Finally, as a substitute for 104 government cleanup, the President may authorize a responsible party to carry out the cleanup act provided such party is "qualified" and the action will be performed "properly and promptly." 42 USC 9604(a)(1).
20. The EPA is required under 105(a)(8) to develop "criteria" for determining priorities among releases or threatened releases throughout the United States for purposes of taking remedial actions and to develop and maintain a "list" of such "national priorities." See 42 USC 9605(a)(8). EPA has set forth a structured program for evaluating sites and placing sites on the National Priorities List (NPL) in *40 CFR 300.66.* Basically, the NPL serves as a basis to guide the allocation of Superfund resources among releases.
21. *See* 42 USC 9613(f)(1) and *40 CFR 300*.66(c)(2).
22. See, e.g., *U.S. v. Shell Oil Co.*, 605 F. Supp. 1064, 1083 (D.C. Colo. 1985) (CERCLA permits imposition of joint and several liability); *U.S. v. A & F Materials Co., Inc.*, 578 F. Supp. 1249, 1256 (D.C. Ill. 1984) (Congress in enactment of CERCLA did not intent to preclude imposition of joint and several liability); *New York v. Shore Realty Corp.*, 759 F.2d 1032, 1042 (Congress intended that responsible parties be held strictly liable under CERCLA). *City of Philadelphia v. Stepan Chemical Co.*, 544 F. Suppl. 1135, 1140 (D.C. Pa. 1982) (legislative history clearly establishes Congress' understanding that it was incorporating a standard of strict liability under CERCLA)

23. 42 USC 9601(20)(A).
24. *United States v. Maryland Bank & Trust Co.*, 632 F. Supp. 573 (D. Md. 1986).
25. *State of Idaho v. Bunker Hill Co.*, 655 F. Supp. 665 (D. Idaho 1986).
26. Excerpted from Harrington, James T.; *The Legal Implications;* Ross & Hardies, Attorneys at Law; Chicago, IL; Copyright 1988.
27. 42 USC 9607(b).
28. 42 USC 9601(35).
29. 42 USC 9601(35)(B).
30. 53 Fed. Reg. 37082 (September 23, 1988). These rules became effective December 22, 1988.
31. Harrington, James T.; ibid.
32. Harrington, James T.; ibid.
33. 33 USC 1344.
34. See 33 CFR 320, 326 and *40 CFR 225, et seq.*
35. 42 USC 4321, *et seq.*
36. 16 USC 1531, *et seq.*
37. 53 Fed. Reg. 20764 (June 6, 1988), effective July 6, 1988.
38. A negligent violation of the *Clean Water Act* may result in the imposition of a fine of not less than $2,500 nor more than $25,000 per day and/or a sentence of up to one year in prison. 33 USC 1319(c)(1). Anyone who knowingly violates the *Act* may be subjected to a fine of up to $50,000 per day and/or imprisonment for up to 3 years. A "knowing endangerment" violation of the *Act* is punishable by a fine of up to $250,000 (or $1,000,000 if the violator is a corporation) and/or imprisonment for up to 15 years. 33 USC 1319(c)(3).
39. The sales of pollution credits has already been reported as involving some $5 billion dollars in transactions prior to the encouragement apparently to be offered under the CAA revisions.
40. 42 USC 9603.
41. 42 USC 9603(a).
42. See 53 Fed. Reg. 27268 (July 19, 1988) (proposed rules); 54 Fed. Reg. 29306 (July 11, 1989) (supplemental proposal regarding *Clear Air Act).*
43. See 53 Fed. Reg. 12868 (April 19, 1988) (proposed rule). However, statistically significant increases of those releases must be reported to the National Response Center. 42 USC 9603(f)(2).
44. 42 USC 9601(14). *See Wilshire Westwood Assoc. v. ARCO,* ____ F.2d ____, 30 ERC 1065 (9th Cir. 1989).
45. 42 USC 9603(c).
46. 42 USC 9603(b) and 9609.
47. See *40 CFR Part 302, Table 302.4*. See 54 Fed. Reg. 33418, 33449 (Aug. 14, 1989) final rule adjusting RQs and amending list of CERCLA hazardous substances); 54 Fed. Reg. 22524, 22538 (May 24, 1989) final rule adjusting RQs for radionuclides).
48. *40 CFR 302.56(a).*
49. 42 USC 9603(c).
50. 42 USC 9607(b)(3).
51. Section 101(35)(C) 42 USC 9601(35)(C).
52. 42 USC 9620(h).

53. See procedures at *40 CFR Part 373.*
54. See 42 USC 11004(b) and 11023(a).
55. Sec. 304.
56. 42 USC Sections 11004(a) and 11002(a)(2). See *40 CFR Part 355,* Appendix A (list of extremely hazardous substances and their RQs).
57. 42 USC Sec. 11004(b).
58. 42 USC Sec. 11004. See *40 CFR Sec. 355.40(a)(2).*
59. 53 Fed. Reg. 27268 (July 19, 1988).
60. Sec. 311.
61. Sec. 312.
62. Sec. 313.
63. 42 USC Sections 11021–11023.
64. 42 USC Sec. 6930(a).
65. 42 USC Sec. 6934. See USC Sec. 6927 (Inspections).
66. *40 CFR Sections 270.41, 270.72.*
67. *40 CFR Sections 264.112(d), 265.112(d).*
68. *40 CFR Sections 270.40, 270.72.*
69. 42 USC Sec. 6991(2).
70. 42 USC Sec. 6991(a).
71. *40 CFR Sec. 280.22.*
72. Ibid.
73. *40 CFR Sec. 280.50.*
74. *40 CFR Sec. 280.61.*
75. *40 CFR Sec. 280.62.*
76. *40 CFR Sec. 280.63.*
77. *40 CFR Sec. 280.64.*
78. *40 CFR Sec. 280.65.*
79. *40 CFR Sec. 280.66.*
80. Ibid.
81. *40 CFR Sec. 280.53.*
82. 42 USC Sec. 6991a(a); *40 CFR Sec. 280.22.*
83. *40 CFR Sec. 280.71.*
84. *40 CFR Sec. 280.106.*
85. The following list is taken from *Underground Storage Tanks: Basic Information for Appraisers;* Research Department; American Institute of Real Estate Appraisers; Chicago, IL; December, 1988.
86. 42 USC Sec. 1321(b)(5); *33 CFR Sec. 153.203.*
87. 42 USC Sec. 1321(b)(5).
88. *40 CFR Sections 761.123* and *761.125.*
89. 42 USC Sections 2606(f), 2607(a), and 2602(9).
90. N.J. Rev. Stat. Sections 13:1K-6, *et seq.*
91. N.J. Rev. Stat. Sections 7.1-7.10.
92. N.J. Rev. Stat. Sections 58:10 23.11, *et seq.*
93. Gen. Stat. 22a-1 34-1 34D.
94. Cal. Health & Safety Code Sec. 25500, *et seq.*
95. Sec. 25507; 19 Cal. Code of Regs. Sections 2620–2703.
96. Sec. 25295(a); 23 Cal Code of Regs. Sections 2650, 2652.

97. 15 USC Sec. 77a, *et seq.*
98. Sec. 11.
99. Sec. 12.
100. SA Sec. 11(a).
101. *17 CFR Sec. 230.405.*
102. *17 CFR Sec. 230.408.*
103. 15 USC Sec. 78a *et seq.*
104. Sections 10, 18 and 20 (15 USC Sections 78 r and t).
105. 17 USC Sec. 240.10b-5.
106. 15 USC Sections 78m(a) and (b).
107. *17 CFR Sec. 229.303.*
108. *17 CFR Sec. 229.303(a)(1).*
109. 54 Fed. Reg. 22427; May 18, 1989.
110. See Thomas A. Cole, SEC No Action Letter [Fed Sec. L. Rep. (CCH) 78,962 at 78,813 (Jan. 31, 1989)] for discussion of the difficulty of quantifying remedial action costs.
111. 42 USC 9606..
112. See Thomas A. Cole, *supra.*
113. Rule 303, Regulation S-K, *17 CFR Sec. 229.103.*
114. The balance of this chapter is primarily extracted from the Harrington document, *The Legal Issues.*
115. Ill. Rev. Stat. Ch. 111-1/2 para. 1001, *et seq.*
116. See *supra* notes 8-9 and accompanying text.
117. N.Y. Envtl. Conserv. Law 27-1301, *et seq.*
118. N.Y. Comp. Codes R. & Regs. tit. 6 375.9.
119. For definition of "hazardous substance" under the *Illinois Environmental Protection Act,* see Ill. Rev. Stat. Ch. 111-1/2 para. 1003.14.
120. N.J. Rev. Stat. Ann. 58:10–23.11f(f).
121. P.A. 85-1347. This law became effective when signed on August 31, 1988.
122. H.B. 3462, 85th Gen. Assembly, 1988 Session. This *Act* becomes effective January 1, 1989.
123. 42 USC 7410(a).
124. 53 Fed. Reg. 34500 (September 7, 1988).
125. The Prevention of Significant Deterioration (PSD) program provides measures to protect air quality that is cleaner than required by national standards. See 42 USC 7470, *et seq.* The PSD program is administered through review of certain new sources and modified existing sources. If subject to review, a source must meet certain requirements before it will be allowed to construct. 42 USC 7475.
126. See 42 USC 7501, *et seq.*
127. See *40 CFR 52.*
128. 33 USC 1342 and *40 CFR 121–125.* The NPDES permit performs two functions — it establishes certain levels of performance the discharger must meet and it requires the discharger to report any failure to meet those levels.
129. 33 USC 1317(b). See also *40 CFR 403.*
130. 42 USC 6901, *et seq.*
131. 42 USC 6922–6925. See also *40 CFR Parts 262–267.*
132. See Ill. Rev. Stat. Ch. 111-1/2 para. 1022.01.

133. A *citizen's suit* is a civil action commenced by any person on his own behalf. For example, CERCLA provides that any person may commence a civil action on his own behalf against any person who violates CERCLA or against the EPA when the EPA allegedly fails to act or to perform a duty under the *Act*. 42 USC 9659(a). See also *Clean Air Act* 42 USC 7604, *Clean Air Act* 33 USC 1365; *Resource Conservation and Recovery Act* 42 USC 6972.
134. See 5 USC 552.
135. See *supra* note 10 and accompanying text.
136. See 42 USC 9604(b).
137. See 42 USC 9603(a) and (c). Notice to the National Response Center is required upon release of a hazardous substance and notice "to the EPA is required of the existence of a storage, treatment, or disposal facility by the owner or operator of such facility."

3 PROFESSIONAL LIABILITY AND RESPONSIBILITY[138]

INTRODUCTION

This chapter covers a multitude of professional issues from the point of view of the consultant. For the purposes of this chapter, a consultant is any individual or firm who seeks to advise an "owner, operator, or transporter" of hazardous substances or environmental risks as those terms are generally understood under the Superfund laws. The consultant is advised to discuss these issues with competent legal counsel and this chapter cannot, should not, and does not profess to be legal advice, only an outline of some of the legal issues facing the consultant in the performance of his/her professional duties. Included in this chapter are discussions of client/consultant, attorney-client/consultant relationships, contract issues, inspection concerns, reporting criteria, and insurance concerns.

PRESERVING THE CONFIDENTIALITY OF CONSULTANT FINDINGS

The client will generally want all information to be maintained as confidential because information obtained through performance of the scope of work may be the basis for negotiating a transaction, may be the basis of an enforcement action, could result in criminal liability for the consultant or client, or involve trade secret information that could significantly impair the client's business if revealed. In addition, the consultant needs to protect the client's confidentiality in order to encourage openness by the client in revealing needed information because it enables the consultant to compete in the marketplace, and may be a required obligation of the consultant in a fiduciary sense.

However, the consultant may be placed in a "Catch-22" position by requirements to disclose discovered information under some circumstances. For example, the consultant may need to disclose data to maintain professional standards of conduct, particularly to protect health, safety and/or the environment from imminent damage or harm. If the client makes a claim against the consultant, the information discovered by the consultant may be needed to provide an adequate defense against the claim, or a court may order the disclosure of information by the consultant.

There are some specific guidelines to assist the consultant in knowing what may be revealed and what may not. In particular, the client has the obligation to clearly identify to the consultant that information which is considered to be confidential, and the client may not declare information to be confidential which is available elsewhere. In addition, there are some types of information that must specifically be disclosed by law. For example, almost all of the work undertaken in performing the Phase III ERA known as the "Management Plan" under AHERA is explicitly required to be publicly available under those rules and regulations. Other rules and regulations may have similar provisions such as the Underground Storage Tank rules, although in this case the first and primary obligation to disclose is incumbent upon the owner or operator.

In parallel to considerations of client confidentiality are issues of the ownership of documents that result from the consultant's work. In general, clients generally prefer to own the work that they paid for, but such an arrangement should be clearly spelled out in advance of the commencement of work. One of the main reasons for carefully establishing ownership is the consultant's legitimate concern that the documents might be misused and create additional liability for the consultant.

ATTORNEY WORK-PRODUCT DOCTRINE[139]

Issues of confidentiality are often discussed in terms of the much misunderstood phrase of "attorney-client privilege." Although an extensive discussion of this concept is beyond the scope of this book, suffice it to say that "attorney-client privilege" does not generally apply to an environmental consultant performing his/her functions for a given client even if that client is an attorney. Another concept, the "work-product doctrine" may, however, apply when the consultant is engaged to perform investigative tasks for an attorney.

The purpose of the work-product doctrine is to protect the lawyer's research, analysis, legal theories, and the like, prepared *in anticipation of litigation,* from disclosure. The doctrine applies not only to the attorney, but to his agents and investigators among others, and probably including the consultants that work at the attorney's direction. There are specific limitations to the doctrine that can have a very strong influence on the practices and procedures of the consultant. Empirical data is usually not protected. It is usually only the opinion portion of the work-product that is protected—the interpretation of the data and conclusions drawn from empirical data for example.

There are a number of procedures for the protection of the confidentiality of the consultant's findings under the work-product doctrine and the attorney is the final arbiter as to what procedures should be employed in a given set of circumstances. In general:

1. The confidentiality of the work should be a specific contractual and material condition of the work to be performed.
2. The consultant's contract must be with the law firm, not the law firm's client.
3. Contractual documents should specifically declare that all work is performed at the direction of the attorney and in contemplation that the consultant will be called as a witness in possible litigation.
4. All communications by the consultant regarding the work performed must be cleared by the attorney.
5. All written reports must be directed to the attorney and prepared first in draft for comment; after the final version has been prepared, the draft is destroyed.
6. All documents must be stamped as privileged, confidential, and kept in separate secure files.
7. Meetings with the client without the attorney being present should be limited and held only with attorney authorization.

The foregoing should not be misunderstood. The consultant is presumed to be a professional in his/her own right and competent counsel will not permit or encourage the consultant to participate in a violation of law or ethics. What it should mean is that the consultant's report should be prepared to carefully document the findings in two separate sections:

1. A report section that sets out objective scientific facts regarding the investigation.
2. A separate section that sets out the consultant's opinions, thoughts and recommendations concerning the objective scientific facts.

Most consultant's have some difficulty in recognizing that even apparently very clear statements contained in environmental rules and regulations are not the same as objective scientific facts. For example, even though the AHERA rules and regulations make reference to Permissible Exposure Levels and Action Levels for the number of asbestos fibers in a particular environment, these are not objective scientific facts and their applicability to a particular situation is the subject of competent legal interpretation. The number of asbestos fibers discovered by a particular scientific technique in a specified environment and at a particular time is a scientific fact and should be reported as such.

PROFESSIONAL CIVIL/CRIMINAL LIABILITY CONSIDERATIONS

Statutory Liability

One of the major areas of potential exposure to liability for the consultant is that of becoming an "operator" of a site under the Superfund laws. Many consultants have a strong background of line management experience and the temptation to make decisions in specific situations may be almost irresistible. The making of decisions with regard to the handling, treatment, or disposal of hazardous substances, or the appearance of having made such decisions, can easily expose the consultant to the possibility of status as an "operator"—a situation that must be constantly guarded against not only by the individual, but by the consulting firm itself.

Many states have, or are establishing, technical registration criteria for environmental professionals or are extending the technical registrations of certain professionals to cover environmental issues. A violation of the technical registration criteria can expose the consultant to suspension or revocation of his/her certificate and monetary penalties. In some cases, suspension or revocation of one principal or officer of the firm could result in the same consequence to all principals or officers of the firm, and it will in any case have an effect on the firm's reputation.

The requirement to disclose danger to public health, safety, and welfare is contained in many laws and the obligation to report is often placed on the individual who discovers the situation. The consultant must take great care to avoid, on the one hand, the disclosure of confidential information, and on the other, exposure to a failure to report under the applicable laws.

The consultant as an employer also has specific duties with regard to the health, safety, and welfare of his/her employees. In particular, health and safety plans are required by a number of statutes, notably the OSHA rules, and failure to prepare plans or provide proper training may result in serious consequences. The OSHA Worker Protection Rule covers employers and employees involved in response actions under CERCLA, hazardous waste operations under RCRA, emergency responses to incidents involving handling, processing and transportation of hazardous substances, hazardous waste operation sites designated for cleanup by state or local governments, and work under AHERA. Most states have implemented a plan adopting the federal OSHA standards.

Common Law Liability

One of the major issues with respect to common law liability is that of the standard of care that should be exercised by a professional, particularly with respect to environmental investigations where there are no well-established standards of professional performance or ethics. In general, a professional standard would recognize the skill, care and diligence that members of the profession ordinarily exercise under like circumstances. "Specialists" in a particular field may be held to a higher standard of care. Unfortunately, there is little in the way of established guidance in this area and a number of additional issues must be specifically dealt with on an case-by-case basis until such standards come into existence and are generally recognized.

One of the largest areas of exposure may result from an inadvertent guarantee of specific results that may in turn create a higher standard of performance than can normally be achieved. Compliance with government minimum standards may not be sufficient and more than one method may meet the standard of care. The problem will be to establish exactly what standards of care will be employed and then to carefully document that those standards were, in fact, adhered to,

Another area of common law exposure is the tendency to comment on the work of other "professionals" which may constitute defamation. Defamation may be a fault amounting to at least negligence on the part of the publisher of the information. In general, defamation contains the following elements:

1. A false and defamatory statement concerning another.
2. An unprivileged communication to a third party concerning the work of another.

This does not mean that total silence on the part of a consultant concerning the work of another is required in all circumstances. There are potentially applicable privileges that may protect the expressions of one consultant on the work of another, such as privileged communications in furtherance of a common interest, communications in official proceedings (such as a court), clearly labeled expressions of opinion (provided that the opinion can be objectively supported), and expressions of the truth (if truth can be objectively established).

An emerging area of liability exposure is that of professional malpractice. There are, as yet, few cases against environmental consultants although this is likely to change because of the number of environmental assessments being performed and the magnitude of the losses that can occur. The technical registration rules of professional conduct set standards of professionalism, and failure to comply will be deemed malpractice or negligence. The problem is that the standards of professionalism are being adapted on an ad hoc basis from other fields of professional performance and therefore are not clearly defined or even necessarily appropriate. Some examples of professional malpractice are relatively straightforward, such as performing work outside the scope of employment.

The duty to disclose is an area particularly well-suited to malpractice. The duty to disclose depends on the scope of the work and the extent of the hazard, as well as state and federal laws which place an affirmative duty to disclose the release of hazardous substances in excess of specified quantities on the owner, operator, or person in charge of a facility (''facility'' is very broadly defined). In most instances, the client will own or operate the facility and the consultant will have no obligation to report to regulatory agencies, unless the consultant expressly assumes it. If the client refuses to remedy a situation that presents a serious threat to the public health, safety, or welfare, the consultant will face a particularly difficult situation unless this situation has been contemplated in the drafting of the contract between the consultant and the client.

The trend among the states is to expand the group of persons required to report releases making the resolution of this issue very early in the relationship extremely important. If a specific resolution of the reporting issue is achieved, it should minimize the risk of being held an operator or person in charge on the part of the consultant. The consultant should avoid clients who have serious problems, but who refuse to report.

In general, the setting forth of a precise scope of work should alleviate the problem. One means is to clearly state that the condition will be reported to the client, stating that the client is responsible for reporting and reporting is outside the scope of the consultant's work. In addition, the consultant should not sign manifests or become involved in day-to-day operations. The contract between the consultant and client should contain appropriate indemnifications concerning reporting. Even given all of this, however, the regulations may require disclosure regardless of the scope of work where there is a serious threat. A failure to comply in this case may constitute gross negligence or misconduct or result in claims by third parties.

The consultant must carefully avoid conflicts of interest or even the appearance of conflicts. Implied in every contract is a covenant of good faith that requires the consultant to perform his/her functions in the client's best interest. In particular, the consultant should take care not to acquire an interest antagonistic to the client and has a positive duty to withdraw if such an interest occurs (such as a disagreement on reporting). The consultant does have a right to protect his/her own interest, but should not remain a consultant to the client if the right to protect the consultant's interest conflicts with the duty to function in the client's behalf.

The consultant may also be an agent of the client and as such may be accepting fiduciary responsibility. As a fiduciary, the consultant is charged with protecting and preserving the client's assets and operations from harm and the consultant would be well advised to carefully discuss the very broad implications of fiduciary responsibility with competent counsel.

Most consultants are not attorneys, but conversely, many attorneys are not environmental law experts. The consultant must be very careful not to become engaged in the illegal practice of law in this situation. It is the attorney's place to give the client legal advice and the consultant must avoid the appearance of giving such advice improperly. Unfortunately, there are few clear rules to employ to avoid this situation. In general, the consultant's services should be limited to gathering and interpreting data; it should not include giving advice regarding implications or ramifications under the law. The consultant should avoid using legal phrases or buzz words that could have unintended legal implications.

Criminal Liability

There are expansive provisions of criminal liability under many of the environmental laws to which the consultant may be exposed; for example, as an aider or abettor of an illegal activity or as a co-conspirator to cover up a violation of the laws. This exposure extends not only to the consultant in the field, but possibly to the "responsible corporate official" of the consulting organization.

The potential sources of liability include the federal and state water pollution statutes, the *Toxic Substances Control Act* (TSCA), federal and state Superfund laws, and federal and state hazardous waste laws. For example, TSCA includes severe criminal penalties for deliberate failure to disclose the presence of an imminent threat to human health to those who may be affected by the threat.

In addition, a number of general criminal statutes have been utilized in environmental prosecutions including conspiracy, the making of false statements, the fraud statutes, and the false claims statutes. For the professional, again, the best advice is to make clear to the client in all appropriate communications that the consultant will not participate in a failure to disclose conditions that threaten potential harm to individuals or the environment and to adhere to that policy.

Along these same lines, there are violations which can occur during an agency or criminal investigation including the making of false statements, obstruction of justice, witness tampering and perjury. It is difficult to imagine a consultant being willing to risk a criminal violation, but many clients are unaware of the potential consequences of a violation of the environmental laws and the consultant should be prepared to clearly enunciate the criminal violation nature inherent in many of those laws to the client, preferably through the client's own attorney.

Disclosure of Professional Opinion in Disagreement with Client

Obviously the foregoing leads to the need to consider disclosing information and professional opinion in disagreement with the client. This raises issues of when to disclose, how to prepare for disclosure, and what to disclose. Of equal importance is the issue of when not to disclose. Given a lack of widely accepted professional standards and ethics for environmental consulting, this issue will have to be dealt with on a case-by-case basis, but the general outlines of a policy can be formulated in advance. The policy should be set forth clearly to the client in the contract as a matter of record and thereafter strictly adhered to by the consulting organization.

The general disclosure policy should take into account the specific requirements of the regulations governing the circumstances of the discovery of an environmental risk, such as the required reporting times included in CERCLA/SARA for the discovery of an uncontrolled release or threatened release, *and* the specific obligations of the various parties as to who is required to disclose.

For example, the UST rules and regulations require that a release must be reported within a specified period of time by the owner or operator. The consultant should therefore carefully notify the client of the discovery of the release in a timely fashion, stating the required format of notification and the responsible agency to which the notification should be given. The contract itself should make clear to the client that if the release is not reported within a specified time period then the consultant will be obligated to report the release directly and, if necessary, withdraw from the client's service (remember the duty to withdraw if an antagonistic interest is acquired).

The point is, the environmental consultant would not be hired unless there were a reasonable possibility of the discovery of an environmental problem. Given this fact, it is simple prudence to address the issue of the disclosure of environmental problems at the time of contract negotiation and to provide to the client a clear policy as to when the consultant will be obligated to disclose a discovered problem without the client's consent and the procedures that the consultant will follow both before and after such a disclosure.

CONTRACT ISSUES

The development of a contract between the consultant and client should consider a number of issues, both specific and general. One of the general issues revolves around the question of who should, in fact, draft the contract. In general, it should be kept in mind that the party who drafts the contract is assumed to know the most about the subject matter of the contract, and it is therefore normal for any ambiguity in the contract to be interpreted against the interest of the party who drafted it. Clear, concise language that covers all of the salient points in an unambiguous fashion, but leaves room for dealing with unexpected contingencies in the actual execution of the work, should be the ideal. As usual, the ideal is very difficult to achieve.

Contracts, for reasons that should by now be obvious, are very advisable. A properly thought out and drafted contract should tend to minimize or avoid misunderstandings between the parties. If nothing else, it provides a basis upon which to sue if everything goes sour, but ideally, it will provide such a clear understanding between the parties that suits will not arise. The contract can provide additional protections to both parties where needed, but these protections must be reasonable and legal.

Uniformity of Contract Provisions

There is a major problem in drafting environmental consulting contracts—the lack of uniform standards of professional practice, which must be carefully guarded against and will result in voluminous contract documents. Many attorneys are accustomed to drafting contracts for such services as architectural or civil engineering work where the standards of professional practice are well developed, and will consequently use such phrases as "state of the art" or "Phase I Environmental Audit," not realizing that these terms do not have generally accepted meanings in environmental consulting work. This means that these, and many other terms, phrases, and conditions of the work to be performed must be carefully spelled out in detail in the contract documents.

Without widely accepted standards, a consistent approach within the industry or within a firm may be difficult to achieve. The development of published standards for the work of a given firm, supplying those publications to the client, and carefully ensuring that the client understands the standards of the firm prior to signing a contract, is at least one method of dealing with this issue. This will normally cut down on the amount of negotiation required to reach an agreement between the parties. It will also tend to raise the level of sophistication of even the simplest contracts and reduce the chance of overreaching by one party. Note, however, that the use of published standards of practice, even those published only for the firm, determines in part the level of performance of the work for the execution of the contract that may, or may not, be applicable to a given situation. Careful review of standard contract documentation is required by both parties in light of the specific circumstances in all cases.

Contracting With a Law Firm as the Client

Contracting with a law firm as the client is substantively different from dealing with a non-law firm client. While contracting with a law firm can provide greater protection from disclosure for the ultimate client—and sometimes for the consultant—the consultant must always be aware of the fact that the law firm does not represent the consultant. A law firm can substantially assist in the solution to performance/contract questions, and if it is an experienced environmental attorney, often reduce significantly the consultant's concerns over disclosure.

Generally, the law firm as the client, will require that reports and opinions (written and oral) be given only to the law firm. This may create some tensions for the consultant on the site due to the natural desire of the law firm's client (the owner or operator of the site, for example) to know what is going on, and the consultant should be prepared for this. Ideally, the law firm will have already carefully briefed its client on how the work will be conducted.

In general, all actions must be taken at the direction of the law firm. The consultant must carefully avoid performing work at the direction of the law firm's client even though they may be in charge of the site. The attorney will have to determine what work is to be undertaken, what methods are to be used, and how the samples, data, reports, and other information are to be handled, not the attorney's client. The protocol should be to refer the client to his/her attorney for discussions of work to be performed that is not already covered by the attorney's instructions to the consultant.

The law firm will expect exclusivity; that is, the consultant should not have any other clients who are involved in the specific matter. The obvious reasons are to protect confidentiality and guard against potential conflicts of interest.

Contracting With a Non-Attorney Client

In many respects, contracting with a non-attorney client is a more difficult situation for both practical and legal reasons. The client may be considerably less sophisticated in environmental consulting, generating both problems in practical understanding and communications between client and consultant, and an additional burden placed on the consultant by the common law due to the fact that the consultant is more sophisticated (for lack of a better word). In general, a contract is ideally made between two equally knowledgeable parties who are equally balanced with respect to the ability to negotiate with one another. To the extent that these conditions are not true, there is a tendency in the law to construe the terms of the contract in favor of the weaker or less sophisticated party.

With a non-attorney client, the definition of the scope of the work to be performed must be drawn with extra care to ensure that the client understands exactly what is to be done, how the work will be performed, how the scope of the work may be altered as the work progresses, and to carefully inform the client on the issue of disclosure. Particularly with an unsophisticated (in environmental matters) client, the consultant must take great care to explain and document all of the conditions surrounding the work to be performed. Again, as a reminder, ambiguity will generally be construed against the drafter and against the more sophisticated party to the contract.

The First Contract Issue

The very first consideration on the part of the consultant should be the answer to the question: ''Should I enter into any relationship with this potential client?'' Obviously, this will involve considering whether the client is reputable and financially sound, and if there is a risk, whether it's worth the reward. In considering the client's reputation, special attention should be given to the possibilities surrounding the disclosure issue, keeping in mind that disclosure of an environmental problem may have significant negative financial consequences for the client. If the client is at all likely to resist disclosure, the risks to the consultant may overshadow the financial rewards.

Scope of Work

The scope of the work to be performed is probably the single most important part of the contract. The goal is to be as specific as possible while leaving room for flexibility. This part should contain a very careful description of the field work and analyses to be performed and, particularly, what is *not* included, e.g., sampling costs, laboratory costs, alterations of the scope of work depending on information discovered, and responsibility for reporting to agencies. Also, if special authorizations are required for the activities (drilling permits, local licenses of all types, prior notice to agencies, etc.), who is responsible for obtaining them and who bears the costs.

In general, clients prefer that the cost elements be broken down. In addition, the client will want a realistic time schedule and timely notification if the time schedule will have to be changed for some reason. The relationship between the client and consultant will involve a large element of mutual trust and respect to be successful, and one of the indicators is reliable adherence to cost and time schedules.

The contract should contain a realistic and reasonable description of the expected output and specify the time deadlines for drafts and final reports (if necessary, build in flexibility for lagging lab reports). The specifications should clearly establish who will receive the output on behalf of the client. With respect to conclusions and recommendations that can be expected, the consultant should make clear the realistic expectations associated with the various levels of investigation. A Phase I ERA will not provide the information required to develop precise engineering cost estimates for remediation. An implication that the work will result in a higher level of output than is justified will simply result in problems for both parties.

The issue of disposal of waste and contaminated equipment should be carefully considered in the scope of work. Disposal policies will usually depend upon client's status, e.g. whether the client has an EPA generator ID number. The risks to the consultant of disposal should be considered as it may establish the consultant as an agent of the client, or possibly an operator, and additional indemnification by the client to protect the consultant may be necessary.

Clients will often seek to reduce the costs of environmental work by asking a consultant to use previous inspection reports. This matter should be carefully considered, particularly if the quality of the earlier work will be difficult to establish. Again, the consultant is reminded of the implications of the negative default principle and its particular application to sampling and testing. Many of the older investigation reports may not have observed all of the appropriate sampling and testing protocols to establish that a suspected substance is not a hazardous substance and acceptance of an incorrect conclusion may be possible. The acceptance of a positive finding, that is that a substance is a hazardous substance, does not contain as many risks for the current consultant as the acceptance of a negative finding. In general, the use of previous inspection work should be qualitative as a guide to assist in the current inspection, but it should probably not be accepted for quantitative use unless the quality of the work clearly warrants this.

Consultants are not normally hired to simply investigate, but to provide expert advice, particularly in the areas of hazard assessment, remediation priority setting, remediation option selection, and remediation cost estimation. Each of these areas requires the expert knowledge of the consultant and the exercise of considerable judgment in areas of risk assessment and estimation that have, as yet, relatively few guidelines. In fact, one of the major purposes of this book is to provide a number of quantitative tools for the consultant to use in precisely this area. As a contract issue, the consultant should carefully establish the mechanics of how such decisions will be made, who has ultimate decision-making responsibility, and particularly the consultant's position as adviser as opposed to being the decision maker.

Clients often ask for consultants to recommend contractors for the work to be performed. The consultant should be aware that the less sophisticated the client, the greater the consultant's exposure if the recommended contractor does not perform the work correctly, and be guided by this thought in making selection recommendations. In general, it is probably better to list generic selection criteria than to name contractors unless the consultant is very certain of the contractor's performance history. If names are submitted, it would probably be best done in the form of a list of reputable contractors as a starting point for the client's selection process without favoring any one of the contractors on the list.

Although not generally well publicized, many of the environmental rules and regulations allow the financial concerns of the client to be taken into account in the selection and timing of remediation options. The contract should clearly establish that the consultant is not in a position to determine if the budget provided with which to do remediation work is appropriate to the magnitude of the problem or the ability of the client to fund the work. The methods developed in this book for the selection of remediation strategies are based on a budgetary figure, supplied by the client, as one of the constraints in determining remediation strategy. It should be clearly and unequivocally the client's responsibility to determine if the amount provided is appropriate under the circumstances.

The Professional's Daily Records

Even if the scope of work does not specify, everything should be documented and records kept. This particular will apply to decision making; why the decision was made; what standard practices were involved in making the decision; if standard practices were not available or appropriate, what factors were relied upon; who was talked to (name, phone numbers, addresses, date and time); what records were generated or reviewed, and the like.

One of the best and most useful means of accomplishing the above is the professional's personal daily log of activities. Such a daily log should meet several basic requirements. First, it should be a bound book with prenumbered pages in which the professional records events as they occur. The notes in the book should never be erased, if a mistake occurs then a single line, possible with a notation, should be drawn through the incorrect data but it should not be totally obscured. This daily record should be the personal property of the individual and kept as a record of daily activities for a considerable period of time (possibly as long as 20-30 years) and not surrendered except under unusual circumstances. The professional should probably consult an attorney for further advice on the keeping and preservation of such a daily log of activities as it can be extremely important later.

Limitations on Liability

Warranties and Disclaimers

Warranties are intended to provide a level of assurance from one party to another concerning the work to be performed and the methods to be used. The objective is to assure the party receiving the warranty that the results will be appropriate and satisfactory with respect to the objective for which the work is carried out. Warranties will traditionally make reference to the applicable professional standard of care, something which should be carefully considered for reasons already discussed. Keep in mind that an ambiguous warranty, like any other ambiguous wording in a contract, will generally be construed against the drafter, and that improperly or carelessly worded warranties can subject the consultant to a higher standard of care than may be necessary, desireable, or even possible.

Disclaimers are of only limited effectiveness in limiting liability and are generally not favored by the law. Under certain circumstances, they are valid and inforceable, and the factors that are likely to be considered in making this type of determination are:

1. The relative bargaining position of the parties. Specifically, does one party hold a position of unusual strength in bargaining with the other over the terms of the warranties?
2. The nature and subject matter of the agreement. Is this an agreement that is unusual with respect to the work to be performed for one of the parties? This has been referred to as the relative sophistication of the parties in the earlier discussions.
3. Is the language clear and conspicuous or is it ambiguous with all that implies for the party who drafted the language? Does the languague include the word "negligence" and if so, how does the warranty limit or define negligence? Generally, the parties may not disclaim liability for harm caused by intentional or reckless conduct.
4. The warranty must be bargained for by the parties. The warranty is not a given, and evidence of bargaining can assist in establishing that the parties clearly understood the terms and conditions of the warranty. The bargaining process should also assist in establishing the value of the warranty to the parties.

Similar to, but not the same as warranties, are the client's acknowledgments of the conditions surrounding the work as a limiting condition on the work to be performed. Such acknowledgements, provided that the client understood the terms and conditions of the acknowledgment, may be more effective than a warranty in certain circumstances.

Warranties and acknowledgments do not relieve the consultant of the duty to exercise due care. They may ensure that the client understands that the result is not "guaranteed." The provision for warranties and acknowledgements, by the process of bargaining for their inclusion and their language content, may be used to support the argument that the transfer of risk accomplished by an indemnity provision was "brought home" to the client.

The federal and state environmental statutes may have a strong influence on the enforceability of warranties and disclaimers. In particular, a warranty or disclaimer that is contrary to the environmental statutes may be worse than meaningless; it may be misleading. The reporting requirements, for example, cannot be avoided through a disclaimer or warranty, but the giving of a warranty in this area may lead one or both parties to believe that the other will be responsible for dealing with the issue, and neither do, exposing both to the problem.

Disclaimers of liability to third parties are particularly difficult, especially now that environmental investigation reports are becoming a relatively standard part of property transactions. Despite language to the contrary, it is likely that if a consultant provides an environmental report to a specific client, that client will very likely use that report to demonstrate to a third party—totally unknown to the consultant—the status of the property. In addition, some states do not require privity; that is, third parties such as labor organizations may claim a right of access to the consultant's report or damage based on the withholding of information contained within the report. These issues should be carefully discussed with competent counsel before entering into a contract.

Indemnity and Hold Harmless Clauses

Indemnity and hold harmless clauses are generally viewed by the law in much the same manner as warranties and disclaimers, without a great deal of favor. They are, however, enforceable under certain circumstances. The factors to consider are generally the same as provided above for warranties and disclaimers.

Indemnifications and hold harmless clauses are generally not enforced to indemnify against intentional torts, criminal acts, or conduct deemed to be contrary to special statutory enforcement policies. State and federal environmental laws can greatly influence the value of an indemnity. Examples of fair indemnification provisions would be when each party indemnifies the other for their negligence or when the client indemnifies for existing contamination. Normally, indemnifications are limited to the value of the contract or insurance policy.

INSURANCE

In general, there are four types of insurance of direct concern to both the consultant and the client; (1) workers' compensation, (2) automobile liability, (3) commercial general, and (4) errors and omissions. Each of these types must be considered with respect to the work to be performed.

Workers' compensation insurance generally covers occupational diseases and work-related disabilities. It may be the only form of insurance that will cover a worker dealing with hazardous substances and generally has state-mandated limits of coverage.

Automobile liability insurance is essentially what its name implies; coverage against liability arising from or connected with an automobile. This coverage can sometimes be of a very general form, but should be in place to cover the travel of the employee related to the work to be performed.

Commercial general (used to be called comprehensive general) covers bodily injury and property damage resulting from the actions of the insured. However, most commercial general policies contain absolute pollution exclusion clauses and as such, will not normally cover liabilities arising from damages related to hazardous substances. In addition, commercial general policies do not normally cover errors and omissions on the part of a professional. Most commercial general policies are also limited to a claims-made basis of coverage which can severely limit the value of the policy.

There are two very general and broad classifications into which policies will fall relating to when a claim may be filed to insure coverage: (1) occurrence, and (2) claims made. Briefly put, occurrence policies provide coverage for any covered incident that occurred during the period when the policy was in force without regard to when the claim for damage is filed. This type of policy is extremely rare today although it was quite common in the past. The claims-made policy, on the other hand, only covers incidents that both occurred during the policy period *and* the claim is filed within a specified period of time. Claims-made policies are the common form of general liability policy today and unless other specific provisions are elected, require that the claim be filed within the same time period as the policy is in force.

Claims-made policies are often modified by amendments called "tails" that may extend the *filing* period backward or forward in time. For example, a claims-made policy with a one-year filing tail does not extend the period of coverage during which the incident occurred, but does allow an additional year after the policy period in which to file a claim for damages arising from an incident that did occur within the base policy period. A backward tail extends coverage for filing to incidents that occurred before the policy period and is a very rare form.

Associated with commercial general liability policies may be a so-called "umbrella" policy that provides coverage after the primary policy coverage has been exhausted. Such policies rely on the primary policy to provide coverage for damages arising from a covered incident up to the limit of the primary policy and then take over if the amount of the claim exceeds the primary coverage maximum.

Errors and omissions (professional liability) coverage is probably the most sought-after and difficult to obtain form of coverage in the environmental field. This type of insurance is intended to cover damages or losses caused as a result of an error or omission in professional practice. It is normally offered in a claims made form and the client and consultant need to be aware that most such policies contain pollution exclusion clauses and will not generally provide the coverage needed by either party. There are a number of reasons for the unwillingness of insurance companies to offer this type of insurance to the environmental professional, but one of the major ones is a lack of recognized standards for the performance of the work, a condition that is expected to continue for some time in the future. Both the client and the consultant should be aware that, if obtainable, this form of insurance may carry a premium equal to the total amount of the coverage, implying that it may be both less expensive and more efficient for both parties to look at other forms of assurance that the costs arising from mistakes are covered.

It is normal for clients to be named as additional insureds on commercial general and errors and omissions policies as a means of assuring the client that claims will be paid to them if valid. In addition, naming the client as an additional insured will provide direct assurance to the client that the insurance coverage desired is in force.

The key issues to keep in mind concerning insurance coverage are to carefully read the policy to understand precisely what is covered, the terms and conditions that will maintain the coverage, and the financial and claims payment record of the company that issued the policy.

Reading an insurance policy is often best left to experts in the field, but as in real estate, the promises of the agent mean virtually nothing; it is what is said in writing in the policy that counts. Policies often appear to say one thing in the preamble or title that is completely set aside by the actual language of the policy itself. In addition, specific clauses in the policy may preclude important areas of coverage. For example, a commercial general policy might exclude claims made by occupants or tenants of the property, or it might preclude *anticipatory damage claims* (claims made based not on actual physical harm already experienced, but damage or harm that is anticipated as a result of some event).

No matter how adequately the policy addresses the issues of concern, it is worthless if the issuing company does not have the assets with which to pay the claims or has a record of claim denials for arbitrary reasons, or is in a position where it cannot be forced to pay a claim (for example, all of its assets are in a country that will not allow a suit to force payment). These, as with the reading and understanding of the policy itself, are issues best left to experts for determination, but a determination is critical to both the client and the consultant.

REFERENCES

138. Much of this material originates with notes on this subject supplied by Mr. Roger Ferland of the law firm of Streich, Lang, Weeks, and Cardon of Phoenix, AZ. The narrative use of Mr. Ferland's notes is strictly the responsibility of the author.
139. FRCP 26(b)(3).

4 ENVIRONMENTAL RISK MANAGEMENT POLICY

INTRODUCTION

In many senses, this chapter is for the education of senior management as well as for the direct line manager of environmental risks. Experience has shown that a number of Environmental Risk Management Programs (ERMPs) are inadequate to the situation because senior management, not line management, has failed to fully appreciate the issues.

AHERA was chosen as a model for the development of an environmental risk program simply because it is one of the best explicit models of a long-term environmental management and control system existent in the law today. AHERA includes excellent environmental risk management principles in implementing rules and regulations.

Quite often these statements come as a surprise because AHERA is generally perceived by the public as an immediate removal law. AHERA explicitly states in the preamble to the rules and regulations at *40 CFR 763* that it is assumed that a school will locate, identify, and maintain control of the asbestos inventory until it is removed during the course of normal renovation or prior to demolition. Although removal and disposal prior to major renovation or demolition is not prohibited, it is not required nor encouraged by AHERA.

The major reason for introducing public relations as the first major topic in this chapter is that public relations sets out one of the fundamental management issues for environmental risk. Recent legislation, promulgated rules, and court cases point toward the existence of a defined right to know of any individuals or entities that may be affected by an environmental risk; a right to know of the existence of that risk and the action being taken by the controlling entity regarding that risk.

Except for limited considerations revolving around such issues as trade secrets or national security, the bottom line is that anyone who may potentially be affected must be informed fully and completely. Under AHERA, this right is implemented through the notification of concerned parties[140,141] of the existence and location of the management plan and the guarantee of the right of nearly unrestricted access[142] to the detailed documents concerning the operation of the management plan. AHERA further defines the requirements of the management plan in a way that any action involving the inventory—even many that only involve the inventory in a subordinate fashion—must be documented and that documentation must then become a part of the management plan.

Whether the right to know is established by AHERA, the SARA Community Right-to-Know provisions, the OSHA Hazard Communications Rules, the *Illinois Responsible Property Transfer Act,* the *New Jersey Environmental Cleanup Responsibility Act,* or good management practice, is immaterial. The point is that notification and access concerning the environmental risks are essentially an established public right and must be addressed as one of the most important management policy issues.

PUBLIC RELATIONS — INTRODUCTION

The management of environmental risks requires an understanding that often the major part of the problem will result from a difference between perceived and actual risks in the mind of the public. The scientific, engineering, and business facts of a situation often have very little to do with the concerned publics' perception of the risks. The actual costs borne by the entity for the control of—or a failure to control—environmental risks may be substantially influenced by the publics' perceptions.

The NIMBY (Not In My Back Yard) syndrome is very often based on the publics' perception of the risks and the ability of the controlling entity to effectively manage those risks, even though the arguments are presented on the basis of apparent facts.[143] Critical to the success of an ERMP are two key elements: respect for the potential impact of a risk, and respect for the publics' perception of both the possible impacts and the ability of the entity to successfully control the risk.

The use of "publics" as a plural noun is deliberate. There are a number of publics that will be involved in environmental risk management, ranging from employees, owners, lenders, and management to concerned citizens, the media, the courts, and the regulators. Each of these have their own point of view that must be addressed to some extent in the development of an ERMP. Some of the relationships are governed by the laws as have already been mentioned and will be discussed further. Other relationships are based on proven practice with respect to employees and the general public.

PUBLIC DISCLOSURE AND PUBLIC RELATIONS

Legal requirements, ranging from the ''Community Right-to-Know'' provisions of SARA to Proposition 65 in California, place the direct and explicit burden on organizations to disclose information about potential hazards resulting from risk sources. Simultaneously, customers, workers, communities and those having a financial stake in the organization are asking questions about risks and acceptable levels of risk.

Activist groups are encouraging these regulations for both strategic and tactical reasons. From a strategic point of view, they believe that disclosure is consistent with the public's ''right to know.'' From the tactical point of view, they believe that disclosure will cause organizations to alter their behavior and significantly reduce potential risks.

PUBLIC PERCEPTIONS

Technology, and technological advances—especially as they regard hazardous materials—are viewed by the public as mixed blessings. People enjoy the benefits of technological advances, but feel that some of them may threaten disastrous harm to workers, the community or the general environment; a position that has some amount of truth to it. Underlying these feelings are a number of complex processes, ranging from the discredited philosophy of ''stonewalling'' the issues, or outright denial of their existence, to a fundamental lack of understanding on the part of the public (and even the scientific community) as to the mechanisms by which risk sources can become threats to human health and/or the environment.

With respect to public perceptions, several generalizations may be found in both the academic and popular literature.

Facts and Logic

Facts, or the use of logic based on those facts concerning risk sources, do not always reduce or eliminate public concerns. Studies show that the perceived risks from highly publicized causes of death (natural disasters, airplane accidents, cancer) are generally overestimated by the general public. More commonplace causes of death (auto accidents and heart diseases) are underestimated.

The implication of this is that disclosure processes cannot rely on the assumption that facts will control or alleviate negative reaction without regard to how compelling those facts may appear to be. The emotional aspects of risk disclosure can be powerful and overwhelming. To say that properly trained and equipped workers are safer in an asbestos-containing environment than they are smoking tobacco is factually true. However, it is not likely to convince the spouse of the worker that smoking presents the greater danger.

Basic Risk Attitudes

The reinforcement by friends, family, fellow workers and respected public officials cause the basic attitudes to be very difficult to change. The perceptions of risk, built and reinforced by a number of complex factors, shape the way in which new information and new risks are interpreted and understood.

This implies that one of the primary objectives of risk communications is to establish trust in the organization's commitment to contain and control risk, that is, its creditability. This objective is fundamentally different from the one often perceived by most organizations; that of changing the public's opinion about the size of the risk. Efforts to tell the public why the risk of a well-known hazard is small, generally backfire or fail. Efforts to convince the public that the organization is committed to the implementation and operation of systems designed to control or reduce risk tends to result in a public attitude that the organization is genuinely concerned and is sharing, not minimizing, their concern.

Zero Risk

Often public reaction seems to indicate a strong desire for a "zero risk" solution. However, there is little real evidence to support this idea. Generally, the public is seeking a reasonable trade-off between benefits and risks.

A demand for zero risk may be a result of failure to establish a bond of trust between the organization and the public, and/or a failure to clearly explain the benefits that are gained in return for a reasonable level of risk. Neglect of either of these factors may result in demands expressed in terms of zero risk.

Credibility of Information

When risk information is presented by or on behalf of self-interested parties, the public will respond with caution. Information presented by an apparently disinterested party is likely to receive a more positive response.

The organization's credibility can only go so far, and it is often necessary to accept the risk estimates of an apparently disinterested third party as part of communicating the organization's concern and the value of its control programs. This may mean the involvement of the major disinterested parties in the design and development of the ERMP in order to obtain their endorsement on its benefits and performance.

Emotional Symbols

When dealing with the environmental risks, scientists and engineers speak in terms of probabilities (e.g., the expected number of incremental cancer deaths resulting from occupational exposure to asbestos). While these are the scientific facts, they will be swamped by the emotional result of one well-publicized death resulting from the hazardous substance of concern.

Emotional symbols are essential parts of communications. The building owner who maintains his/her offices in the same building containing the "worst" asbestos problems is making a positive statement about the risks involved. The presence of an environmental auditor in a "moon suit" is also making a statement, but one that may be highly undesirable and very difficult to counteract. On the other hand, insistence that workers who will deal with hazardous substances routinely go through serious training and retraining and are disciplined for failure to observe proper safety precautions makes a strong and continuing positive statement.

KEY ELEMENTS OF RISK PERCEPTION

There appear to be two primary factors in the perception of risk by the public: control, and the ability to understand. In general, the public tends to underestimate the risks where they feel that they are in charge and can make decisions on whether or not to take the risks and how the risk factors will be dealt with. The ability to understand the risks implies that risks from sources perceived as unknown, unseen, or unfamiliar are more likely to generate negative reactions. Risks from familiar or well-known sources, particularly ones that can be seen without difficulty, do not usually generate strong negative reactions. A simple but clear example would be the fear most people would have in approaching a working sawmill blade. Those who work around that same giant blade, even having seen or experienced the damage it can do, have a much milder reaction. The difference is familiarity.

The implications are that communications involving hazardous materials must try to use techniques that create a sense of both familiarity and control. A feeling of control can be developed by encouraging dialogue and a willingness on the part of the organization to respond to public concerns. A feeling of familiarity can be developed by education and prompt, reliable, and accurate information.

GUIDANCE FOR COMMUNICATIONS

Risk Statistics and Public Concerns

A common technique is to compare a known risk to a less well-known risk. This might take the form of a statement that driving to the meeting contained a greater risk of death or injury than . . . .

Few such comparisons will work to actually assure the public that a risk is not worth worrying about and often these types of assurances will seem flip and insincere. It is generally useless to compare a familiar and controlled risk (e.g., driving) to an unfamiliar risk, or one over which the concerned public has little or no control.

The most useful risk comparisons are those that compare the organization's performance to some objective or reputable third party's standards of performance. This places the organization's position in a context of familiarity for the concerned public and does not require the organization to justify the very existence of the risk factors or the standards of comparison. This is particularly powerful if the organization is not only performing better than the standards, but is committed to further improvement of its performance.

In the event that the organization is not exceeding the accepted standards, it must be prepared to demonstrate what it is doing to improve its performance and when it reasonably expects to exceed the standards.

Recognition of Emotions and Risk Perceptions

The public may be generally said to know what scares them. By understanding what scares the public and accepting the fears as real and legitimate, the organization can develop specific responses to those fears. Generally, the public will range from a total lack of concern to an unreasoning level of concern. The communications program must seek to convince those in the middle of this range that, while there is a basis for their concern, the risks are understood and under control. There is probably no way to convert the truly fearful; the objective is to convince all others of the organization's concern and ability to deal with the risk factors in a responsible and meaningful fashion.

A major area where emotions govern more than any other is when the familiar suddenly appears to become threatening. When this occurs, sympathy, not reasoned facts, is the primary mechanism of addressing the problem. Later, after the public has gotten over its initial emotional reactions of denial, anger and depression, the issue can be dealt with based on facts; particularly facts that show an aggressive program to squarely face the problem and solve it in the best and most responsible fashion available.

The discovery of asbestos in a popular and heavily occupied office building can generate exactly this type of situation, and the communication program must be initially based on a willingness to listen to the expressions of fear that may be generated. Later, the facts of the situation and the action program to deal with the problem can be communicated.

Benefits and Risk Communications

Individuals tend to make risk-benefit decisions unconsciously rather than explicitly. Similarly, a community's willingness to tolerate the risks will be influenced by the public's attitudes with respect to the contributions of the source of the risk to the community's welfare. This attitude will only rarely be expressed explicitly and the organization must therefore establish a history of explaining general community benefits in order to have a reasonable balance. To attempt to define the benefits explicitly at the time of risk-benefit decisions on the part of the community will generally not work.

An essential part of a successful risk communications program is the ability to start from a position of established and well-understood economic and social benefits to the community on behalf of the organization. These economic and social benefits should not be discussed in connection with the risk.

Effective Communications

Risk communications studies have identified several trends:

- The media is interested in strong statements, not couched statements.
- The media is more interested in the politics of a risk issue than in the science of a risk issue.
- The media is most likely to cover the human side of the story, not the statistical side.
- Materials for distribution to the media should include:
  - Strong visual aids.
  - Clear backgrounders on the organization and its benefits to the community.
  - Understandable information about the organization's toxic/hazardous substances; how these substances are used; how they are handled, and emergency procedures.

Building Trust and Credibility

A basic rule of crises communications is to devise a message as soon as possible which answers the question "What are you going to do about it?'' The more a communicator can focus attention on solving a crisis problem, the more likely he is to convince audiences that the crisis will pass and the organization can be trusted.

The most credible message the public can hear from an organization is not that the risk is small, but that the organization is working hard to manage and minimize it. The key messages should include:

- Efforts taken by the organization to reduce risk over time.
- Efforts the organization intends to take to reduce risks in the future.
- Efforts the organization is currently taking to manage risks.

Build Understanding and Confidence

No group is more important in meeting the Community Right-to-Know challenge than an organization's employees. If the employees are doubtful and uncertain about the safety and health situation, no amount of external communication will build community confidence. Employees are both the formal and informal ambassadors to the community. This implies the need for an internal communications program that should use normal internal channels.

Communicators should design internal programs which bring employees up to speed on the disclosures which they will ultimately hear about from the outside. Where appropriate, enlist key employees as spokespersons for the organization on the commitment to control dangerous substances in the workplace.

Third Parties and Risk Communications

Developing third party support and validation for the basic risk messages of the organization is essential. This support should ideally come from:

- Fire and police officials
- Medical authorities
- Environmentalists
- Relevant academics
- Political leaders
- Union officials
- Regulators

These individuals should be personally briefed on the basic information which will be released under Community Right-to-Know and on the key messages. Communicators and organization officials should launch a program of briefings of key outside experts who would carry weight with the media and the community as credible sources of guidance on risk issues.

Communications with the Public

Much of the driving force of the emotionality of risk perceptions has to do with the issue of control. Risks which involve high levels of personal control tend to be more acceptable and generally underestimated. For risks imposed on a community by an organization, the essential problem is a public sense of a lack of control—a sense that what is happening behind the organization's walls is unknown and is creating new health and safety risks for the individual and the individual's family.

The mechanisms of Community Right-to-Know are, in essence, new control mechanisms which, on the positive side, will give the public a greater sense of confidence that the community does indeed control its own destiny. To that degree, the success of the law could ultimately be measured by the fact that disclosure finally leads to reduced fear because it offers the chance to create the crucial elements of familiarity and increased control.

But to reach such a happy outcome, it will be essential that the organization's stance be one of openness and a willingness to let the public inside the gates to look around and be satisfied by what they see. This requires that the organization respond positively to the spirit of disclosure and be willing to listen and respond positively to public concerns.

Equally important, the organization must never seem to shirk its responsibilities for the risks it is effectively imposing on the community. Even if the risk seems small, the organization must implicitly recognize one fundamental reality: It is in fact adding to the risks faced by that community.

Communicators should constantly check and recheck their response time to community inquiries. Communications techniques which emphasize openness (tours, community forums, briefings, etc.) should be part of any external communications program. In addition, organizations must be ready to join with other organizations in their areas to deal with community risk concerns which transcend but nonetheless involve their own facility.

THE OVERALL OBJECTIVE: PUBLIC UNDERSTANDING AND TRUST

The final objective of risk communications is to create a level of public comfort. Both research and experience suggest that this objective is best achieved not by attempting to create greater public understanding of the size of a risk, but by creating greater confidence that an organization can be trusted to manage the risk.

Creating such trust requires time, patience, and a great deal of work:

- You must build your equity position in public esteem for your contributions to the community.
- You must be ready to address with sympathy any outpourings of outrage that may be the starting point of a dialogue, but need not be the end point.
- You must always be willing to share the public's concern rather than deny its legitimacy or importance. The organization which seems to understand how the public feels about a risk is the organization most likely to be trusted.

In the final analysis, the best risk communications will not come from words but from deeds, from the actions that people see an organization take which adds value to community life and which protect the community from harm. Organizations that win public confidence through such deeds will turn out to be the most effective communicators on risk issues. They will be saying to the public that they are not different but that they share everyone's hopes and fears. They will not be separate from but will be seen as an essential part of the community, and therefore what they do will be acceptable to the public because, above all, they will be one of "us."

THE POLICY REQUIREMENTS OF ENVIRONMENTAL RISK MANAGEMENT

Any ERMP must begin with a deliberate set of decisions on the part of the controlling organization, decisions that are required in order to develop and implement a program. Good management practice, many environmental laws, and simple self-defense require that these policies be in writing and that they begin with a statement of the fundamental purpose and objectives of the program.

Program Objective Statement

The objective of the ERMP was set out in general terms as "maximization of risk reduction at minimum feasible cost." The prime objective is to reduce risk to human health and/or the environment given the constraints of technology, economics, and local circumstances as they exist for the particular entity in control of the environmental risk.

The absolute elimination of environmental risks is not required either by the strategic objective of the ERMP, by the law of the land, or is such an objective generally within the realm of scientific or technological possibility. Conversely, failure to reduce risk is not acceptable.

The reduction of risk is placed into three separate but related contexts:

technology, economics, and local conditions. If it is technologically feasible to reduce risk, the reduction must be considered. Given the economics of the controlling entity, risk must be reduced to the lowest level. Given local conditions, the risk must be reduced to the lowest level. It is the relationship between the last two statements that can cause the greatest problem for the controlling entity.

If the local conditions are such that risks to human health and/or the environment are high or will increase without the application of some positive action on the part of the controlling entity, action must be taken and the economic concerns become subordinate. It is not the economics that govern, but the local conditions as they relate to the level of risk.

The controlling entity must balance the requirement to reduce risk with the technological capabilities of the available risk reduction methods, the need to reduce risk based on the local conditions, and the entity's ability to fund the risk reduction. The formal recognition of this balance will become the governing policy for the ERMP and should be stated in as clear, concise, and positive manner as possible.

NOTIFICATION

Notification of the existence of an environmental risk has both strategic and tactical purposes. On the strategic level, the belief is that the right to know is fundamental to the community that may be affected. On the tactical level, the regulators and environmental advocates intend that notification will generate a form of self-policing action by the party responsible for the risk. From a management point of view the right to know may as well be an assumed fundamental condition of environmental risk management, but a condition that has been pointed out as having strong positive benefits as well as negative ones.

Notification shall be defined in the broadest possible context. Notification includes formal written notices, press releases, formal and informal informational briefings, and training ranging from simple "don't touch this" to complex formal health and safety training.

Who to Notify?

The first major decision must concern who to notify. Certainly the requirements are partially governed by the laws, such as the requirement for annual hazardous release reporting under SARA, the provision of a Manufacturer's Safety Data Sheet under OSHA, or the access rights to the management plan under AHERA.

A careful analysis of who might be affected by the existence of an environmental risk is discussed later, but upon completion of such an analysis, the notification policy should define when and how those potentially affected publics will be notified. The argument for open, honest, and straightforward communications of environmental risks has already been made, but management should make the explicit policy decisions required to implement such communications in the very beginning of the development of a ERMP.

The major publics to be notified would certainly include all potentially affected employees, surrounding property owners and residents who may be affected, guests or "short-term workers," and the users of products who may be affected by environmental risks contained within the products or their application. This would naturally be in addition to the notifications of regulatory bodies as required under the applicable laws.

How to Notify?

As many organizations have already found out, simply providing information to the USEPA under the Community Right-to-Know provisions may be very counterproductive. Simple compliance with the established rules and regulations is usually not adequate to the purposes of the organization. A careful analysis of how notification should take place must include a recognition of the differential risk exposure of those to be notified as well as the timing and content of the notification. In addition, the concerns of liability defense require that some means of documenting the provision of notice be considered.

The AHERA experience provides several important points to be considered in the notification policy. The first is a recognition of the impact of differential risk on notification. Specifically, those individuals at greatest risk require a different level of notification than those with only a relatively small likelihood of risk exposure. All individuals having a risk exposure have a right to a certain minimum level of notice.

Differential risk considerations require that those segments of the general public having a real but removed or indirect relationship to the risk be notified on a regular basis and that notification be carried out in a manner that has a reasonable assurance of reaching all or nearly all of those concerned. At a minimum, the notification under AHERA to the general public must inform the public of the existence of the risk and how additional information on the risk may be obtained.[144] This notification should also inform those segments of the actions taken or planned that are intended to reduce risk.

At a higher level of differential risk, AHERA requires that formal informational and rudimentary training sessions take place for those whose duties place them in the vicinity of the risk source, but who are not normally expected to be in direct contact with the risk source or the control mechanisms for the risk source. Such individuals are required to be informed of the presence of the risk, the specific locations of the risks in relation to their duties, the nature of the risks including the health and/or environmental effects that may result from the realization of the risks, and proper individual and collective conduct in the presence of the risk.[145]

Policy decisions concerning this type of notification should be made with respect to exactly who is likely to be active in the presence of the risk and therefore require this minimum level of notification, how frequently the notification is to take place, the exact content of the notification curriculum (which may be different depending on the activities of the individual to be notified), and how assurance of notification may be obtained. Note that the phrase ''active in the presence of the risk'' refers to the type of activity of the individual in relation to the condition of the control mechanism for the risk.

An additional consideration may relate to the type of notification to be provided to those who are ''passive'' in the presence of the control mechanism, but are in immediate proximity to the risk. The classroom teacher provides an excellent example of this issue. The teacher is normally not in a position to disturb most of the ACBMs in the building. However, if the ACBM is disturbed (say by an earthquake), what are the appropriate actions for the teacher to take? In many respects, the development of a risk management program should allow this and similar situations to be dealt with in an environment of reasoned consideration and objectivity as opposed to the type of environment that would exist after the earthquake had taken place.

At a still higher level of differential risk, policies should be developed with respect to how the notification of the AHERA ''short-term worker'' will take place,[146] and the implications of the very existence of a short-term worker in the presence of an environmental risk. AHERA has some major flaws as a model for environmental risk management and one of them occurs here. Suppose that the policy adopted with regard to notification of the short-term worker includes placing notice of the existence of environmental risk on all purchase orders and the requirement that all short-term workers be formally notified of the risk prior to entry to the premises. While these policies may serve to notify the individual, do they serve the overall objectives of a ERMP?

Consider the implications of having a short-term worker on the premises who has been fully informed and who is working in close proximity to the control mechanism for the risk. Is this individual properly trained and equipped to avoid disturbing the control mechanism and thereby cause its failure, or, if the job requires working with or deliberately disturbing the control mechanism, is the individual able to do so properly? The bottom line question is: How can the facility management be certain that such an individual will not generate a major problem? Having a new phone line run above the ceiling may present major issues of asbestos disturbance and the building owner and/or operator will almost certainly bear primary responsibility.

At the highest level of differential risk are those whose function requires that they work directly with the risk control mechanisms or the risk itself. These individuals must receive the highest level of training, going far beyond simple notification. Again, AHERA recognizes these individuals and requires formal training for them.[147]

Formal training in proper work practices and personal/public protection, even with periodic retraining, has long been recognized as insufficient by itself. Explicit policy attention should be given to how proper practices and procedures are to be consistently employed and what disciplinary mechanisms are to be used to ensure that the tools, techniques, practices and procedures will be consistently utilized. There is no difference between the release of a hazardous substance caused by an unforeseen event and that caused by negligence, except that the liability burden for negligence is greater. Discipline, to be effective, must have two components; an explicitly stated and well-understood policy, and proof that the policy was enforced.

Obviously, the development of notification policies involves a number of major considerations from risk and liability management through labor relations and operating cost impacts. The need for the training is not really an issue because so many laws require it that such training is rapidly becoming universal. The content and extent of training and the enforcement of the practices and procedures are the primary areas for policy.

When to Notify?

The purpose of notification is to reduce risk and this objective should determine when notification should take place, particularly with respect to retraining and updating the information. People need to be reminded of both existence of the risk and appropriate behavior in the presence of the risk. AHERA requires that new employees be notified within 60 days of employment and that specified existing employees be notified at least annually.[148]

Given the types of activities in which an individual is employed and the environmental risk likely to be encountered during these activities, a series of questions concerning the time of notification and renotification should be answered. In particular, can an individual perform his/her functions safely from the first moment on the job without proper notification and/or training? Should the individual require retraining when job classification or duties change? Should the individual be required to go through formal retraining if the individual's performance is judged inadequate? When should new information be provided; on a regular basis or when significant new information becomes available, or both?

In answering these questions, management must take into account not only the direct costs of providing training and notification, but the likely costs associated with inadequately trained or notified personnel. The direct cost may prove to be substantial, but the likely costs of failure of the risk management system resulting from inadequate or improper training can be monumental.

What Information to Make Available, and in What Format?

Telling someone that they may contract an absolutely fatal disease and that the materials from which the disease may be contracted are present in the room they are working in can have very unfortunate consequences. However, this is exactly the minimum requirement. The issue is not what to provide in the way of the minimum of information, but what the total amount of information to be provided should be and the format for that information.

A recent article in the *New England Journal of Medicine*[149] contained details of a study of the workers in the filter manufacturing operation for Kent[150] cigarettes where asbestos was used in the manufacturing process. Of the 33 individuals studied, two-thirds of them are dead from causes directly related to asbestos or have contracted a debilitating asbestos-related disease. This information is readily available to anyone with an interest in the subject and the source would be considered to be nearly irrefutable.

The facts of the matter are that the article is talking about occupational exposure; that is, exposure as an integral part of the work environment at elevated levels of asbestos and over a prolonged period of time. The statistics for non-occupational exposure to asbestos indicate that smoking is 21,900 times more likely to cause premature death than asbestos in a well-maintained building.[151] Given this information as background, how can an individual in an asbestos-containing building be assured that they will not suffer the same fate as the Kent workers?

A major difference between environmental risks in the occupational versus the non-occupational settings is common. It is the source of the misconception of risk on the part of the general public. The explanation of the difference and the enunciation of the risk management program elements that insure the safety and health for both occupationally exposed and non-occupationally exposed individuals must be clear, concise, and provided by authorities perceived to be credible by the public to which the information is addressed.

An earlier section provided extensive background information on this topic, but several points deserve emphasis. First is the credibility of the source of the information. If the source of the information is not deemed credible, the notification process will generate more problems than it is designed to prevent.

To the extent that hard and fast guidelines for the notification of individuals exist, they may be summarized as follows:

1. Do not lie in any form or fashion.
2. Be the first to provide the information.
3. Provide all of the information. More information is better than less, if less might be perceived as withholding information.
4. Provide the information through authoritative sources.
5. Be a part of the concerns of the public. The fears are real and in a context of less knowledge than you possess, they might have been yours.
6. Clearly demonstrate the actual risks to the public being addressed and the differences between the risks evidenced in the media/literature/elsewhere and the actual risk situation in effect locally, if any such differences legitimately exist.
7. Provide verifiable information concerning:
   a. The actions already taken to reduce the risk and their impact on risk.
   b. The actions planned to further reduce the risk.
   c. An explanation of what obvious actions have not been taken and why in objective terms (what alternative was used?).
8. Provide a means through which the individual may actively participate in controlling the risk.

Most of these guidelines are straightforward, but numbers (4) and (8) deserve additional thought and planning on the part of management. The best information distribution systems should make those most likely to be concerned with and negatively react to the risk situation an integral part of the development of the ERMP from the beginning. The inclusion of the thinking and concerns of those individuals indicate that they should also be a part of making the tough decisions concerning how the program will operate. The ERMP should honestly allow for the inclusion of input from opposing points of view in defining and executing the program. Exclusion of a viewpoint is substantively different from agreements based on reasonable compromise of concerns.

THE MANAGER: AUTHORITY AND RESPONSIBILITY

Someone must be placed in charge of the program who will have the responsibility, authority, and knowledge to deal with the issues. These elements are all required parts of the AHERA statutes, where the manager is known as the *designated person*, and whose training and responsibilities are required to be a part of the written record.[152] The designated person will be referred to throughout this text as the Environmental Risk Manager (ERM).

The identity and skills of the ERM are critical to the success of an environmental risk reduction program, but they are meaningless without a clear definition of the responsibilities and authority of the position itself. AHERA is very strong in terms of defining the ERM's responsibilities, but deserves major criticism for failure to clearly require a definition of the authority of the ERM to act.

This is a major flaw because of the very nature of the environmental risk management problem. Contrary to popular opinion, very few organizations deliberately set out to create environmental hazards. Most of the significant environmental hazards have resulted from a failure to appreciate the outcomes of an action, a lack of knowledge about the consequences, or as a result of an unplanned event. The ERM must have the authority to deal with these situations and the very nature of the unplanned event requires that the authority be carefully defined in advance. The charter of authority and responsibility granted to the ERM must deal with both the results of actions taken given the existing state of knowledge about likely impacts and changes in that fund of knowledge and to deal with emergencies.

Traditional management practice provides a number of examples of how the charter of authority and responsibility may be developed and this fund of experience is partially included in the AHERA rules. Specifically, the ERM under AHERA is charged with the implementation of the management plan.[153] Further, the performance of the program is to be monitored through a system of internal audits (called *"periodic surveillance"*[154] in AHERA), that are to take place no less often than every six months. In addition, the overall performance of the program is to be audited *(AHERA recommends that this be accomplished by an accredited Building Inspector/ Management Planner who has no implementation responsibility for the management plan)*[155] no less often than every three years.[156] This system is similar to that of financial performance analysis.

The charter of authority and responsibility granted to the ERM must then require that the ERM implement an approved management plan of action governing the inventory of environmental risks, maintain the inventory and risk management control systems, conduct periodic internal audits of the system performance, and have the authority to deal with any environmental risk issues. The performance of the ERM and the management plan is to be monitored through the use of classic auditing procedures as they would be applied to the environmental risk inventory management program.

In a very real sense, the charter of authority and responsibility for the ERM should be defined through the approved management plan. This plan will be shown to contain the details of what is to be done to reduce risk, how the risk reduction is to be achieved, the resources to be made available for achieving the risk reduction, and regular plan performance auditing.

RISK EVALUATION METHODS

The existence of risk implies uncertainty with respect to future events. In a policy sense, this truism translates into a need to decide how risk will be evaluated in order that consistency may be introduced into the decision-making process. In the case of environmental risk, the decisions fall into several areas, some of which may not be obvious.

Risk Rate of Return

In the financial sense, the first area is in defining the risk rate of return to be applied to investment decisions, particularly decisions that involve large capital expenditures. In a later chapter, a method for determining a risk rate of return will be developed that involves both objective and subjective considerations. The objective part of the calculations can be dealt with as required, but the subjective considerations should be handled as an explicit management decision.

With respect to environmental risk, the major areas of subjective concern relate to civil litigation and the anticipation of regulatory changes. The question of the probability of becoming the subject of civil litigation and the likely financial consequences of a negative outcome of such litigation as well as the cost of defending the entity, can significantly impact the risk rate of return. Management's evaluation of this issue should result in the addition of a component to the risk rate of return as determined by the objective methods. The likelihood of regulatory changes that will act to increase or decrease the costs of risk management should also result in an additional rate of return component.

In a sense, both of these additions to the risk rate of return are a form of ''self-insurance'' premium designed to cover the financial impacts of a given situation on the entity. The insurance carriers do not provide environmental liability insurance because of the lack of an ability to predict the outcomes from these two areas. The assignment of a value to this component is difficult, but developing even an approximate value can provide two major benefits. First, the discipline of explicitly evaluating these subjective areas of risk can lead to uncovering significant risks not previously considered. Second, accumulation of a self-insurance reserve fund based on this evaluation can act to stabilize the financial affairs of the entity in the event of a problem, a highly desirable result from the point of view of all stakeholders in the entity.

Assignment of a Risk Value to Human/Environmental Impacts

Consideration of environmental or human impacts resulting from risks can confuse even the initiated. Threshold Limit Values (TLV), Permissible Exposure Limits (PEL), Action Levels (AL), environmental fate of hazardous substances, linear versus non-linear human and global response curves, and other similar issues are highly technical in nature and subject to continual change and re-evaluation with new information. Added to these issues is the problem of synergistic response, where one or more risk elements combine to result in multiples of the independent risk levels. The point is that many of these issues simply cannot be rigorously defined at this time, but decisions must still be made.

Added to these problems are significant issues of how to determine the level of risk exposure, a subject of significant technological, legal and environmental controversy. Should the level of exposure be primarily based on objective scientific measurement of some quantity, or the subjective evaluation of a situation by an expert? These are issues that should not be left solely to the "expert," if for no other reason than the fact that the "experts" cannot agree. Management must select from among the possibilities those methods and measures to be used in developing and evaluating the ERMP to obtain consistency in decision-making information.

This text, recognizing the importance of the Negative Default principle, adopts a conservative position. It also recognizes the inherent complexity of the total environmental risk management subject area. In doing so, the following approach has been adopted.

First, in the case of multiple risks, the risk element having the most severe impact on human health and/or the environment will be used as a proxy for all of the risks impacting a given situation. This is consistent with the standards established for evaluation through the USEPA Hazard Ranking System (HRS) found at *29 CFR 300* Appendix A. In certain circumstances, this may either overrate or underrate the "real" risk, but given the present state of knowledge on this subject, it is the most appropriate assumption.

Second, evaluation of the risk potential will be based only partially on quantitative methods. The primary basis of evaluation will be codified subjective judgments. Devices to codify the judgment, such as decision trees and fault trees, will be extensively used.

The third major area is with respect to performance measurement. Specifically, what will be the quantity used to measure program performance? It has already been stated that monetary units should not be used either explicitly or implicitly to avoid the major issue of placing a monetary value on human health or environmental damage. In addition, the use of monetary values for performance measurement is unnecessary given the newer management decision-guidance tools such as multi-objective linear programming. The risk reduction objective naturally leads to the use of a risk reduction measurement unit. In most of the balance of this text, the units of risk measurement will be the number of thousands of hours per year of risk exposure time, the *relative* rank of the risk measured in terms of the number of incremental deaths per hundred thousand of population for the exposure setting (occupational or non-occupational), and possibly a weighting factor for the type of activity engaged in by the individual with respect to the condition of the control mechanism.

Monetary units will be used to evaluate the costs for remediation. When coupled with the probability of the occurrence of an event using fault tree logic, this will lead to the risk rate of return and the likely measurable cost to be associated with a given course of action. You should not allow this application of monetary values to confuse the basic objective of measuring performance through risk reduction, it is simply the inclusion of the issue of probable remediation costs as a recognized factor in ERMP evaluation. From the point of view of the mathematics of multi-objective linear programming, the use of incommensurable units (e.g., dollars and thousands of hours of risk reduction per year) simultaneously in the solution of a decision-guidance problem is completely acceptable.

This text assumes that the basic objective will be the progressive reduction of risk given the limits of technology, local conditions, and economics. Risk reduction will be measured throughout this text in Risk Reduction Units (RRU) that will usually be in units of thousands of hours per year of exposure time. There are other methods of performance measurement and for the evaluation of risk potential and management should determine which methods shall be employed.

THE MANAGEMENT PLAN FOR ENVIRONMENTAL RISK REDUCTION/CONTROL

The management plan for environmental risk reduction and control is the key to the success of the ERMP just as sound planning is the key to the success of any business operation. The environmental risk management plan is not substantively different from a regular management plan except for one major feature—the benefits are measured in units of risk reduction instead of dollars.

In general, for reasons to be demonstrated throughout the balance of this text, the management plan should be developed in three major parts. Part I should contain the overall policies and procedures that will govern the execution of the plan throughout the organization and over time. This part will not change radically from period to period. Part II should contain the plan details specific to a particular situation or facility and accurately summarize the details of the execution of the ERMP. Part III consists of all of the detailed supporting records of plan performance.

AHERA is substantively different in its requirements from those set out above in that it requires that all three parts of the plan be contained within a single unit. So long as AHERA does not specifically apply to the entity, this form of plan organization is to be avoided. It is equivalent to requiring that all of the detailed documentation of the daily operations of an entity be included in the annual performance reports and generates an unmanageable situation. If AHERA is to be modified, this is one major area because of the burden it imposes on all concerned.

In addressing these issues, the management plan should be viewed as a set of operating expectations, the comparable operating results compiled on a regular basis, and the details behind the results summarized as set of environmental balance sheets, profit and loss statements and annual reports, along with supporting documentation.

Definition of the Environmental Risk Inventory Specifications

An old axiom of management states that only those things that can be measured can be managed. The definition of the environmental risk inventory to be managed is therefore of critical importance, especially given the lack of precision contained in the actual rules and regulations that often govern environmental risk management.

Management policy decisions are required in terms of both the quantitative issues surrounding the development of the risk inventory and the basic philosophy upon which inventories will be developed. It is the latter issue that is of primary concern here.

One major application of the Negative Default principle will be in terms of the extent to which resources will be devoted to the quantitative definition of the risk inventory. Specifically, how much of the available resources are to be used in defining the exact extent of the risks versus the amount of resources to be put into the management of risks only assumed to be present?

Phase I environmental assessments are specifically designed to answer the question: "Is there a reason to suspect the presence of an environmental risk?" Phase II environmental assessments are designed to answer the question: "Can the suspicion of a risk be confirmed or denied?" Phase III environmental assessments are designed to answer questions concerning the extent of the risk, the current and likely future condition of the control mechanisms, the availability of transport mechanisms, and the level of risk exposure.

Placed in this context, a major difference in the amount of resources required to simply quantify the inventory at various levels of assurance should be obvious. The policy question is then: "How much of the resources are to be devoted to a higher level of assurance, knowing that if a risk is suspected, action must be taken as if it actually existed in its worst possible form lacking better information to the contrary?"

A less obvious policy issue is contained within this decision. What level of assurance of the non-existence of a risk will be acceptable? There is no scientific possibility of absolutely determining that risk does not exist. In addition, the greater the level of assurance that is sought, the higher the costs of achieving that level of assurance.

Although precision is difficult, a properly conducted Phase I, as defined later, that does not uncover a substantive reason to suspect the existence of a risk probably provides a 60–70% assurance that no risk exists.[157] Phase II assessment techniques may raise the level of assurance by 10% or so in return for an order of magnitude cost increase over the cost of a properly conducted Phase I. Phase III assessments may reach a level of 90–95%, but provide substantive quantitative information of primary importance to the detailed execution of an ERMP. The Phase III may cost an amount in excess of several orders of magnitude over the cost of a Phase I.

The balancing of assurance versus resources is not a simple task and the incremental frontend costs of initiating a risk management program can be substantial. Generally, the performance of a Phase I assessment is preparatory to conducting a Phase II or III.[158] The real purpose of a Phase I assessment is to limit the scope of more extensive investigations by uncovering information that will narrow the analytical work. On occasion, the Phase I may provide a satisfactory conclusion, generally when no reason to suspect the presence of a risk can be found or when obvious and readily definable risks are encountered.

Management must determine the level of assurance that it will require and devote the resources necessary to achieve that level of assurance.

BUDGETING

Budgeting is explicitly a disciplined method for thinking through all of the alternatives and the allocation of resources necessary to implement the alternatives selected. To this end, the following major budget components are generally recommended.

Operations and Maintenance (O&M) Budget

The daily ongoing task of maintaining the control mechanisms can best be dealt with and controlled through a specific budget. This budget may have a strong fixed cost component with an essentially infinite lifetime resulting from the engineering nature of specific environmental risks. For example, it may be necessary to monitor groundwater contaminant levels surrounding a site for decades into the future even if the site is no longer in active use.

The O&M budget should cover the costs for daily maintenance of control mechanism activities including provision for special individual, public and environmental protection devices employed during the maintenance activities. Environmental risk operations and maintenance programs are specialized activities that not only require specific equipment and work practices, but are also frequently governed by specific record-keeping and reporting requirements resulting both from statutes and from self-defense considerations on the part of the entity.

One of the basic reasons for placing all O&M activities involving environmental risk into a special budget is to provide visibility and control. O&M, properly conducted, is frequently the best risk reduction strategy in an absolute sense and is usually the least expensive strategy. By its very nature as a long-term activity, O&M will require constant attention to be effective, attention that can be focused through the O&M budget.

Training/Notification Budget

The concerns of training and notification have already been addressed. The training and notification budget should implement the policy decisions regarding those concerns. Who shall be trained/notified, how often, by whom, what are the contents of training/notification actions, and similar considerations all have budgetary consequences and should be specifically funded and tracked to insure proper implementation and adequacy.

Planned Response Action Budget

If specific response actions are planned to deal with specific problems, they must be funded. This is a budget that often achieves a level of visibility out of proportion to its risk reduction potential because of the large sums of funds expended and the pressure to achieve a "quick fix." A large proportion of environmental risks do not have a "quick fix" and funds allocated to planned response actions should be evaluated to ensure that the actions are being taken for the right reasons.

As an example, consider the case of asbestos in a building. The experience of the past several years has demonstrated that removal of asbestos has left the building more contaminated with airborne asbestos fibers than it was before removal 60% to 80% of the time. This increased risk has been purchased for incremental (to the costs of removal at renovation or demolition) prices in the range of $10 to $50 per square foot, a cost that must be capitalized under current tax law. In return, the property has been returned to its approximate original market value. This tradeoff often does not appear to make good sense except as a reaction to public perception; perceptions that may have been changed through proper notification and understanding of the objective requirements of the law and capabilities of technology.

Emergency Response Action Budget

A specific budget to deal with possible unplanned events should be developed. Delays in taking immediate action to at least contain the release of a substance have demonstrated an almost exponential increase in costs with delay time (e.g., the *Exxon Valdez*). Management should allocate funds and authority to the ERM for as nearly instantaneous action in critical situations as possible.

Delay in taking actions for any reason can have a large number of negative consequences in terms of human health, environmental damage and public relations. Any restriction to the ERM's ability to immediately respond to an emergency should be very carefully considered prior to acceptance by management. This is especially the case as time goes on and nothing has occurred to justify the existence or size of the budget; the impulse to cut funding should be strongly resisted. The probability of the occurrence of an environmental problem generally *increases with time,* not decreases, as control mechanisms age and people begin to take the proper functioning of control systems for granted.

RECORDKEEPING POLICIES AND PROCEDURES

The recordkeeping policies are very important to the ERMP both for the obvious reasons associated with performance measurement and the less obvious reasons associated with the ability to demonstrate to the various publics that the program is well-founded and functioning properly. The records of the ERMP must be considered to be documents available for public examination and sufficient precedent probably exists to establish this idea in the law.[159] As previously stated, this text recommends that the records of the ERMP be kept in a manner similar to those of a financial reporting system.

The analogy of a financial reporting system is very realistic and many of the principles of financial reporting can be directly applied. The major exceptions revolve around the facts that dollars are not the only, or even primary, performance measure, and the same level of accuracy as may be reasonably expected in a standard inventory cannot be expected for the environmental risk inventory. As will be shown later in this text, these considerations lead to some changes in the accounting and reporting methods, but the analogy will still hold. The adoption by management of an "open books" policy, and the preparation of reports recognizing the very real possibility of examination by those outside the organization should be fundamental to record-keeping policies and practices.

PERFORMANCE AUDITING

The analogy to financial reporting systems has already been drawn and the auditing principles for financial systems are directly applicable. Management should recognize that the negative leverage capabilities of environmental risks may be orders of magnitude greater than the negative leverage associated with all of the other operations of the entity. A seemingly minor environmental risk in terms of the number of dollars associated with the material or the quantity of the material may result in an environmental impact that could totally destroy the net worth of the organization. No other common risk has this magnitude of negative leverage capability, where several drums of chemical worth $500 can result in a multi-million-dollar remediation and liability exposure.

As a result, management should establish the most stringent internal and external auditing standards on the ERMP. These standards should be more carefully reviewed and governed by senior management than any of the traditional financial management audit practices, but they will be similar to classic financial auditing with emphasis on the objectivity and competence of the auditor and the thoroughness of the audit practices and procedures.

REFERENCES

140. 763.84(c)(4), 763.93(e)(1) and (10), and 763.93(g)(2) require notification of all potentially affected parties or their parents/guardians on no less than an annual basis of the existence and location of the management plan.

141. 763.84(c),(d),(e), and (f) require notice of all asbestos-related activities occurring within a facility. The notification extends to all workers, building occupants or their legal guardians, and to such individuals as "short-term workers" (e.g., telephone repairpersons). All such persons must be notified of the presence of asbestos in the building and the location of and conditions of access to the government management plan.

142. 763.93(g)(1), (2), (3), 763.93(g)(4) require that notification of the location of the management plan be given at least annually and that access be granted to anyone requesting access within no more than five working days of the request, and that a copy of the plan be provided to anyone requesting a copy at "reasonable cost."

143. For example, one of the most often cited arguments used to support a NIMBY position is that of a reduction in property value. To date, property value impacts beyond the immediate line-of-sight have only very rarely been demonstrated, although this may be more a consequence of the available analysis techniques than of reality. See the discussion of Topovalue Mapping later in this book.
144. 763.84(c), (d), (e), and (f) requires that notice of the asbestos management plan be communicated to all workers, building occupants or their legal guardians at least on an annual basis. 763.95 requires warning labels to be placed in specific maintenance activity areas.
145. 763.93(e)(4) requires formal written notification. 763.93(e)(10) requires a record be kept of the notification steps. 763.92(a) requires that custodial and maintenance staff be given at least two hours of ''awareness training'' consisting at least of the information noted.
146. 763.84(d) ensures that short-term workers (e.g., telephone repair workers, utility workers, or exterminators) who may come in contact with asbestos in a school are provided information regarding the locations of ACBM and suspected ACBM assumed to be ACM.
147. 763.92(a) requires that all custodial and maintenance staff receive at least two hours of ''awareness'' training. Operations and Maintenance workers are further required to receive an additional 14 hours of specific training in work practices and procedures, safety, and health. The requirements for this training are contained in 763.92.
148. 763.92(a) requires awareness training of new workers within 60 days after commencement of employment. Retraining of specified individuals is required annually by the Model Contractor Accreditation Plan published in detail as Appendix C to Subpart E, *40 CFR 763* in the Federal Register of April 30, 1987 and is included in AHERA by specific reference.
149. Talcott, James A., M.D., et al; ::Asbestos-Associated Diseases in a Cohort of Cigarette-Filter Workers:'' *The New England Journal of Medicine;* November 2, 1989; Vol. 321 No. 18.
150. Kent is a registered trademark of the Liggitt & Meyers Tobacco Company.
151. ''Summary of Symposium on Health Aspects of Exposure to Asbestos in Buildings;'' Harvard University; John F. Kennedy School of Government; Energy and Environmental Policy Center; Cambridge, MA; August, 1989.
152. *40 CFR 763/84(g).*
153. *40 CFR 763.93(i)* ''Each management plan must contain a true and correct statement, signed by the individual designated by the local education agency under 763.84, which certifies that the general, local education agency responsibilities, as stipulated by 863.84, have been met or will be met.''
154. *40 CFR 763.92(b).*
155. *40 CFR 763.93(f).*
156. *40 CFR 763.85(b).*
157. NOTE: In a recent audit of so-called Phase I assessments conducted for a lender, 73% of the Phase Is on file failed to uncover substantive risk situations. The quality of work in performing Phase Is, at least for the sample examined, is very low and probably provides a level of assurance of less than 19% when stating that no risks exist.

158. NOTE: If a Phase III will be required, the Phase II questions will be answered by definition and the Phase II is therefore automatically included in the Phase III.

159. AHERA, for example, at 763.84(f) states that the governing body of the facility shall "Ensure that management plans are available for inspection . . . ." Other environmental rules and regulations contain similar requirements in various forms ranging from "Environmental Impact Statements" through the requirements placed on Manufacturer Safety Data Sheets and SARA Community Right-to-Know substance release reports.

5 THE ENVIRONMENTAL RISK INVENTORY SYSTEM (ERIS)

INTRODUCTION

Acknowledgement of the point of view that environmental risk management is a long-term and continuous process implies the need for an information system that will allow management control and the performance measurement of that management. The model of an inventory to be managed and the value of which is to be enhanced is explicitly adopted throughout the balance of this text.

The idea is simple and the fact that an inventory of environmental risks has a negative value as opposed to the positive value of a classic inventory does not invalidate the concept. The Securities and Exchange Commission (SEC)[160] has adopted a policy requiring publicly held firms to provide a value assignment for potential environmental risk liabilities. To this point in time neither the SEC or the Financial Accounting Standards Board (FASB) have clearly defined how such a liability may be "valued," but a value cannot be attached to something unless that something has been identified and, to the most reasonable extent possible, quantified.

Inventories of environmental risks are implicitly or explicitly required under RCRA and AHERA. The need for a quantified listing of environmental risks with a "value" attached to those risks is fundamental to the management process. However, the "value" for management purposes is different from the connotations of "value" with respect to the SEC requirements. "Value" for management purposes is measured in Risk Reduction Units (RRUs) and this definition of "value" removes the worst problems facing the SEC and the FASB, the dollar value of probablistic outcomes, and the dollar value of human and/or environmental damage. Such a definition is consistent with the approaches taken in a number of federal and state laws and, for the purposes of this book, most notably the Hazard Ranking System (HRS).

THE INVENTORY MODEL

An Environmental Risk Inventory System (ERIS) must quantify, to the extent reasonably possible, at least the following elements:

1. The type of environmental risk.
2. The quantity and extent of the risk sources.
3. The identity of the control mechanisms governing the level of risk.
4. The current and future condition of the control mechanisms.
5. The identification of the transport mechanisms that might operate to move the risk from its present position to a position where environmental damage can occur in the event of control mechanism failure. This implies that knowledge of the spatial location of the risk is critical.
6. The quantification of the likely damage to the targets of the environmental risk.

Given a knowledge of these six inventory elements, plus a knowledge of the outcome of the various alternatives to prevent realization of the risk, a management plan may be developed, implemented, and its performance measured.

The Format of the Inventory

An inventory of environmental risks must inherently recognize the spatial distribution of the risks that characterize the inventory. It is rare that an environmental risk will be confined to a single point in space or time. Asbestos for example, is generally distributed throughout a building and a fundamental element of the level of risk associated with the inventory lies in this distribution. If the inventory were isolated to an area such that no individual could be exposed to the risk, then there is by definition, no risk. On the other hand, the same inventory located in places where exposure is a virtual certainty, represents the highest level of risk attainable.

Because location is a critical element, it must be a key criteria in the inventory process requiring the extensive use of maps and floor plans as an integral part of the inventory record-keeping and record maintenance systems. These maps and floor plans must, however, convey the information represented to a number of possible users and be maintainable in a simple and accurate way, or their primary application—the reduction of risk—will not be fully realized. The following specifications for floor plans are recommended. (There are others for risks not located inside structures that will be discussed later.) These are illustrated in Figure 5.1 through Figures 5.5.

DEF: Homogeneous Area: An area of risk with similar composition and primary risk control mechanism.

1. There should not be more than one homogeneous area shown on a single drawing.
2. The drawing should display the standard two dimensions of a floor plan by limiting homogeneous areas to a single floor of the structure.
3. Drawings should be limited to a size such that they will clearly reproduce the information shown on them on an $8^1/_2 \times 11$ sheet of paper.
4. The drawings should allow unfamiliar users to quickly orient themselves to their physical location within the structure. Room numbers shown on the floor plan for example, should be those in common use throughout the structure by the occupants of the structure. Major structural features such as permanent walls, stairwells and elevator shafts should be clearly denoted.
5. A standard set of symbols for the risk inventory should be adopted, clearly defined, and consistently used. At a minimum, the symbols should denote the following information:
   a. The location and identity of the homogeneous area.
   b. The location and identity of any analytical samples taken of the substance.
   c. The identity of the primary control mechanism for the substance.
   d. The location and identity of any salients within the homogeneous area
   e. The location, orientation, and identity of any photographs taken of the homogeneous area or salients.
   f. A special symbol denoting areas on the floor plan that *have not been inspected for environmental risks.*
6. The general location of the major transport mechanisms for the substance should it escape control.

Backing up the graphical information on the homogeneous area should be a series of supporting documents along the following lines.

1. A narrative description of the risk source and particularly the current condition of the control mechanisms for the source and a description of likely failure condition for the control mechanism.
2. A description of the likely risks that might be encountered if the unexplored areas are entered and the precautions to be taken upon entry, if any.
3. A description of the transport mechanisms available to the homogeneous area and the extent to which those mechanisms would be likely to spread the risk source in a reasonable period of time after control mechanism failure.
4. A log of any analytical samples taken of the risk source and a copy of the laboratory results of such samples.

5. A log of the salients in the homogeneous area giving a brief description of the salient and, if applicable, the current condition of the control mechanism if that condition is different from the average condition of the control mechanisms for the area.
6. A collection of photographs of the salients. The photographs should identify the subject salient and the date of the photograph.

For convenience, the above group of inventory data shall be referred to hereafter as the Homogeneous Area Inventory. The Homogeneous Area Inventories should be organized in a form providing convenient access to those who will need the information for the operation of the ERMP. A suggested form of organization would be to collect the most current version of all of the Homogeneous Area Inventories into an indexed group by wing of the building, the wings grouped by floor, and the floors grouped by building.

As activities occur which change the characteristics of the inventory, the Homogeneous Area Inventory records should be updated and the prior records archived, thus providing a complete audit trail of the results of any changes in the inventory. Subsidiary to the Homogeneous Area Inventory records and supporting the information on any changes in the inventory will be a series of other records such as work orders, personal activity logs, incident reports, and the like. These will be discussed in some detail later.

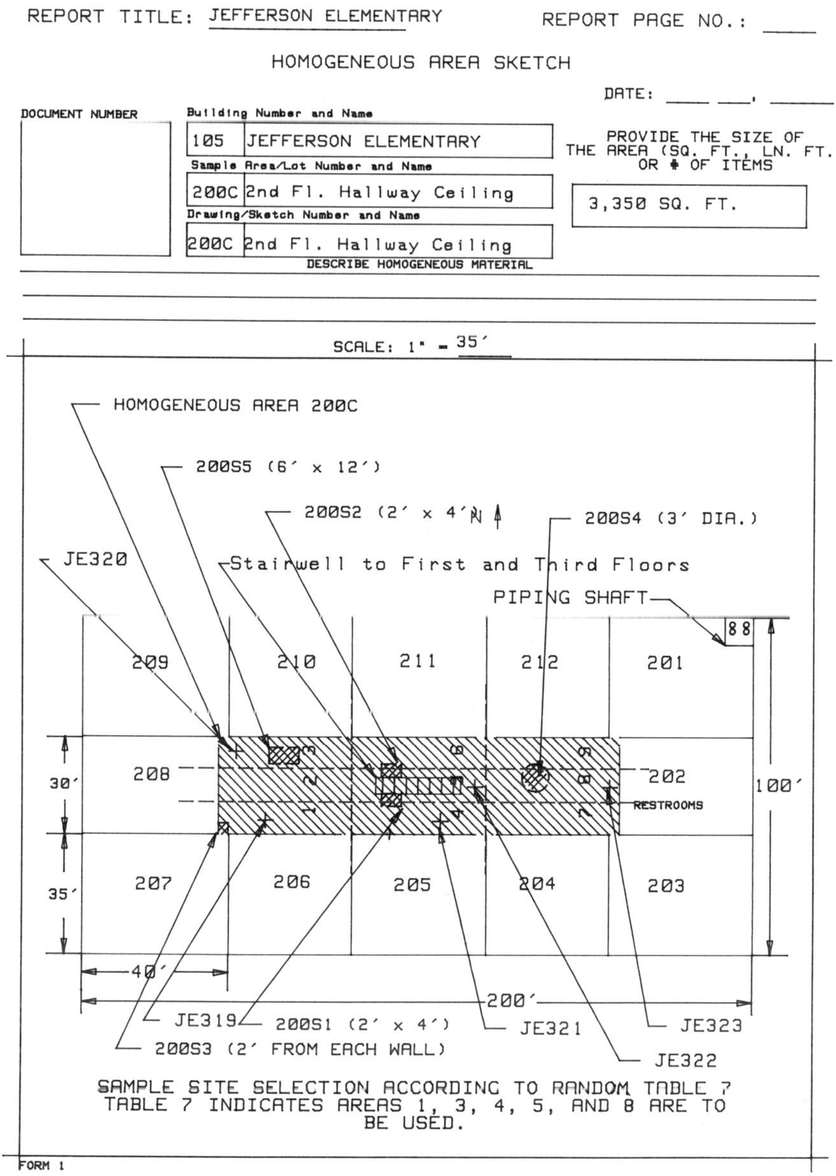

Figure 5.1 Homogeneous Area Floor Plan.

REPORT TITLE: ____________________ REPORT PAGE NO. _____

ROOM CROSS-REFERENCE AND ASSESSMENT

PAGE 1 OF 2

PAGE _____ OF _____

DATE: __________ ____, 19___

Building Number and Name

Building Location

ROOM NUMBER/NAME

HOMOGENEOUS AREA OR SALIENT ID	HOMOGENEOUS AREA/SALIENT DESCRIPTION	ACBM			CONDITION				
		ASSUMED ?	CONFIRM ?	FRIA. ?	MATER. TYPE	DAMAGE COND.	POT. DAMAGE EST.	EQUIVALENT AREA	C.C. NO.

MANAGEMENT PLANNER'S ROOM HAZARD ASSESSMENT

MANAGEMENT PLANNER'S SIGNATURE CERT. NO. & STATE

FORM9A

INSPECTOR'S ROOM CONDITION ASSESSMENT:

BUILDING INSPECTOR'S SIGNATURE CERT. NO. & ST.

Figure 5.2 Room Cross-reference and Assessment Form Page 1.

REPORT TITLE: ________________ REPORT PAGE NO. ____

ROOM CROSS-REFERENCE AND ASSESSMENT
GENERAL ROOM DATA
PAGE 2 OF 2

ROOM USEAGE DESCRIPTION

PERSON-RISK HOURS OF OCCUPANCY PER YEAR

	ACTIVE	PASSIVE	TOTAL PERSON-RISK HOURS/YEAR
AVG. NO. OF PEOPLE/HR	____a1	____a2	
AVG. NO. OF HOURS/DAY	____b1	____b2	(a1 x b1 x c1) x 5 = ____
AVG.NO. OF DAYS/YR	____c1	____c2	(a2 x b2 x c2) = ____

DAMAGE FACTORS

PHYSICAL DAMAGE	WATER DAMAGE	DETERIORATION
SIGNIFICANT ☐	EXTENSIVE ☐	HEAVY ☐
DAMAGED ☐	MODERATE ☐	MODERATE ☐
NO DAMAGE ☐	SLIGHT ☐	SLIGHT ☐
	NONE ☐	NONE ☐

DISTURBANCE FACTORS

PROXIMITY TO REPAIR ITEMS	ACCESSIBILITY	ADJACENT ROOMS
LESS THAN 1 FT. ☐	WITHIN REACH ☐	GYMNASIUM ☐
1 TO 5 FT. ☐	BARELY REACHABLE ☐	MUSIC ROOM ☐
MORE THAN 5 FT. ☐	NOT REACHABLE ☐	MECHANICAL ROOM ☐
		OTHER ☐

AIR FLOW FACTORS

BARRIERS	VENTILATION	AIR MOVEMENT
PERMANENT ENCLOSURE ☐	YES ☐	HIGH ☐
ENCLOSED ☐	NO ☐	MODERATE ☐
ENCAPSULATED ☐	INTAKE ☐	LOW ☐
NONE ☐	EXHAUST ☐	NOTE IF: AIR PLENUM ☐

FORM 9A

Form Page 2, Room Cross-reference and Assessment

REPORT TITLE: ____________________ REPORT PAGE NO.: _____

SAMPLE LOG and SAMPLE NOTES

DOCUMENT NUMBER	Building Number and Name	
	Sample Area/Lot Number and Name	
	Drawing/Sketch Number and Name	

THIS IS A CHAIN-OF-CUSTODY DOCUMENT. PLEASE FILL OUT COMPLETELY AND SIGN ACCORDINGLY

PAGE _____ OF _____

Sample Number	Receiv-er's Initials	Photo. Number	Description of Sampled Material	SAMPLE SITE LOCATION

NOTE: IF MAILED, HAVE CARRIER PROVIDE SIGNED PICK-UP AND DELIVERY RECEIPTS TO BE ATTACHED TO THE ORIGINAL OF THIS DOCUMENT.

Receiving Analyst's Name	Receiving Analyst's Firm
Receiving Analyst's Signature	Date Samples Received

Inspector's Name	Inspector's Firm
Inspector's Signature	Date Samples Collected

FORM 7

Figure 5.3 Sample Log.

REPORT TITLE: ____________ SALIENT LOG REPORT PAGE NO.: ____

DOCUMENT NUMBER

Building Number and Name

Sample Area/Lot Number and Name

Drawing/Sketch Number and Name

DATE: ____ ____, ______

PAGE ___ OF ___

Salient ID Number	Sketch Number	Photo. Number	Sample Number	Description of Salient	Access Notes	Recommendations

FORM 4

Figure 5.4 Salient Log.

PERIODIC SURVEILLANCE REPORT

RETURN COMPLETED FORM TO ASBESTOS PROGRAM MGR.

Page ___ of ___

DATE: ____ ____, ______

DOCUMENT NUMBER

Building Number and Name

Building Location

ROOM Number and Name

IF THE STATUS OF THE ACBM HAS CHANGED, THEN PHOTOGRAPH THE AREA AND RECORD THE PHOTOGRAPH NUMBER IN THE SPACE PROVIDED. NOTIFY THE ASBESTOS PROGRAM MANAGER CONCERNING THE CHANGE.

Sample Area/Lot or Salient ID	Sample Area/Lot or Salient Description	LAST MAT. COND.			CHANGE?		NEW PHOTO. NUMBER	NOTES
		T	DC	PD	YES	NO		

Signature of Person Completing Report

Title of Person Completing Report

* REFERS TO MATERIAL TYPE AND DAMAGE CATEGORIES
. T = MATERIAL TYPE AS:
. S = SURFACING
. M = MISCELLANEOUS
. T = THERMAL SYSTEMS

DC = DAMAGE CONDITION
. ND = NO DAMAGE
. D = DAMAGE
. SD = SIGNIFICANT DAMAGE

PD = POTENTIAL DAMAGE CATEGORIES
. NPD = NO POTENTIAL DAMAGE
. PD = POTENTIAL DAMAGE
. PSD = POTENTIAL SIG. DAMAGE

FORM 18

Figure 5.5 Surveillance Report.

THE GENERALIZED ENVIRONMENTAL RISK INVENTORY SYSTEM

Figures 5.6 and 5.7 present detailed and summary inventory reporting forms for a generalized environmental risk inventory system. The information presented by these forms and their application will be the subject of a large portion of the balance of this book, including the development of the outlines of methods for standardizing the reporting of environmental risk management performance and strategy development methodologies.

The basic arrangement of the inventory summary clearly displays the fact that both the risk, measured in dimensionless risk units, and the financial costs for dealing with the risks are important. Subsidiary to the summary information is the detailed risk inventory data shown on Figure 5.7.

The major divisions of the risk inventory detail will correspond to the definition of the risk system set out in the next chapter, specifically; the risk source, the primary control mechanism, the transport/secondary control mechanism, and the target. Each of these items must be separately evaluated in order to arrive at a risk "value" for the inventory. In general, for each risk source there are assumed to be five allowable forms of the transport mechanism that may result in damage to different targets.

Also subsidiary to the summary is a separate set of detailed worksheets that will analyze the various alternatives for dealing with a given component of the risk system. There are six generic alternatives for dealing with any risk:

1. Operations and Maintenance (O&M) programs;
2. Repair of the primary control mechanisms;
3. Encapsulation of the contaminated area;
4. Enclosure of the contaminated area;
5. Isolation of the contaminated area; and
6. Removal of the risk source.

Each has its own costs, associated risks, and associated efficiencies. The objective of management will be to balance the risk reduction objective with the available resources to develop a strategy that will provide maximum risk reduction within the constraint of available resources, just as management now seeks to achieve maximum return on investment and utilizing similar tools.

ENVIRONMENTAL RISK INVENTORY SUMMARY

REF DATA		RISK SOURCE DESCRIPTION	ENVIRONMENTAL RISK VALUE		CONTROL COST PER YEAR	REMEDIATION			PRIORITY NUMBER	NOTES
REF DWG	HOMO AREA		HUMAN TARGET	SENS. ENVIRON.		DESCRIPTION	COST	REF NO.		
		TOTALS								

ERINVTRY1/2

Figure 5.6 Environmental Risk Inventory Summary.

ERIS DETAILED INVENTORY WORKSHEET

REF DATA: REF DWG	REF DATA: HOMO AREA	RISK SOURCE DATA: DESCRIPT.	RISK SOURCE DATA: HRS QUAN	RISK SOURCE DATA: TOX/PERS	RISK SOURCE DATA: RS CS	PRIMARY CONTROL MECHANISM ANALYSIS: DESCRIPT	PHYSICAL COMPONENT: CURR COND	PHYSICAL COMPONENT: t 1	PHYSICAL COMPONENT: t 2	PHYSICAL COMPONENT: t 3	PHYSICAL COMPONENT: HRS SCRE	HUMAN COMP.: AWAR NESS	HUMAN COMP.: TRAI NING	HUMAN COMP.: EQUI PMNT	HUMAN COMP.: DISC IPLN	PCM RS	TRANSPORT/SECONDARY CONTROL MECHANISMS: DESCRIP	PROB USE	SURF	GRND	AIR	DC	F/E	TM RS	CRS	TARGET ANALYSIS: HUMAN: CURR NO.	HUMAN: t 1	HUMAN: t 2	HUMAN: t 3	HUMAN: ERIS SCRE	SENS ENVR SCRE	HUMAN RISK SCORE	SENS ENVR RISK SCRE	ROUTE	NOTE NO.
																																		SURF	
																															NA		NOT APP	GRND	
																																		AIR	
																																		DC	
																																		F/E	
																																		SURF	
																															NA		NOT APP	GRND	
																																		AIR	
																																		DC	
																																		F/E	
																																		SURF	
																															NA		NOT APP	GRND	
																																		AIR	
																																		DC	
																																		F/E	
																																		SURF	
																															NA		NOT APP	GRND	
																																		AIR	
																																		DC	
																																		F/E	
																																		SURF	
																															NA		NOT APP	GRND	
																																		AIR	
																																		DC	
																																		F/E	
																																		SURF	
																															NA		NOT APP	GRND	
																																		AIR	
																																		DC	
																																		F/E	
																																		SURF	
																															NA		NOT APP	GRND	
																																		AIR	
																																		DC	
																																		F/E	

ERINVTRY1

Figure 5.7 ERIS Detailed Inventory Worksheet.

DEFINITION OF TERMS

Lawmakers and rule-writers are generally isolated from the real world and the concerns of practical implementation in specific cases. This unavoidable and probably necessary situation results in issues of language that those responsible for implementation of a ERMP must be aware of and deal with from the outset. The point is that for a given situation, management must define and will define by implication if not explicitly, the operative definition of critical terms utilized in the rules and regulations. Several of these are addressed here as examples.

Salient

There is a term coming into common use in the practice of environmental risk management that has seen little general application and the concept behind it has been proven to be extremely beneficial to the ERMP.

> *A salient is a limited area of significantly different material, or material control mechanism condition, within a homogeneous area.*[161]

The salient embodies the concept of managing a homogeneous area through management of the exceptions to the homogeneity. For example, consider a large room with a uniform ceiling. To manage the ceiling is a difficult concept and it is particularly difficult in tracking the performance of the management program. On the other hand, if there is a single discolored spot in the ceiling and the objective of the management program is to have no exceptions to the uniformity of the ceiling color, management tracking becomes greatly simplified. The discolored spot is a salient as defined above.

Current

Many of the environmental rules and regulations require that "Each local education agency *(the name given to the risk owner by AHERA)* shall maintain and update its management plan to keep it current with ongoing operations and maintenance. . . ."[162] The question becomes how current is current? Does this mean that records should be posted to and made a part of the management plan on a daily, a weekly, a monthly, or an annual basis?

Failure to define the meaning of the term "current" in the context of the particular situation at hand allows others to define the term by default. Depending on the motivation, the definition provided by others may be more restrictive than good practice and the specific situation justify. Defining "current'' to imply daily or annual posting would probably represent the two extremes. Most situations would likely result in a monthly posting of a summary of the detailed records to a section of the management plan as it was defined earlier.

Exact

The term ''exact'' is frequently used in environmental rules and regulations. For example, one use of ''exact'' in AHERA is to define the ''. . . exact locations where each bulk sample is collected, ...''[163] Other rules and regulations require the recording of ''exact'' quantities of a disposed material. Taken in an absolute sense, there is no scientific possibility of complying with the requirement, but in the context of a particular situation exact may take on a practical operative meaning. In a large building, the recording of the location from which a sample was taken to the accuracy of standard architectural prints may be well-justified and appropriate. If the material were plutonium, ''exact'' may need to be defined in terms of micrograms of the substance. In some cases, the accuracy of measurement is specified by the applicable law, but in many cases it is not and management should adopt a definition of measurement accuracy applicable to the risk and local conditions.

Any

The term ''any'' appears in a number of contexts in environmental laws and regulations. In AHERA, it most notably occurs in the periodic surveillance protocol as ''record . . . any changes in the condition of the material.''[164] Taken in an absolute sense, this requirement could be interpreted as meaning that if a material's color changes to a darker shade of grey over a six-month period of time, the change must be noted.

The recordkeeping policies and practices and an operative definition of the term ''any'' can be dealt with simultaneously in this case by adopting a ''process definition'' of the term ''any.'' For example, the process of determining if any change in condition has occurred might be addressed as follows:

> The person conducting a surveillance shall take a copy of the floor plan, updated to reflect any salients as of the last surveillance or reinspection, a copy of the Salient Log and the Salient Photographs, and compare these datum to the current conditions observed in the space, noting on the Surveillance Report whether or not the salients have changed in appearance, or if any new salients have appeared. In the event that a new salient has appeared, or the condition of a pre-

existing salient appears to have changed from that recorded on the last photograph of the salient, then the Surveillance Logs, Salient Logs and Floor Plans shall reflect this, new photographs taken, and the Photograph Log updated. All documentation reflecting any of the above changes shall be turned in to the ERM.[165]

Similar process definitions of a term or phrase can be adopted that define both the term or phrase and the recordkeeping system practices and procedures.

All

In AHERA, the rules require that the inspection of a building for ACBMs be conducted as follows (in part) ". . . to identify all locations of friable and nonfriable ACBM."[166] Clearly, given the limitations of inspection and sampling techniques the term cannot be taken in the absolute sense.

Given that definitions of "exact" and "any" have been adopted, what remains is to define how the inventory shall document areas of uncertainty. It is necessary to clearly show in the records that an uncertainty exists. The Negative Default principle has as a corollary the concept that if no information is shown, the presumption is that there is no problem. This is perfectly acceptable, if true, but if the situation has simply not been investigated a major problem may result if something happens because the situation was not dealt with cautiously as the Negative Default principle would require. Management should require as a part of the record-keeping practices and procedures that the unknown be clearly and explicitly identified as being unknown in order that appropriate caution may be exercised.

Assume

Most of the environmental regulations require that a situation be assumed to be dangerous to human health and/or the environment in the absence of scientific proof to the contrary; a requirement that has been labeled here as the Negative Default principle. Underlying this statement is a major issue—when to assume versus when to demonstrate that a possible assumption is or is not valid.

The AHERA rules clearly state that the assumption must be made that a material is asbestos-containing given (1) a reasonable suspicion that the material may contain asbestos, and (2) absent scientific proof that the material does not contain asbestos. The first point is—does a reasonable suspicion exist? Management should establish a policy describing what will constitute reasonable suspicion given the existing rules and regulations. The second point is—what will constitute a reasonable standard of proof that the assumption is not valid? This question is similar to those faced in production management in determining the level of confidence for an acceptable quality control program. One hundred percent testing is not possible, but what level of testing will be required to provide a reasonable level of assurance? Analytical techniques can be very expensive and the basic decision criteria as to when they should be used can reasonably be based on consideration of several factors.

First is the question of how likely a series of analytical tests are to provide either needed additional information for the risk management system or to prove that the assumption is not valid. In the case of a commercial building built in 1969 in Los Angeles without a sprinkler system, the probability that the fluffy material sprayed on the decking is not asbestos-containing is so small that testing for purposes of management control is likely to provide little or no additional information. On the other hand, if the building had been constructed in 1980, the chances are probably 50/50 that the material contains asbestos and testing may save considerable sums in operations and maintenance costs and in countering the concerns of tenants.

The second consideration is the likelihood that the control mechanism for the material may be frequently in danger of being disturbed by individuals or that the consequences of a control mechanism failure may be significant. In either case, testing may provide information needed to deal with the substance in the appropriate fashion. Spending large sums to remediate a suspect substance utilizing the full range of protective measures when the substance is not, in fact hazardous can be a costly mistake.

REFERENCES

160. Reported in the "Law" section of *The Wall Street Journal,* November 11, 1989.
161. *Model Management Plan and Inspection Report;* p. 54; Hazardous Materials Institute, Inc.; Columbus, Ohio; 1988.
162. 763.93(d).
163. 763.85(a)(iv)[A].
164. 763.92(b)(2)[ii].
165. *Model Management Plan and Inspection Report;* Hazardous Materials Institute, Inc.; p. 55; Columbus, Ohio; 1988.
166. *40 CFR 763.85(a).*

6 PRIMARY CONTROL MECHANISMS

INTRODUCTION

The risk source (see Figure 6.1) is often mistaken as the object of environmental risk management when in fact, it is not. The real object of management attention is the primary control mechanism for the risk source, its current and likely future condition, its efficiency, and its maintenance. The existence of the risk source implies the need to control it in a manner that generally maintains its usefulness without unnecessary compromise of human health and/or the environment.

It will prove to be useful to think of risk sources, control mechanisms, transport mechanisms, and risk targets as a cohesive risk system. Each component, and the connections between the components are the proper concern of environmental risk management. Because the risk source itself is of secondary concern to environmental risk management, discussions of risk source identification and evaluation will be delayed to a later chapter.

PRIMARY CONTROL MECHANISM RECOGNITION

Often the most difficult part of the evaluation of a risk system is simply the recognition of the primary control mechanism. Primary control mechanisms may be in the form of specifically engineered containment or safety devices, in the form of naturally occurring barriers, or in the form of the training and education of an individual. Control mechanisms may therefore be very obvious or extremely subtle and both extremes can cause recognition problems. For example, many will have seen ornate decorative plaster work in public buildings built during the period 1870–1930. To a 99% level of confidence, that plaster work contains both lead (in the paint) and asbestos

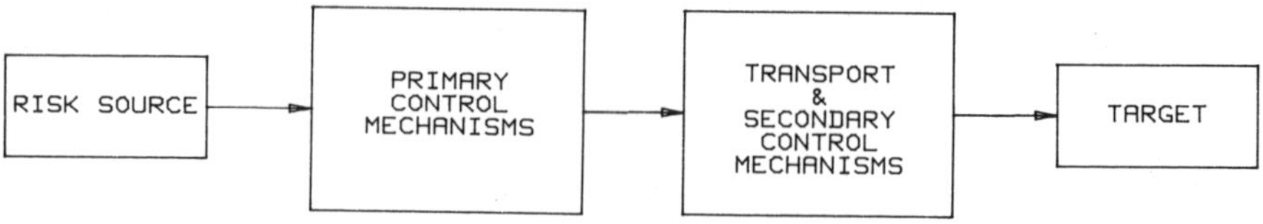

Figure 6.1 The Risk System.

(in the plaster itself), and the control mechanisms are the physical integrity of the plaster matrix and the paint membrane. The point is that these primary control mechanisms are so obvious as to often be overlooked.

A subtle and often overlooked control mechanism is the education and training of the individuals who are commonly in the presence of the risk system. Each of us deals on a daily basis with electricity to the point where serious conscious attention would have to be given to an analysis of how our individual actions have been adapted to minimize risk from this particular source. We have been trained from an early age not to stick a hairpin into an electrical outlet, or a finger into a socket, or a screwdriver into a circuit box to the point where to perform any one of these actions would require a strong effort of will to overcome the training. Recognition of these forms of control mechanisms and the evaluation of their efficiency, is an explicit and important part of environmental risk management.

Primary control mechanisms almost always consist of two major components, the physical control component and a complementary human control component. These are generally not independent, but function as a unit.

The physical component may be thought of as performing one or more of three major functions:

1. The isolation of the risk source from contact with human beings or the environment.
2. The containment of a risk source within a controlled area such that contact with the environment or individuals will only occur under low-risk conditions.
3. The maintenance of the risk source in a benign condition relative to the environment or human beings.

The human component generally operates to maintain the efficiency of the physical control components and in certain situations, as the direct control agency when physical systems will not perform adequately. The human component should never be overlooked in risk analysis for two reasons:

1. The individual involved in the primary control mechanism is also generally the individual at greatest risk if a failure occurs.
2. The individual is the most likely cause of a physical system failure either through direct action or indirectly by overriding the physical system inappropriately.

The analysis of virtually all control mechanisms may be substantially aided by the introduction of a series of logic tools and analysis devices. The major ones of concern are the fault tree, the decision tree, and methods for subjective analysis.

FAULT TREE ANALYSIS

Fundamentally, a fault tree is a decision analysis tool based on a flow chart of possible events and the probability of the occurrence of events along any path through the flow chart. The basic rules for the construction of a fault tree are simple.

First, a flow chart that displays all of the possible events leading to a control mechanism failure is constructed. Consider a simple example, an automatic flow cutoff system in a pipeline. The flow chart would look something like Figure 6.2.

As can be seen on the flow chart, there are six possible places where the system can fail. Let us assume that a failure of any of the individual system components is .01 (1%), and the possibility of an operator manual override decision being correct is .4.

The second rule of fault tree construction is that at the level of each possibility, the probability of failure plus successful operation must sum to one. The third rule is that the sum of the overall probabilities of success or failure of the system must sum to one. Placing the assumed event probabilities into the flow chart the determination of the final probabilities is shown in Figure 6.3.

For the given information, the probability of the successful operation of the pipeline cutoff system to prevent a release of the risk source into the environment, given that a leak has occurred, is .38 (38%). This is the combination probability of the automatic system functioning correctly (95%), times the probability that the operator made a correct manual override decision (40%). The system therefore has a 62% probability of failure to prevent a release as a result of the failure of one of the components, or an incorrect operator decision to override the automatic cutoff. Note that raising the reliability of the automatic components to 99.9%, but leaving the operator override probabilities alone, will only increase the overall probability of the system functioning correctly to 40%.

FAILURE PROBABILITY DEVELOPMENT

For an engineered control mechanism, the failure probabilities can generally be obtained with a reasonable degree of accuracy, from component manufacturers or system simulation tests for example. In addition, it is possible to define the change of failure probabilities with time. The probability of failure generally increases along a learning-curve shaped path.

The difficulty in estimating failure probabilities comes with control mechanisms that have not been engineered for the specific purpose; for example the asbestos-containing matrix in spray-on fireproofing or a naturally occurring clay lens to contain a soil contaminant. In an ideal world, tables of experience data could be relied on to estimate these failure probabilities. Unfortunately that type of data is not currently being collected and disseminated. Other means must often be relied upon to develop failure probabilities, or decision analysis methods that do not rely on the assignment of probabilities must be employed.

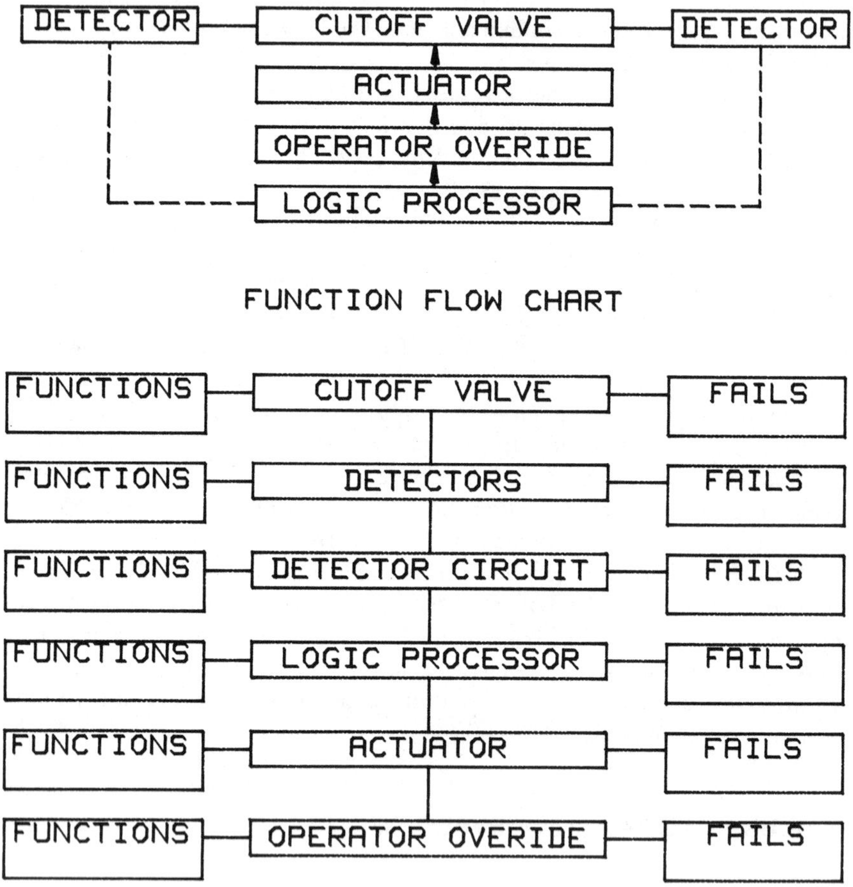

Figure 6.2 Event Flow Chart of an Automatic Pipeline Cutoff.

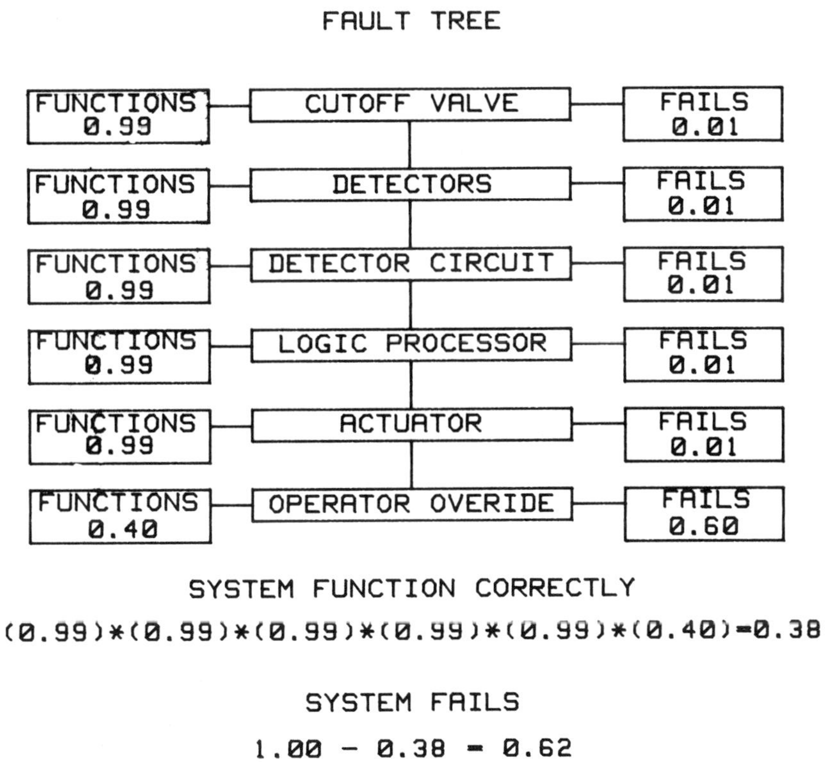

Figure 6.3 Fault Tree for an Automatic Pipeline Cutoff.

At the opposite extreme from the development of failure probabilities through testing and simulation, lies a technique that is surprisingly useful and accurate—the Delphi Technique. The Delphi Technique was developed in the 1950–1970 period to quantify the subjective knowledge of experienced individuals. Basically a small group of highly experienced individuals in a particular field are gathered together in a structured discussion and analysis session on a particular topic or set of topics. During the course of the discussion, the individuals are asked to rank their subjective opinions in a format that will require the numeric assignment on a scale of 1 to 10. After all of the individuals have expressed their opinion, the numeric scores can be utilized to develop a probability of the occurrence of a specific event.

Consider the automatic cutoff example above. The Delphi Technique could be used to develop the probability of incorrect operator overrides in several action scenarios. For example, if the automatic system has incorrectly shut down the pipeline in the past, the opinion of the Delphi group may be that the probability of an incorrect operator override is higher than if the system is not prone to this type of failure. Since most safety systems function in exactly this manner, the lesson of the automatic cutoff scenario is that significant resources should be employed to decrease the probability that the operator will incorrectly override the system.

The Delphi Technique and similar probability of occurrence development tools utilizing a basically subjective approach are particularly useful, and in fact may be the only tools that will work when complex human interaction with the control mechanisms are a natural and unavoidable part of the risk system. Other areas of use are when the system is extremely complex and the resources for a quantitative failure analysis are not available, or the underlying information on which such a failure analysis might be based do not exist.

Another important area of application of subjective group analysis is at the very beginning of complex environmental risk evaluations. A group analysis can be used to form an overall picture that is less likely to miss important failure modes.

DECISION TREES

Decision trees are logic devices that map out the path of choices leading to a specific conclusion. Their primary purpose is to insure that all individuals charged with making a decision follow a uniform approach to that decision. In essence, decision trees codify the subjective judgment of an individual.

Although it has often been criticized, AHERA guidance documents set out a decision tree approach for determining the current and future threat of a given asbestos-containing building material. This approach has been criticized primarily because of its lack of quantitative precision and the natural tendency of the observer to react to a risk decision conservatively when quantitative information is lacking.

Within limits these criticisms are valid, but two primary areas of application of the decision tree approach are overlooked when making them. The AHERA decision tree, with its large element of reliance on subjective judgment, deals with both the current condition of the control mechanism, and with its likely future condition. There are currently no reliable methods for quantitative measurement of the current condition of most of the asbestos control mechanisms, and the extremely large number of unquantifiable variables that will enter into the future condition render a quantitative assessment virtually impossible. Both of these problems are generally ignored or overlooked by those who criticize the approach.

On the other hand, those who support the approach often overlook the problems associated with an overly conservative subjective analysis. The argument offered in defense of the conservative approach is that it will result in the removal of the materials and thereby render a safer environment. Again, this argument is far too simplistic because it overlooks the fact that experience has shown 60–80% of removal operations have left the building more, rather than less contaminated. Further it overlooks the fact that removal may be absorbing scarce resources that should be employed in higher risk remediations or other critical applications.

Decision trees may result in a simple conclusion (Yes or No for example), or a ranking system for assigning a numeric priority to the conclusion may be appended to the decision tree. Figure 6.4 shows a comprehensive version of the AHERA decision tree with appended ranking system.

The decision tree of Figure 6.4 includes an analysis of both the risk source (is it asbestos containing or not), the control mechanism type (friable or nonfriable), the current and likely future condition of the control mechanism, and the existence or non-existence of major transport mechanism (air flow) factors. The exact method of application of this particular decision tree will be dealt with in a later chapter; the important point here is that this decision tree displays a set of reasonably complete logic paths leading to one of a specific set of possible conclusions, providing a uniform approach within which the observer can arrive at a conclusion.

TIME DEPENDENCE OF FAILURE

Physical control systems that have been specifically engineered to perform a function are generally the simplest to identify and to evaluate. These may be in the form of automatic flow cutoff systems, containment systems, specifically constructed barriers, or isolation systems that prevent unauthorized access to the immediate presence of the risk source. All control mechanisms will have some probability of failure and evaluation will concentrate on the possible failure modes and the likelihood of failure.

Inherent in all control mechanism failure analysis is a common trap, the effects of time on the probability of failure. The tendency to ignore time as a major factor is quite common, but it is almost always a major analysis error. As mentioned earlier, the probability of failure of a system generally increases with time along a learning-curve shaped path with an initially low probability of failure after the system has been burned in, then a rapidly increasing probability of failure to a relatively high but reasonably stable level at a later date. This shape of the time dependence of failure is shown in Figure 6.5

The application of fault trees to the quantitative analysis of failures in an engineered system has already been illustrated, but generally the resources required to perform such a comprehensive analysis are not available and other less quantitative methods must be employed. This is not necessarily bad because the quantitative approach has two serious problems.

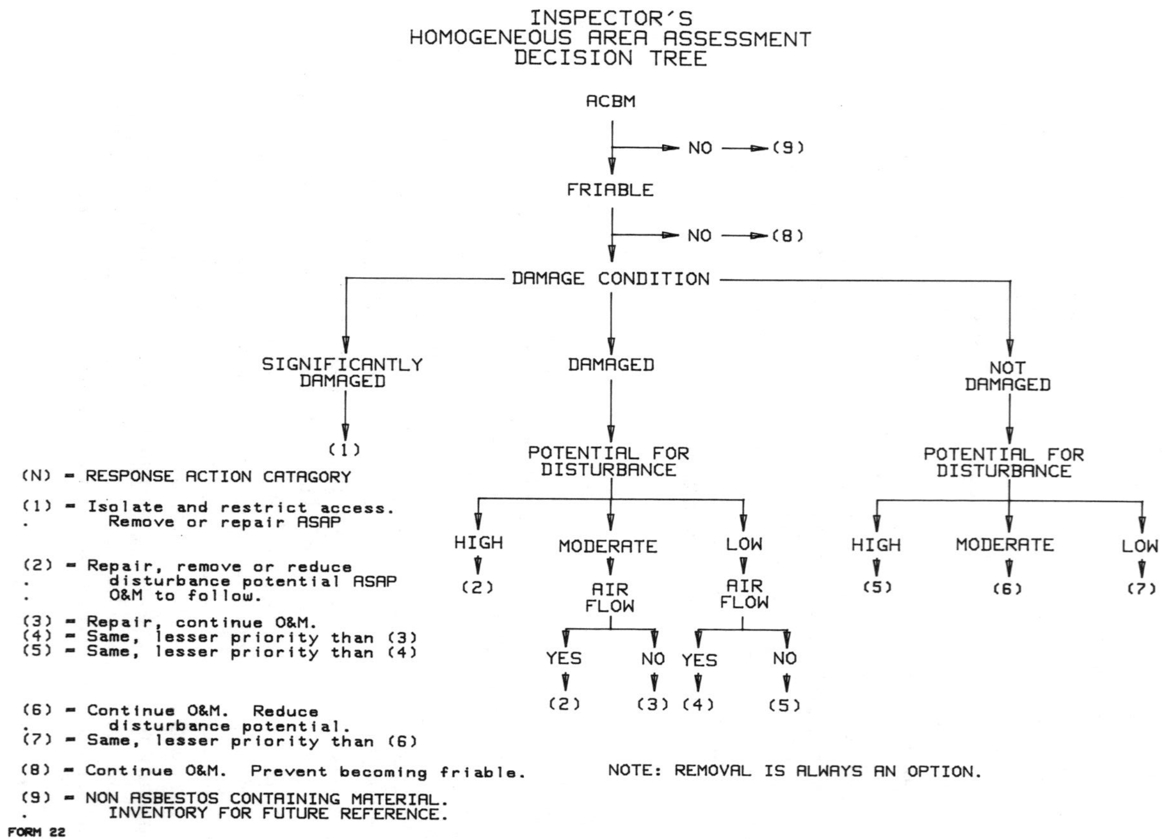

Figure 6.4 AHERA Decision Tree for Control Mechanism Condition Assessment with Appended Result Ranking.

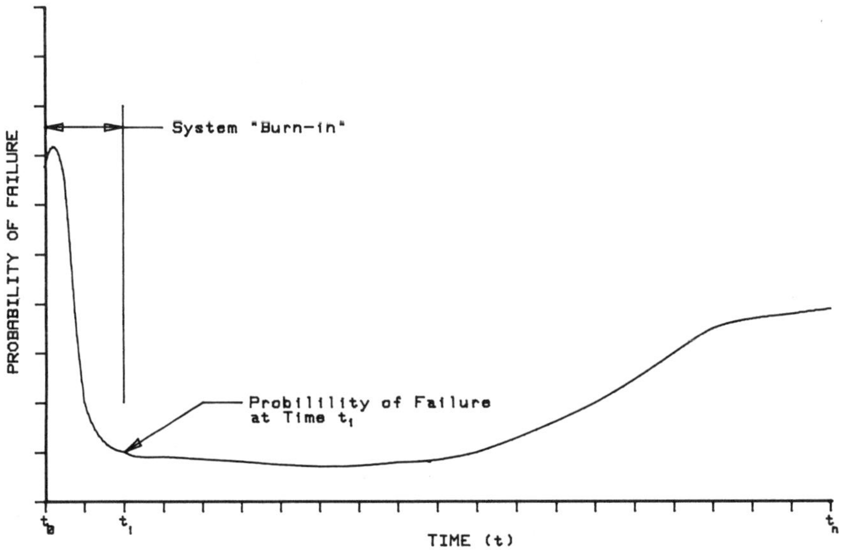

Figure 6.5 Time Dependence of the Probability of Failure for a Complex System.

TYPE II FAILURES

In Figure 6.3 the failure probabilities of an entire system, as determined by the reliability of its individual components, is portrayed. Excluding the human component for a moment, the overall probability of failure for the physical components of the system may be shown to be .0490 (4.9%). In the case of each physical component, failure has been defined as the failure of the component to function as designed, resulting in the overall system failing to operate properly. This will hereafter be referred to as a Type I failure; a failure of the total system to function as designed.

Suppose now that the reliability of each of the physical system components is increased to 0.999 (99.9%). In this case the overall probability of the failure of the physical system to function as designed will have decreased to .0050 (0.5%). Does this mean that the system, including the human component, is more reliable? Unfortunately, experience has shown that this is not necessarily the case! In fact, if the increase in reliability is accompanied by an increased level of sensitivity of the system components to detect a possible problem, the opposite may be the case, the overall system reliability may decrease. The problem results from an increased likelihood of Type II failures.

In determining the reliability of the physical part of the system, it was the reliability of the individual components in functioning as designed that was the measurement of importance. When placed in a complex system, the components may function perfectly and shut down the system when an actual leak has not taken place. As component sensitivity in detecting an abnormal condition increases, the possibility that such a shut-down will occur will increase. The shutting down of the system when an actual problem does not exist is a Type II failure.

Often the shutting down of a system even when an actual failure has not occurred is proclaimed as an additional advantage of system design; a "fail-safe" system. The fail-safe system design is both proper and acceptable so long as it is recognized that the human component will respond to this design with an increased probability of a human Type II failure.

Most systems are designed to have operator intervention capability, the assumption being that the operator should be able to determine, through additional input from outside of the system itself, that the override capability should be used because a false failure has occurred. A human Type II failure is then an override decision in the presence of an actual failure. There are a number of instances of this situation occurring in both everyday life and with respect to several well-publicized environmental incidents such as Three-Mile Island, the Kill Waterway spill, and others.

Reliance on physical systems must be tempered with the knowledge that the more sensitive the physical system, the more likely it is that the human operator will learn to override the system without adequately determining whether the system is experiencing a Type II failure and instead assuming that a Type II failure has occurred. This tendency to commit a human Type II must be explicitly recognized and addressed in the analysis and management of the system.

In general, the more complex the system and the longer the period of time over which it is required to perform, the greater the likelihood of a Type II error becoming the source of a failure. This concern applies to both mechanical and human control systems.

THE DANGER OF THE APPEARANCE OF UNJUSTIFIED PRECISION

Just as there is a tendency to want to develop precise numeric estimates of Type I and Type II failure probabilities, there is also a danger in giving the appearance of precision in probability estimates through the assignment of numeric values. When precision is warranted, the use of numeric values in the traditional sense is appropriate. When precision is lacking, the use of numeric values can be extremely misleading and should be avoided.

In later sections of this book, an estimate of the probability of failure for primary control mechanisms, the probability of the use of a specific transport mode, or the current or likely future condition of the human component of the primary control mechanisms will be required in order to complete the risk analysis. Although a numeric rating system will be employed to determine the magnitude of a risk, the analyst should use a simple non-numeric recording system such as G/F/P (Good, Fair, Poor) or H/M/L/0 (High, Medium, Low or zero) as opposed to a numeric value. These non-numeric values can be converted to a simple weighting scale when needed, but avoid the explicit appearance of precision when such precision is unjustified.

HUMAN CONTROL MECHANISMS

The human being is not generally thought of as being a primary control mechanism, which is unfortunate because they often are the active component in controlling a risk. To appreciate this point, consider the case of electricity as a risk source. Electricity represents a very highly controlled source of potential energy that is normally confined to the distribution system designed to carry the energy—the wires, transformers and control panels. It is, however, a potential energy source of no value until it is converted into useful work—useful work normally controlled by people. Despite all safety precautions, it is the individual that will ultimately decide whether the application will be harmful or not. For example, so-called ground fault detectors designed to interrupt the flow of energy in the presence of a short have been found to malfunction as much as 50% of the time in construction applications. The safety training and implementation of that training to prevent the realization of the risk is primarily under the control of the individual.

The existence of the human being as a primary form of risk control mechanism introduces a series of difficult to quantify issues for the analyst. At the beginning of the discussion of physical control mechanisms, it was pointed out that such mechanisms operate in one or more of three modes. The human being as a control mechanism interfaces with these modes to form a complete system by determining how well the physical systems function. Consider a simple example—lead paint. So long as the membrane containing the lead is intact, the physical containment system for the lead functions with a high degree of reliability. The human being interfaces with this control mechanism by the act of observing whether or not the membrane is intact, or by taking deliberate action that will disrupt the membrane under either controlled or uncontrolled conditions.

Under AHERA the existence of an asbestos-containing material requires that the physical control mechanism for the asbestos be kept under regular periodic surveillance.[167] The entire point of the surveillance is to determine if the physical control mechanisms for the ACBM are still functioning correctly, and if not, to report that fact in order that action may be taken to prevent the realization of the risk. The human part of the control mechanism for the ACBM is, in fact, the real governing factor on whether or not the risk will be realized because without the correct assessment of the condition of the physical control mechanism, the entire system may fail. Whether the physical control mechanism has deteriorated in performance is not sufficient; the human component must function to detect the change in performance.

At refineries, nuclear materials plants, Superfund sites, in automatic cutoff systems, and school buildings, primary emphasis must be placed on the human element. Physical control mechanisms can fail, but the frequency of failure and the magnitude of the consequences of failure are in large part governed by the human mechanisms in place to oversee or operate the physical mechanism. This concept requires that the inspector place a major emphasis on the analysis of Type II failure modes, training and education parameters for the individuals involved, and execution performance.

APPLICATION OF THE HRS TO PRIMARY CONTROL MECHANISM ANALYSIS[168]

The application of the Hazard Ranking System (HRS) begins with the analysis of primary control mechanisms. This application was not envisioned by the designers of the HRS and its application here has been augmented to provide a higher level of visibility into the probable sources of primary control mechanism failure and to form the basis for the calculation of an environmental risk inventory value.

Current Condition

The HRS assumes that a release has already occurred; that is, that the primary control mechanism has already failed. To provide needed management insight into the probability of such a failure occurring as a basis for resource allocation, the environmental risk inventory control system must explicitly look to the current and likely future condition of the primary control mechanism to assess the probability of such a failure.

The assessment of the current condition of the control mechanism should, unless scientifically more precise methods are available, be conducted with the end result of a Good/Fair/Poor condition category assignment in mind. This simple conclusion, that the current condition of the primary control mechanism is good, fair or poor, will normally be sufficient to assist in the allocation of resources and greater precision is often unwarranted or unjustified.

Probable Future Condition

The probable future condition should again receive a simple value assignment, specifically High/Medium/Low probability of failure. With regard to a future failure the analyst would be very hard pressed to demonstrate that a more precise probability statement is justified without engineering data.

The analysis of the probability of a future failure should take place for at least three different points in future time (t_1, t_2, t_3), assigning a H/M/L value to each point in time. This assignment must be recognized for what it is, a best-educated guess, not a precise prediction of the future, but its importance should not be overlooked. Suppose that two risk reduction projects are judged to have nearly equal current impact and that resources exist for implementing only one of the projects. The project selection may then be determined by the estimate of the future condition of the control mechanism, the one degrading the fastest becoming the project of choice. Crude as this may appear, some information on which to base a decision is superior to none if the information is consistently available, consistently arrived at, based on reasonable expectations and experience, and developed in a reasonably unbiased manner.

HRS Factors

Two major HRS factors can provide insight into the level of risk:

1. Whether or not there has already been a control mechanism failure as denoted by the existence of an observed release incident of direct contact or fire/explosion.
2. The containment characteristics.

Observed Release

The HRS provides a series of observed release weighting factors as follows. These are provided for each of the five major transportation routes (to be discussed in a later chapter). An observed release is defined as the detection of the risk source at a statistically significant level above background outside of the primary physical control mechanism boundaries. In the case of Fire/Explosion transport mechanism, the observed release is defined as being direct evidence of the occurrence of a fire or explosion. For our purposes, if there is evidence of an observed release or direct evidence of a fire or explosion, a weight of 45 will be assigned. If such evidence does not exist then a weight of zero is to be used.

Containment

Surface Water Route (Table 6.1)

Table 6.1 Containment Values for Surface Water Route

Assign containment a value of 0 if: (1) All the waste at the site is surrounded by diversion structures that are in sound condition and adequate to contain all runoff, spills, or leaks from the waste; or (2) intervening terrain precludes runoff from entering surface water. Otherwise, evaluate the containment for each of the different means of storage or disposal at the site and assign a value as follows:

	Weight
A. Surface Impoundment	
Sound diking or diversion structure, adequate freeboard, and no erosion evident	0
Sound diking or diversion structure, but inadequate freeboard	1
Diking not leaking, but potentially unsound	2
Diking unsound, leaking, or in danger of collapse	3
B. Containers	
Containers sealed, in sound condition, and surrounded by sound diversion or containment system	0
Containers sealed and in sound condition, but not surrounded by sound diversion or containment system	1
Containers leaking and diversion or containment structures potentially unsound	2
Containers leaking, and no diversion or containment structures or diversion structures leaking or in danger of collapse	3
C. Waste Piles	
Piles are covered and surrounded by sound diversion or containment system	0
Piles covered, wastes unconsolidated, diversion or containment system not adequate	1
Piles not covered, wastes unconsolidated, and diversion or containment system potentially unsound	2
Piles not covered, wastes unconsolidated, and no diversion or containment or diversion system leaking or in danger of collapse	3
D. Landfill	
Landfill slope precludes runoff, landfill surrounded by sound diversion system, or landfill has adequate cover material	0
Landfill not adequately covered and diversion system sound	1
Landfill not covered and diversion system potentially unsound	2
Landfill not covered and no diversion system present, diversion system unsound	3

Ground Water Route (Table 6.2)

Table 6.2 Containment Value for Ground Water Route

Assign containment a value of 0 if: (1) all the hazardous substances at the facility are underlaid by an essentially non-permeable surface (natural or artificial) and adequate leachate collection systems and diversion systems are present; or (2) there is no groundwater in the vicinity. The value "0" does not indicate no risk. Rather, it indicates a significantly lower relative risk when compared with more serious sites on a national level. Otherwise, evaluate the containment for each of the different means of storage or disposal at the facility using the following guidance.

	Weight
A. Surface Impoundment	
Sound run-on diversion structure, essentially non-permeable liner, (natural or artificial) compatible with the waste, and adequate leachate collection system	0
Essentially non-permeable compatible liner with no leachate collection system; or inadequate freeboard	1
Potentially unsound run-on diversion structure; or moderately permeable compatible liner	2
Unsound run-on diversion structure; no liner; or incompatible liner	3
B. Containers	
Containers sealed and in sound condition, no liner or moderately permeable liner	1
Containers leaking, moderately permeable liner	2
Containers leaking and no liner or incompatible liner	3
C. Piles	
Piles uncovered and waste stabilized; or piles covered, waste unstabilized, and essentially non-permeable liner	0
Piles uncovered, waste unstabilized, moderately permeable liner, and leachate collection system	1
Piles uncovered, waste unstabilized, moderately permeable liner, and no leachate collection system	2
Piles uncovered, waste unstabilized, and no liner	3
D. Landfill	
Essentially non-permeable liner, liner compatible with waste, and adequate leachate collection system	0
Essentially non-permeable compatible liner, no leachate collection system, and landfill surface precludes ponding	1
Moderately permeable, compatible liner, and landfill surface precludes ponding	2
No liner or incompatible liner; moderately permeable compatible liner; landfill surface encourages ponding; no run-on control	3

Table 6.3 ERIS Primary Physical Control Mechanism Worksheet

Observed Release	______
Containment:	
Surface Water Route	______
Ground Water Route	______
Direct Contact	______
Fire/Explosion	______
Total Raw Score	______

Air Route There is no containment value provided for the air route by the HRS.

Direct Contact Containment refers to whether or not the material itself is accessible to direct contact. If the material's surface is accessible to contact, such as in surface impoundments, containers (whether sealed or unsealed), piles, tanks, or landfills with a cover depth of less than 2 feet, or has been spilled directly on the ground or other easily contacted surface, then a weight of 15 should be applied. Otherwise, meaning the material cannot be directly contacted easily, a weight of zero is to be applied for this factor.

Fire/Explosion If the materials on the site are segregated in such a way as to prevent the possibility of the formation of an explosive or flammable mixture, or if no such materials are present, then a weight of 1 is to be used. If this is not the case, or if gases are detected in an uncontrolled area, then a weight of 3 is to be assigned.

Physical Component, Primary Control Mechanism ERIS Score

The sum of the weight values as obtained from the tables above will be the total raw ERIS score (Table 6.3). The maximum possible score is 69.

HUMAN COMPONENT ANALYSIS

The analysis of the human component of risk control depends on at least four key elements with respect to the individual's involved: (1) Awareness, (2) Training, (3) Equipment, and (4) Discipline. The assessment of the condition of the human component relative to these four elements can provide significant insight for resource allocation to reduce risk. This component is not explicitly considered by the HRS.

Awareness

The degree of awareness is both a major concern of a number of environmental and occupational laws and critical to whether or not a risk reduction system will function. The analysis of the level of awareness should include the degree to which individuals have been informed of the presence and nature of the risks, of appropriate behavior in the presence of the risks, and at a higher level, the possible consequences and form of Type II errors. Evaluation of the composite group level of awareness will normally take place on a Good/Fair/Poor rating scale. In judging awareness, the analyst is cautioned to look carefully at the level of awareness of the "short-term" worker who may be in an ideal position to damage the integrity of the entire control system.

Training

The level of training of the individuals who will come into direct contact with the risk source or the primary control mechanism will be critical to system performance. Key—although by no means the only—elements for judging the level of training should be the existence and implementation of formal training programs, the nature of the provider of the training, the retraining and information updating program to maintain currency of training, the existence and accessibility of operating manuals and MSDSs, the applicability of the training to the work performed, and the like. Again, the composite level of training should be assigned the Good/Fair/Poor type of rating.

Equipment

The availability, appropriateness, and sufficiency of the equipment provided to both deal with the ongoing maintenance of the physical control systems and the emergency response equipment for dealing with a failure must be evaluated. The rating should again be on the Good/Fair/Poor scale. Naturally, in all cases the detailed evaluation notes to support all ratings should make clear any deficiencies noted by the analyst.

Discipline

Without proper workforce discipline, the awareness, training and equipment levels will be meaningless. The analyst should carefully investigate to determine whether or not the work practices are properly and consistently implemented on the individual level and the consequences when they are not. Consequences for inappropriate behavior are meaningless if they are not implemented. Again, the composite rating should be on the Good/Fair/Poor scale.

NUMERIC VALUE CONVERSION OF G/F/P (Table 6.4)

In order to assign a risk "value" to the inventory, the G/F/P ratings should be converted to a number. Because higher numbers imply higher risk, it is suggested that the following conversion table be used.

Using these ratings will allow the Primary Control Mechanism evaluation portion of the ERIS Detailed Inventory Worksheet to be completed. The remaining column concerns the Primary Control Mechanism Composite Score (PCM-Score) which will be calculated based on the other values recorded as follows.

1. Multiply each of the individual scores for the physical control mechanism current condition, the ERIS score for the physical component, and the individual scores for the human component together. The maximum possible product for this calculation will be 215,625.
2. Normalize the actual product by dividing it by 215,625 and multipling by 100, yielding the PCM-RS value to be recorded on the ERIS Detailed Inventory Worksheet. The maximum PCM-RS score will then be 100.

Table 6.4 G/F/P Rating Conversion

G/F/P Rating	Numeric Value
G	1
F	3
P	5

PRIMARY CONTROL MECHANISM ANALYSIS										
	PHYSICAL COMPONENT					HUMAN COMP.				PCM RS
DESCRIPT	CURR COND	t 1	t 2	t 3	HRS SCRE	AWAR NESS	TRAI NING	EQUI PMNT	DISC IPLN	
	G/F/P (1/3/5)	H/M/L (5/3/1)	H/M/L (5/3/1)	H/M/L (5/3/1)	TABLE 6.3 (Ø<-69)	G/F/P (1/3/5)	G/F/P (1/3/5)	G/F/P (1/3/5)	G/F/P (1/3/5)	

Figure 6.6 ERIS Detailed Inventory Worksheet Primary Control Mechanism Section.

REFERENCES

167. *40 CFR 763.92 [b]* Periodic Surveillance [1] At least once every 6 months . . . shall conduct periodic surveillance in each building . . . that contains ACBM or is assumed to contain ACBM. [2] Each person performing periodic surveillance shall: [i] Visually inspect all areas that are identified in the management plan as ACBM or assumed ACBM. [ii] Record the date of the surveillance, his or her name, and any changes in the condition of the materials.

168. Details on the Hazard Ranking System may be found in Appendix III to this book and at *29 CFR 300 Appendix A—Uncontrolled Hazardous Waste Site Ranking System: A User's Manual.*

7 TRANSPORT AND SECONDARY CONTROL MECHANISMS

INTRODUCTION

No matter how well or poorly controlled a given risk source may be, damage to a given target is not possible until the risk source and the target are in immediate proximity to each other. Immediate proximity usually means direct contact, but with the increased recognition of global environmental risks, indirect contact with the ultimate target must also be considered. Transport mechanisms are the means by which a risk source is brought into proximity to the target or into a position where it may influence the target indirectly.

Transport mechanisms may be thought of in the following major forms.

- Air Movement
- Surface Water Movement
- Ground Water Movement
- Fire/Explosion
- Direct Contact
- Atmospheric Transportation

Each of these normally have two major components, the mechanisms or medium of movement, and barriers to the movement or secondary control mechanisms. For example, if the medium of movement is subsurface (ground) water, then the secondary control mechanisms will be resistance to the rate of movement of water through the ground (permeability and rate of transfer), and the distance over which the substance must move to reach a target.

Under the USEPA Hazard Ranking System, the first five are denoted as hazard migration modes or route factors. In AHERA, only the first transport

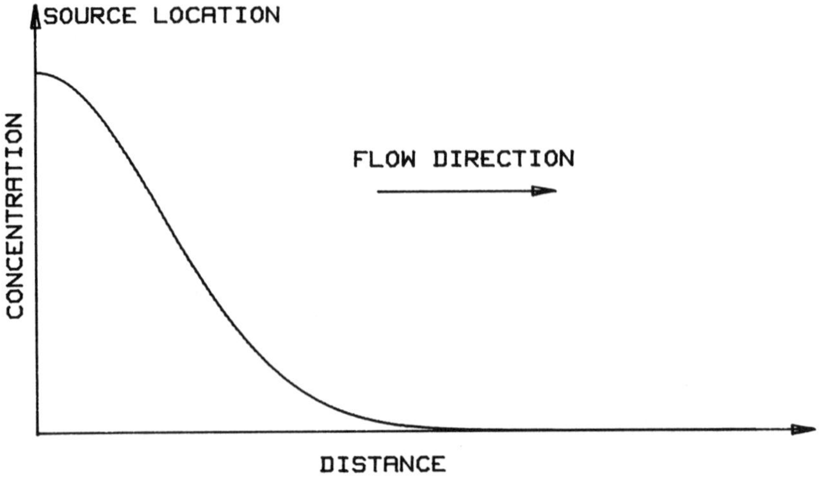

Figure 7.1 Distribution of a Risk Source with Distance
Risk Source has a Higher Density than the Medium.

mechanism, air movement, is considered because of the nature of the risk source. To this point, atmospheric transportation is not directly considered by existing federal legislation although the proposed *Clean Air Act* reauthorization bill and several international treaties directly address the issue in acid rain and atmospheric ozone degradation controls.

TRANSPORT MECHANISMS IN GENERAL

This is not intended to be a text on fluid flow or the environmental fate of a substance, but some understanding of these ideas are required and are reviewed here. Basically, the ability of a given medium to transport a risk source depends on several physical and chemical properties of the source and the medium. The first, and probably the most obvious, is the relative density of the risk source and the medium. Lead will not be transported over a great distance in either water or air for obvious reasons, while oil may be transported over a great distance in water and light hydrocarbons in air. The density of the risk source relative to the density of the medium governs the distance of transport, all other things being equal.

This should not be misunderstood. A risk source that is more dense than the transport medium does not usually drop straight to the bottom of the medium, but is instead distributed with distance from the original location of the source in a non-linear pattern such as that shown in Figure 7.1.

This pattern assumes several other properties of the risk source and the medium. First is that the medium is a fluid moving with a given steady state velocity along the x-axis. Second is the assumption that the risk source does not chemically react with the medium or otherwise change in physical or chemical form upon release into the environment. The consequences of each of these assumptions must be explored, particularly with respect to a violation.

Velocity of the Transport Medium

Violation of the assumption of a steady state velocity of the transport medium may occur in two ways—the velocity is zero, and the velocity changes rapidly relative to the rate of release of the risk source. A velocity of zero implies that the risk source will essentially build up in the immediate vicinity of the point of release until the concentration has reached an equilibrium with respect to the capacity of the medium to retain it. It is this mechanism that allows the buildup of toxic materials at low points in tunnels, clay lenses in the ground, or low areas in the beds of bodies of water, if the risk source is denser than the medium.

This can generate high localized concentrations of the risk source and dangerous conditions for both the investigator and for users of the transport medium. It is also a problem because a test of the transport medium at a point away from the area of concentration may result in an incorrect assessment of the amount of the risk source actually present. The zero velocity distribution is shown in Figure 7.2.

The problem of uneven distribution in a low-velocity medium has been emphasized by recent investigations in the Great Lakes that have identified localized very high levels of PCB, dioxin and heavy metals near potable water intakes for Detroit, Chicago, and other major metropolitan areas resulting from the collection of these substances at low points in the lake beds.

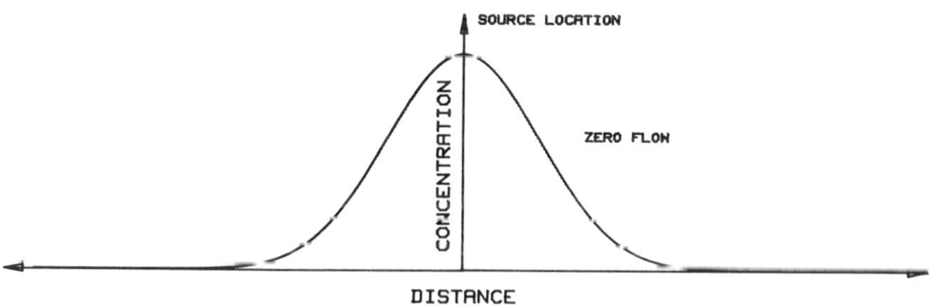

Figure 7.2 Risk Source Distribution, Zero Velocity of Medium.

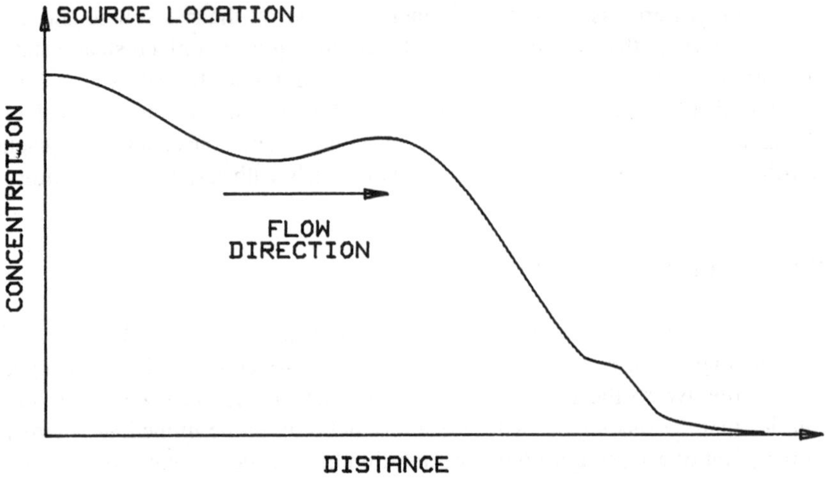

Figure 7.3 Risk Source Distribution, Variable Velocity of Medium.

If the velocity of the medium is rapidly changing relative to the rate of release, the issue of downstream concentration and distribution may become extremely complex. Generally, the higher the velocity of the medium, the greater the distance over which the risk source will be distributed. In addition, changes in velocity may result in a "pulse" in concentrations downstream that may be compounded by the dislodging of localized concentrations deposited when the velocity was relatively low. This possibility is shown in Figure 7.3.

One of the major areas of risk evaluation will be to determine if the concentration of the risk source exceeds acceptable limits. In the case of a variable velocity medium, this issue will rest primarily on whether or not the pulse or localized concentration exceeds acceptable limits and its location relative to a target. Assessment of the risks associated with a pulse is further complicated by the fact that most risk source impacts on targets are measured based on time-weighted averages. Such averages may or may not correctly assess the impact of a short-duration event. This does not mean that time-weighted averages are incorrect measures, although in certain circumstances this may be true, but rather that target impact analysis is in its infancy and subject to rapid development and change.

Continuous Versus Point-in-Time Releases

Continuous releases of the risk source into the environment present additional complications in distribution pattern analysis. All of the distribution mechanisms operate as discussed, but their affects may be compounded with time. In a very simple example, consider the process involved in repairing a car during the period 1910 through 1970. During this period, lead was used as a primary repair material and on a daily basis a small body shop might use 20–30 pounds with 10% being ground off in the smoothing process. Over a period of a year, total wastage might amount to 400–600 pounds of lead. For a shop in the same location for 10 years, 2–3 tons of lead may have been ground off on to the floor with a large proportion being swept out on to the surrounding ground. Such a situation would result in a high concentration of lead in the soils.

Continuous releases of Freon through leaks in refrigeration systems, discharges of pollutants into streams or the air, and the deposition of small quantities in landfills over a period of time may all lead to the same end result—high localized concentrations. The transport mechanism will define the location of the final deposition site, and as indicated earlier, those sites may not be in obvious locations. Unless these locations are identified and brought under primary control, an alteration of the flow pattern of the medium, or the normal working of the flow pattern over time, may result in high concentrations in the immediate vicinity of a target.

A corollary consideration is the opportunity presented for property development to expose the site of a high concentration. The exposure of a site generally will occur in one of two ways; direct exposure, or the disturbance of the primary or secondary control mechanism surrounding the site. Both provide the risk source with new access to transport mechanisms.

The first mechanism is relatively obvious and is the reason for the emphasis on the need to document unknown areas. Especially around heavily urbanized or industrialized areas, careful investigation for the location of concentration sites is very necessary, but the problem may occur in agricultural areas or far downstream. There are a number of recorded incidences of high levels of arsenic in remote locations on farms resulting from a government program of the 1950's that supplied arsenic-based pesticides to farmers who, choosing not to use them, dumped them in otherwise untillable locations. Dredging operations on waterways have uncovered sites dozens of miles downstream.

The second mechanism is probably of greater concern because it is less obvious. There are a number of recorded instances of the construction of a sewer line, for example, allowing a new path for risk source transport leading to the collection of methane gas in the basement of homes distant from the original site. Underground storage tank leaks have had similar problems. Although the control mechanisms for the localized concentration may not be obviously disturbed, those control mechanisms may not be efficient enough to prevent the transport of the risk source along a newly exposed path resulting in the relocation of the risk source.

As is evident from the foregoing discussion, the transport of a risk source from one location to another is a highly complex and, at least at present, a poorly understood phenomenon. Particularly if the release of the risk source occurs on a continuous basis, the development of high concentrations may be very difficult to predict or analyze simply because the processes may be very subtle, hard to define, or difficult to detect.

Environmental Fate of the Risk Source

The assumption that the risk source and the environment or medium do not react in a manner that chemically or physically alters the risk source or its distribution generates a number of very complex issues when it is violated, as it often is. The environmental fate of a risk source is governed by the relative physical and chemical properties of both the source and the environment. An example is the complex transport and interaction of Chloroflurocarbons (CFC) with the atmosphere to result in the depletion of the ozone layer. A number of remediation and environmental risk impact analysis methods depend on the extent and kind of interactions that may take place between the risk source and the general environment.

The introduction of a substance into the general environment adds other factors to the equation of distribution. The old adage of "the bigger fish eat the smaller fish" has very important implications for the distribution of a risk source. In general, all forms of life tend to absorb chemicals from their environment. As the less-developed organisms are utilized by higher forms, the absorbed materials tend to concentrate at a rate approximating the consumption of food materials by the life form. Suppose that each higher form tends to eat ten times its own body weight in lower form foodstuffs and that the risk source is perfectly absorbed into the higher form. Under this scenario, if there are n steps in the food chain, the last step will receive a concentration approximating 10^n of the original ambient environmental concentration. Since man is at the end of most food chains, this transportation process can have very important implications.

A third alteration mechanism occurs when a particular substance that may be low risk itself, operates on the environment in a manner that results in the release of higher risk substances. The acid rain phenomenon is one example. In the mountain West, early and current mining practices have left large areas of mining spoil. Acid rain tends to leach hazardous substances from the spoil as well as directly from the rock and mineral formations, resulting in the contamination of surface and underground aquifers with lead, antimony, bismuth, arsenic and other heavy metals. The acid in the rain water is not a particularly hazardous substance, but the leachate is hazardous. A second example is the action of CFCs as catalysts to a reaction that depletes the supply of ozone, allowing more ultraviolet rays to penetrate to the earth's surface. The greenhouse effect is the result of similar mechanisms.

Again, as with the difference between a point-in-time release versus a continuous release, the cumulative effects can be very hard to predict and may outweigh the point-in-time effects by orders of magnitude.

Secondary Control Mechanisms

Secondary control mechanisms are defined as barriers to the movement of the risk source associated with the transport mechanism. The difference between primary and secondary control mechanisms can be visualized by considering asbestos. The matrix that holds the asbestos fibers in place is a primary control mechanism. The barriers to air movement that may transport the asbestos fibers from one point to another such as walls and doors, are secondary control mechanisms because they are associated with the movement of the air, not with holding the asbestos fiber in its original position.

Secondary control mechanisms often appear in the form of something that retards the velocity of movement of the transport mechanism, such as the tendency of a dam or soil to slow down the movement of water. Secondary control mechanisms may also occur in other manmade forms such as retaining or blast walls, fences, security systems, and the like. Again, the difference is that these secondary mechanisms do not prevent the release of the risk source from its original location, but prevent or retard the distribution of the risk source to a larger area or prevent contact with an area containing a risk source.

One of the major secondary control mechanisms is simply the distance of separation between risk source and target. This factor is explicitly considered in the Hazard Ranking System. The key to distance as a secondary control mechanism is in the measurement of the distance, which is always measured as the shortest distance from the farthest distribution point of the risk source to the target of concern.

Summary of General Transport Mechanisms

By way of an initial summary, consider Figure 7.4. The transport mechanism has been shown to be a function of the medium of transport and the secondary control mechanisms operating in that medium that result in a given distribution pattern in the environment. The final distribution pattern is a function of the rate of release of the risk source into the environment, the relative physical properties of the risk source and the medium of transport (relative density and the velocity of the medium, barriers to transport, etc.), and the environmental fate of the risk source (chemical and physical interactions of the risk source and environment, introduction into the food chain, natural or artificial disturbance of the localized concentration sites, etc.).

> Given a primary control mechanism failure: The final distribution pattern of the risk source is the fundamental reason to understand and identify transport mechanisms.

In the end, the issue is: Will the risk source be placed into sufficiently close proximity to the target or into a position where the presence of the risk source will result in an adverse change in the target's environment? Because of the extremely complex nature of the transport mechanisms and our relative lack of knowledge concerning how those mechanisms function, the objective of environmental risk management will be to maintain the primary control mechanisms in a state that will prevent the entry of the risk source into the transport mechanism. Since this objective cannot always be fulfilled, careful analysis and monitoring of the transport mechanisms to prevent concentrations from exceeding acceptable limits, either globally or locally, becomes the next area of environmental risk management attention.

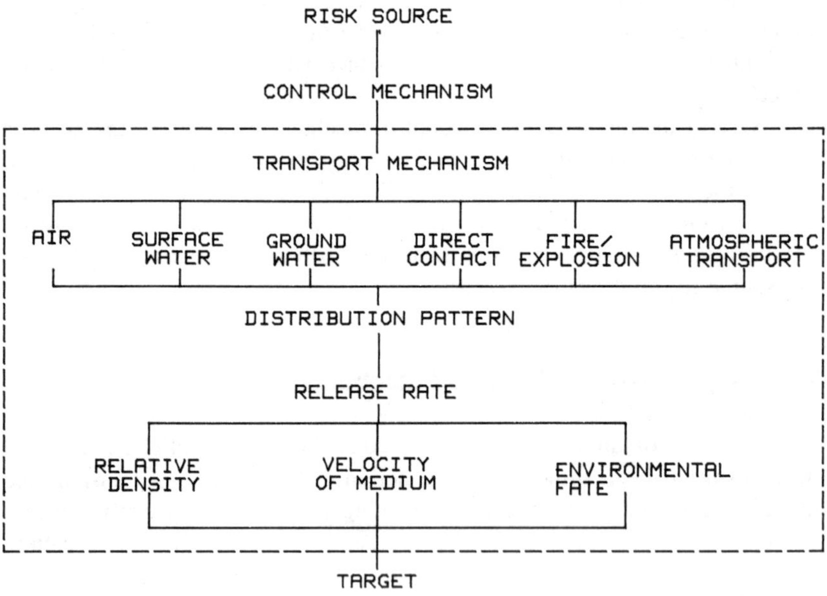

Figure 7.4 General Transport Mechanism Evaluation.

THE HAZARD RANKING SYSTEM AND TRANSPORT AND CONTROL MECHANISMS

The following sections discuss transport mechanisms using the USEPA Hazard Ranking System (HRS) analysis procedures. Because of the complexities and unknowns, it will not be possible to separate out each of the components discussed individually above. Table 7.1 shows the USEPA listing of the major factors to be considered in a hazardous substance release evaluation.

Note that the concern of the HRS is relative to a release, *past tense,* whereas the objective of management is to prevent a release. The difference in viewpoints amounts to the assignment of a probability of occurrence to each of the major factors in the HRS to determine the areas of appropriate management attention.

The use of the HRS is recommended for several reasons. First, the HRS has been defined in law. Second, any other system will, in the case of a serious site problem, probably be forced to meet the standards of the HRS, and the safe approach for the analyst is to meet this criteria in advance. The third reason is simply that the HRS exists, thereby removing a need to create a new analysis system that may or may not be judged adequate in a given set of circumstances. Reinvention of the wheel may be individually satisfying, but it is rarely a rewarding venture in the real world.

There are several things to keep in mind in the following discussion. First, the HRS uses a system of weights for various factors that have been developed through a form of Delphi Technique. These weights presumably reflect the considered opinion of a number of experts after consideration of the factors that influence a particular situation. The HRS has been the subject of considerable criticism because it gives more or less weight to a given factor than a particular individual or group feels appropriate. A version of this weighting system will be used to develop ERIS scores here.

Under the HRS, when several hazardous substances are present that may use a given hazard mode or route, it may be necessary to evaluate all or a major portion of each of the substances or at least the most toxic and/or persistent of them to determine which will give the worst score for that route or mode. It is the worst score that is to be used in defining the character of the site. This is consistent with the ''negative default'' principle set out in an earlier chapter. The ERIS looks at each available transport mechanism for each individual risk source/primary control mechanism combination.

PROBABILITY OF USE: THE FIRST TRANSPORT MECHANISM EVALUATION ISSUE

The first issue in evaluating a transport mechanism is whether or not it is even available for use by the given risk source. Solid lead is highly

Table 7.1 Comprehensive List of Rating Factors[169]

Hazard Mode	Factor Category	FACTORS: Ground Water	FACTORS: Surface Water	FACTORS: Air Route
Migration	Route Characteristics	• Depth of Aquifer of Concern • Net Precipitation • Permeability	• Facility Slope and Intervening Terrain • One-Year 24-Hour Rainfall	
	Unsat. Zone	• Distance to Nearest Surface Water		
	Containment	• Physical State • Containment	• Physical State • Containment	
	Waste Characteristics	• Toxicity/Persistence • Hazardouse Waste Quantity	• Toxicity/Persistence • Waste Quantity	• Reactivity/Incompatibility • Toxicity • Waste Quantity
	Targets	• Ground Water Use • Distance to Nearest Well/Population Served	• Surface Water Use • Distance to Sensitive Environment • Population Served/ Distance to Water Intake Downstream	• Land Use • Population within 4-mile Radius • Distance to Sensitive Environment
Fire and Explosion	Containment	• Containment		
	Waste Characteristics	• Direct Evidence • Ignitability • Incompatibility • Hazardous Waste Quantity		

	Targets	• Distance to Nearest Population • Distance to Nearest Building • Distance to Nearest Sensitive Environment • Land Use • Population within 2-mile Radius • Number of Buildings within 2-mile Radius
Direct Contact	Observed Incident Accessibility Containment Toxicity Targets	• Observed Incident • Accessibility of Hazardous Substance • Containment • Toxicity • Population within 1-mile Radius • Distance to Critical Habitat

unlikely to utilize air transport mechanisms over long distances, but gaseous lead would be very unlikely to utilize groundwater (until after it has precipitated).

Keeping in mind that the HRS assumes that a release has already occurred, while management's attention is properly focused on the prevention of an occurrence, the major difference between the HRS and the ERIS will be the analysts' assessment of the probability of an uncontrolled release and the probability of the use of a particular transport mechanism. The assignment of a probability of use should follow the earlier arguments governing the appearance of undue precision and a High/Moderate/Low/Zero assignment of the probability values will normally be the most appropriate. As with the G/F/P ratings, the H/M/L/0 will be converted to a number for recording as H=5, M=3, L=1, 0=0. This assignment is an integral part of the ERIS evaluation process and should be explicitly made for each of the transport mechanisms for each risk source/primary control mechanism combination.

WATER TRANSPORT

Water is one of the basic commodities required for all human and animal life on earth. From the point of view of hazardous substances, water presents an even more basic set of issues as a transport mechanism than air for the following reasons:

1. Water, because it is a denser medium than air, tends to carry substances in higher concentrations for longer periods of time. Even when streams of water join, they may not mix for long distances. There are some rivers where, for several hundred miles after joining another river, the two streams of water are distinctly visible from the air. This same effect often is seen with hazardous substances in water.
2. Again as a result of its density, water can transport a greater quantity of heavier contaminants than air, thereby acting to increase concentrations at the point of contact.
3. There is relatively little fresh water available for human use and much of that supply is underground where its exact condition is often difficult to determine, but which is surprisingly vulnerable to contamination from the surface.
4. The exact consequences of contamination of a body of water at a given point are very difficult to determine, partially because the exact flow path of the water, particularly subsurface water, is difficult to determine.

SURFACE WATER

Surface water can be divided into three general categories based on the velocity of movement. The HRS defines these as:

1. Contained, that is essentially unmoving as in a small pond with only a small or no outlet.
2. Intermittent flow, or water that tends to have significant movement only when some additional driving force such as heavy rains are active.
3. Flowing, as in a river that flows year round.

The purpose in making these distinctions is to point out the idea that the size of an area affected by a hazardous substance entering a body of surface water is greatly dependent on the flow characteristics of that body of water.

Contained Bodies of Surface Water

Depending on the density and chemical nature of a hazardous substance in a contained body of water, it will float on the surface, sink to the bottom and become part of the sediments (and possibly sink further to pollute an aquifer), or chemically react with or become suspended in the water. So long as the substance remains contained within the water, the possibility of human contact can be significantly reduced by means of secondary control mechanisms (dikes, containment walls, fences, gates, security guards, etc.). However, prevention of contact with endangered species or nearby sensitive environments may not be possible.

In addition, it is possible for local flooding to result in an overflow of a containment pond, spreading the contaminant downstream to highly vulnerable areas such as wetlands, human or agricultural water supplies, and the like. When evaluating a presumably contained body of water, the analyst should look at the following possibilities:

1. That the body is not truly contained. For example, seepage through the bottom to an underground aquifer.
2. Proximity to human habitation and the effectiveness of secondary control mechanisms in preventing contact.
3. Proximity to sensitive environments and/or endangered species, particularly ones that are land mobile and may be attracted to the body of water.
4. The possibility and downstream consequences of overflow or the breaching of the containment.
5. Consideration should be given to extremely low pond levels. The smaller the volume of water for a given volume of contaminant, the higher the concentration of contaminant, therefore the greater the likelihood of health/environmental damage if contact with a target should occur.

Intermittent Flow

Intermittently flowing bodies of water have many of the characteristics of both contained and flowing bodies. There is a strong tendency for contaminants to concentrate on/in the bottom sediments when there is no flow, and then for both the contaminated water and the contaminated sediments to move downstream when there is flow. The analyst should look carefully at the following items in evaluating the intermittent stream.

1. Items (1) through (5) above.
2. The possibility of contaminated stream beds and flood plains downstream as a result of prior flows and the consequences of that contamination. Bear in mind that in the HRS evaluation process the farthest-on point of contamination defines the boundary of the hazardous substances for purposes of evaluation.

Flowing Bodies of Water

Flowing bodies of water present the additional concerns associated with a larger area of distribution if the primary and secondary control mechanisms fail. In general, for all types of surface bodies, an average condition and a worst case scenario should be evaluated.

The worst case scenario may not be conditions of maximum flow. Draught conditions in heavy agricultural regions may present significantly worse situations by concentrating contaminants in a smaller volume of water, thereby increasing the amount of contaminant per unit volume. This has been occurring in the upper Midwest for the past year where the concentration of nitrogen (from fertilizers) and pesticides have several times reached levels at which stream-side communities can no longer draw drinking water from the streams.

For the purpose of developing the ERIS, the transport factors will be developed on a specialized worksheet shown at the end of each section. Note that there will be a single normalized score for each of the applicable transportation routes and it is possible for a score to be zero, indicating that the route is not available to the risk source.

The various weights for the factors are developed through a series of tables to be discussed below; the weights being given additional emphasis through a multiplier to result in a score for the factor under consideration. For example, the 1-year, 24-hour Rainfall factor is given a multiplier of 1, while the Distance to the Nearest Surface Water is given a multiplier of 2 in the HRS. This is a result of the consensus arrived at by the group of experts who assisted the EPA in designing the HRS that—while rainfall is an important consideration—the distance to the water from the site of the contamination

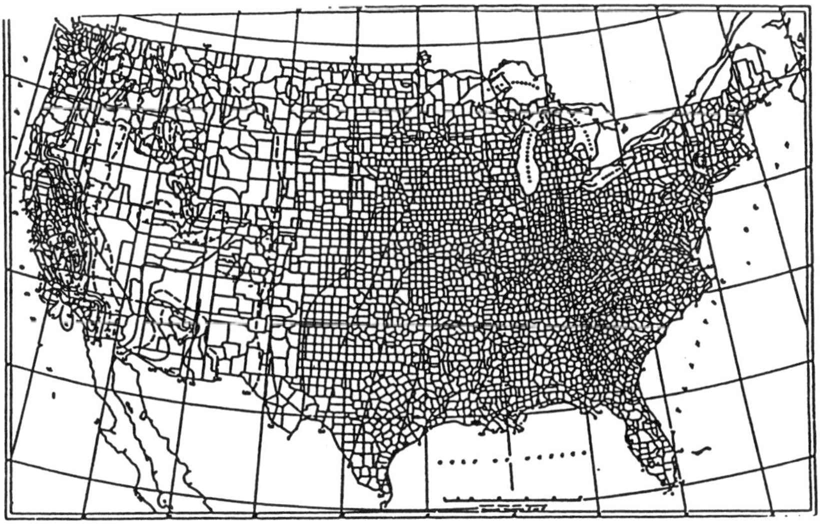

Figure 7.5 1-Year, 24-Hour Rainfall.[170]

is twice as important in the evaluation of the threat of a given site to human health and/or a sensitive environment. In the weight tables shown here, the HRS multipliers have been accounted for.

1-Year, 24-Hour Rainfall

For surface water, the issues of concern are represented by the possibility of heavy rains which may breach control mechanisms and/or increase the flow of streams to the point where the hazardous substance may migrate further downstream. The factor used by the HRS is the "1-Year, 24-Hour Rainfall" (shown graphically in Figure 7.5). This provides information as to the expected maximum amount of rainfall in a given region of the country during a 24-hour period.

Other factors may be used such as the 100-year or 500-year flood potential. The point is to consistently represent the possibility of a sufficient amount of rainfall breaching or compromising secondary control mechanisms and/or spreading the contaminant to previously uncontaminated areas. The rainfall data is used to assign a weight value according to the Table 7.2.

Facility Slope and Intervening Terrain

Particularly for surface water, the slope of the terrain and/or the presence some natural feature may act to block runoff from the hazardous substance

Table 7.2 Weights for Rainfall

Amount of rainfall (inches)	Weight
<1.0	0
1.0–2.0	1
2.1–3.0	2
>3.0	3

Table 7.3 Weights for Facility Slope and Intervening Terrain

	Intervening Terrain				
	Terrain average slope				
Facility slope	<3%*	3 to 5%	5 to 8%	>8%	Water Slope**
Facility is closed basin	0	0	0	0	3
Facility has average slope (<3%)	0	1	1	2	3
Average slope (3–5%)	0	1	2	2	3
Average slope (5–8%)	0	2	2	3	3
Average slope (<8%)	0	2	3	3	3

* This column is to be used both for terrain having a slope of less than 3% or for a site separated from the surface water body by areas of higher elevation.

** This column is to be used when there is a distinct gradient in the surface water streambed as opposed to a gradient (slope) in the surrounding surfaces. This implies a fast-moving stream.

site to a body of surface water available to humans or sensitive environments. This type of secondary control mechanism will influence the level of concern for that hazardous substance. Table 7.3 provides the HRS weight factors for this consideration.

Slope is measured as the amount of drop per 100 feet of horizontal distance and is expressed as a percentage. For example, if the ground drops 4 feet for every 100 feet of horizontal distance, then the slope is 4%. (4/100 = .04 or 4%.) Note that the HRS considers a slope of less than 3% to have the same effect as an intervening terrain feature to retard transport (water does not flow with a high velocity on a flat surface).

Physical State of Soils

The physical state of the soils will obviously have a great impact on the transportability of the substance. To this end, the HRS assigns weights according to Table 7.4. The HRS does not distinguish between Surface or Groundwater Routes with respect to the physical state of the soils.

Table 7.4 Weights for Physical State of the Soils

Physical State	Weight
Solid, consolidated or stabilized	0
Solid, unconsolidated or unstabilized	1
Powder or fine material	2
Liquid, sludge or gas	3

Distance to Nearest Surface Water

The distance from the farthest-on point of existing contamination to a body of surface water is particularly important. Table 7.5 provides weights for this secondary control mechanism.

SUBSURFACE (GROUND) WATER

Subsurface water has many of the same characteristics as surface water, but with the following major differences.

1. Subsurface water tends to move much more slowly than surface water.
2. It tends to cover broader areas because it is not confined to a narrow and well-defined streambed as surface waters generally are.
3. The actual characteristics and extent of flow are far more difficult to precisely define.
4. The action of wells tends to cause flow in directions that may be unexpected based on surface topography or even subsurface flow characteristics elsewhere in a given aquifer.
5. The water is in a denser, three-dimensional media (soil, rock), allowing for several aquifers to overlie each other at different depths and with different flow characteristics.

The HRS identifies the following major route characteristics: Depth to Aquifer of Concern; Net Precipitation; Permeability of the Unsaturated Zone; and Physical State.

These concerns are illustrated and the specialized factors utilized in the HRS are defined in Figures 7.6 and 7.7.

Table 7.5 Weights for Distance to Nearest Surface Water

Distance	Weight
>2 miles	0
1 to 2 miles	2
1000 foot to 1 mile	4
<1000 feet	6

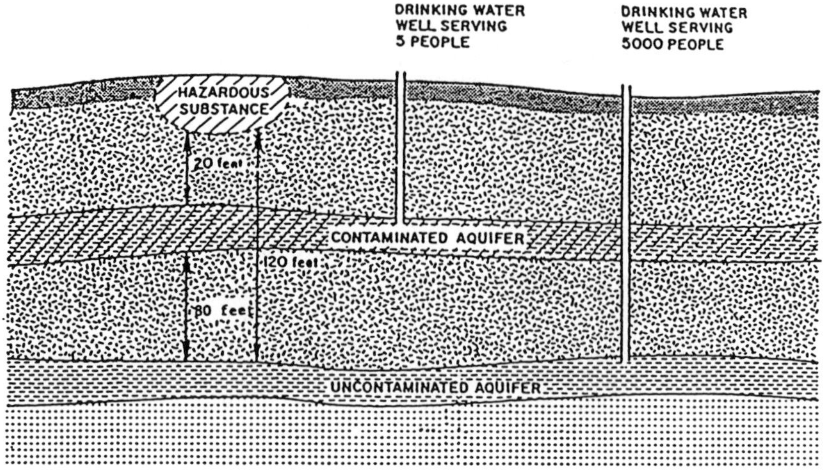

Figure 7.6 Depth to Aquifer of Concern.[171]

Treat target and route characteristics factors consistently. For example, if the upper aquifer is the aquifer of concern, the "depth to aquifer of concern" is 20 feet and the "population served" is 5 persons. If the lower aquifer is "of concern", the "depth" is 120 feet (assuming no known contamination below the indicated "hazardous substance") and the "population" is 5000 persons. If the upper aquifer is contaminated and the lower aquifer is "of concern", the "depth" would be 80 feet (vertical distance between the contaminated aquifer and the aquifer of concern) and the population would be 5000 persons.

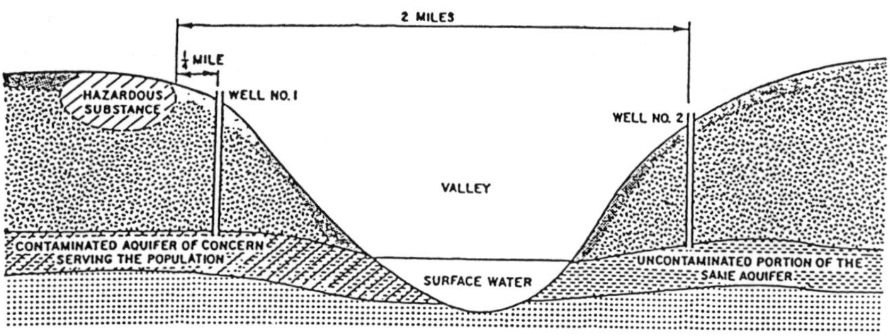

Figure 7.7 Distance to Nearest Well.[172]

In the situation depicted above, the distance between the hazardous substance and the nearest well (No. 1) is ½ mile. If well No. 1 did not exist, the distance to well No. 2 would be immaterial since there is a discontinuity in the aquifer (the surface water) between it and the hazardous substance. Under such circumstances, the factor score would be "0". However, if it could be demonstrated that the contaminant had bridged the discontinuity, the distance to the nearest well would be *[about ½ mile (because the surface water would then be the "farthest-on" point of contamination)].*[173]

Table 7.6 ERIS Surface Water Worksheet

1-year, 24-hour Rainfall	______
Facility Slope and Terrain	______
Distance to Nearest Surf. Water	______
Physical State of Soils	______
Raw Score	______
	×1.11
Normalized Surf. Water Score	______

Precipitation and Evaporation

For groundwater, the issue is not the amount of rain that falls in a given short period of time, since such amounts do not fully penetrate to the aquifers, but rather the net amount of rainfall over a reasonable period of time. The HRS uses the concept of net annual precipitation—precipitation minus lake evaporation—to represent the amount of water entering an aquifer from rainfall. Figures 7.8 and 7.9 show the precipitation and evaporation on an annual basis.

Figure 7.8 Normal Annual Total Precipitation.[174]

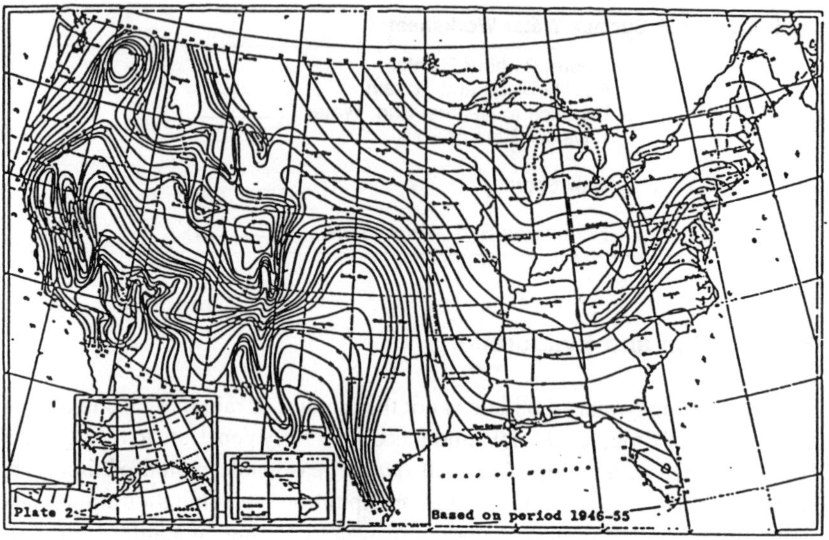

Figure 7.9 Mean Annual Lake Evaporation.[175]

As with the 24-hour rainfall information, more precise information on a given locality can be developed from a number of sources including the nearest weather bureau office and state departments of natural resources. The better the local information, generally the better the assessments, and unless the analyst is preparing a formal Superfund analysis, the use of more precise local data is advised. The HRS is intended to allow the comparison of sites on a national basis. As with rainfall, net precipitation is converted into a weight through the use of Table 7.7.

Permeability

Permeability of the geologic structure is a measure of the rate at which a hazardous substance might migrate through those structures. Given that the

Table 7.7 Weights for Net Precipitation

Net Precipitation (inches) (Rainfall-Evaporation)	Weight
<10	0
−10–+5	1
+5–+15	2
>15	3

Table 7.8 Permeability of Geologic Materials[176]

Type of Material	Approximate Range of Hydraulic Conductivity	Weight
Clay, compact till, shale; unfractured metamorphic and igneous rocks.	$<10^{-7}$ cm/sec	0
Silt, loess, silty clays, silty loams, clay loams; less permeable limestone, dolomites and sandstone; moderately permeable till.	$<10^{-5}>10^{-7}$ cm/sec	2
Fine sand and silty sand; loams; loamy sands; moderatly permeable limestone, dolomites, and sandstone (no karst); moderately fractured igneous and metamorphic rocks, some coarse till.	$<10^{-3}>10^{-5}$ cm/sec	4
Gravel, sand; highly fractured igneous and metamorphic rocks, permeability basalt and lavas; karst limestone and dolomite.	$>10^{-3}$ cm/sec	6

investigation has revealed the type of structures involved that may act as a secondary control mechanism, Table 7.8 gives the weight to be assigned. Note that the permeability relates to the unsaturated zone shown in Figure 7.10.

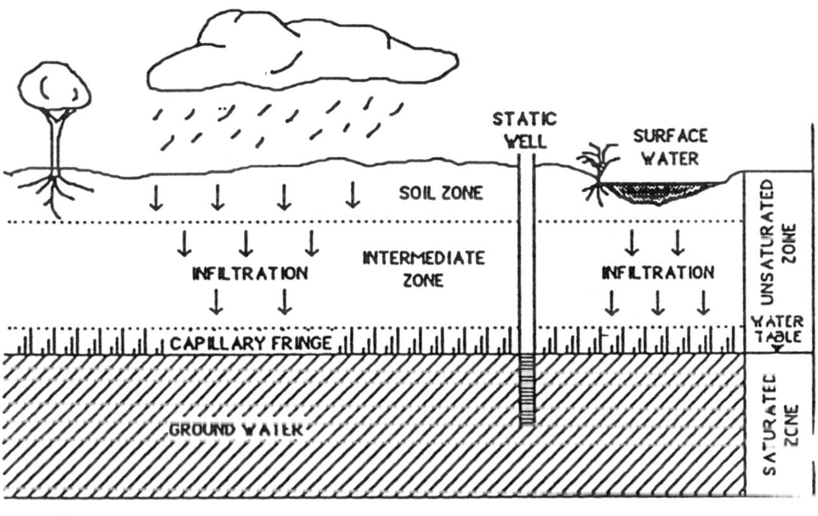

Figure 7.10 Ground Water Nomenclature.

Table 7.9 Weights for Depth to Aquifer

Distance (feet)	Weight
>150	0
76–150	2
21–75	4
0–20	6

Table 7.10 Weights for Distance to Nearest Well

Distance	Weights
>3 miles	0
2–3 miles	1
1–2 miles	2
2001 feet–1 mile	3
<2001 feet	4

Depth to Aquifer

The importance of, and the methods for determining the depth to an aquifer were discussed earlier (see Figures 7.5 and 7.6). Table 7.9 provides the weights for this secondary control mechanism.

Distance to Nearest Well

The relationship between distance from the hazardous substance and the nearest well was also illustrated earlier and the weight factors are provided in Table 7.10.

Physical State

The physical state of the substance is evaluated for groundwater exactly as it was for surface water, the weight factor being obtained from Table 7.4.

AIR TRANSPORT

Exterior Transport

Air transport must be evaluated from three distinctly different points of view. The HRS evaluates whether or not a given site is releasing a substance into the air. If the site is releasing a substance on a regular basis, as opposed to a transitory release caused by site investigation activities and that substance can be detected at a statistically significant level above background, a weight of 45 is assigned. If these conditions are not met, a weight of zero is assigned. The HRS does not deal with the following two points of view at all.

The second point of view is regional, as described in the "bubble" concept earlier in this book. In this case, restrictions on release may be applied to the various sources within the bubble without regard to whether or not the release from a specific source can be detected at any distance from the source.

The third point of view is just beginning to be developed and concerns global releases. As yet, no models have been developed that are accepted into federal regulations although several different models for acid rain on a multi-state basis and CFCs and greenhouse gases on a national and international basis are being seriously considered. In general, releases will probably be treated in the same general manner as they are in the bubble concept.

All three of these exterior release considerations will be based on time-weighted averages of releases over a period of hours, days, weeks, months, or even a year. Only in the case of extremely toxic substance releases are the standards likely to deal with short-duration events. In these cases preventative measures are far more important as a failure to prevent is likely to result in body counting as the measurement of damage (e.g., Bophal).

The basic transport mechanism is the circulation pattern of the local or global atmosphere relative to the substance released. These circulation patterns may be very difficult to identify and evaluate because of their complex and rapidly changing three-dimensional character.

Interior Transport

In some senses, the release of a substance into the air transport system within a structure is simpler because the patterns of air movement are generally more easily established and those patterns are often under mechanical control. The problem will be to identify the exact distribution pattern and the relationship between the distribution pattern and the likely sources of releases. Particular care must be taken whenever very large spaces or spaces with open air return plenums are involved and especially whenever the spaces are heavily occupied.

Reactivity and Incompatibility

The transport mechanism issue considered by the HRS is the reactivity and incompatibility of the substances. These are developed by choosing the highest weight using the three tables which follow. Additional information on the reactivity and incompatibility of substances can be found in Exhibits 8.1 and 8.2.

Table 7.11 ERIS Ground Water Worksheet

Depth to Aquifer	______
Net Precipitation	______
Permeability	______
Physical State	______
Distance to Nearest Well	______
Total Raw Score	______
	×0.91
Normalized Score for Ground Water	______

Table 7.12 NFPA Reactivity Ratings

NFPA level		Weight
0	Materials which are normally stable even under fire exposure conditions and which are not reactive with water	0
1	Materials which in themselves are normally stable, but which may become unstable at elevated temperatures and pressures or which may react with water with some release of energy but not violently	1
2	Materials which in themselves are normally unstable and undergo violent chemical change but do not detonate. Includes materials which can undergo chemical change with rapid release of energy at normal temperatures and pressures or which can undergo violent chemical change at elevated temperatures and pressures. Also includes those materials which may react violently with water or which may form potentially explosive mixtures with water	2
3	Materials which in themselves are capable of detonation or of explosive decomposition or of explosive reaction but which must be heated under confinement before initiation. Includes materials which are sensitive to thermal or mechanical shock at elevated temperatures and pressures or which react explosively with water without requiring heat or confinement	3
4	Materials which in themselves are readily capable of detonation or of explosive decomposition or explosive reaction at normal temperatures and pressures. Includes materials which are sensitive to mechanical or localized thermal shock	3

DIRECT CONTACT

With respect to transport mechanisms, direct contact for our purposes will consider only the issue of accessibility. Accessibility refers to the availability of secondary control mechanisms in place around the hazardous substances—weights to be assigned according to Table 7.16.

FIRE/EXPLOSION

Fire and explosion as a transport mechanism appears to be a relatively simple proposition—either the substance will or will not explode or catch fire. However, the issue is far more complex because both fire and explosion may result in chemical or physical changes of the basic risk source to a far more transportable and possibly more dangerous form. We will deal with the transport mechanism part of fire and explosion from the point of view of the characteristics of the risk source: Ignitability, Reactivity, and Incompatibility.

Ignitability

See table 7.19.

Reactivity

See table 7.13.

Incompatibility

See tables 7.14 and 7.16.

ENVIRONMENTAL RISK INVENTORY DETAIL

The data from the primary control mechanism analysis and the transportation/secondary control mechanism analysis are entered into the Environmental Risk Inventory Detail worksheet section shown below.

Again, there is a need to calculate a Transportation Route Composite Score for each route. This is simply calculated by multiplying the probability-of-use numeric rating by the normalized score obtained from the earlier ERIS tables. The maximum value for this composite score will be 100 and is the reason for having normalized all of the values in the earlier tables to 20.

Table 7.13 Incompatible Materials*

In the lists below, the mixing of a Group A material with a Group B material may have the potential consequence as noted.

Group 1-A	Group 1-B
Acetylene sludge	Acid sludge
Alkaline caustic liquids	
Alkaline cleaner	Battery acid
Alkaline corrosive liquids	
Alkaline corrosive battery fluid	Electrolyte acid
Caustic wastewater	Etching acid liquid or solvent
Lime sludge and other corrosive alkalies	other corrosive acids
Lime wastewater	Spent acid
Lime and water	Spent mixed acid
Spent caustic	Spent sulfuric acid

Potential consequences: Heat generation; violent reaction

Group 2-A	Group 2-B
Aluminum	Any waste in Group 1-A or 1B.
Acid and water	
Berylium	
Chemical cleaners	Calcium
Lithium	
Potassium	
Sodium	
Zinc powder	
Pickling liquor and metal hydrides	Other reactive metals

Potential consequences: Fire or explosion; generation of flammable hydrogen gas.

Group 3-A	Group 3-B
Alcohols	Any concentrated waste in Groups 1-A or 1-B
Water	Calcium
	Lithium
	Metal hydrides
	Potassium
	So_2Cl_2, $SOCl_2$, PCl_2, CH_3, $SiCl_3$
	Other water-reactive waste

Potential consequences: Fire, explosion, or heat generation; generation of flammable or toxic gases.

Group 4-A	Group 4-B
Alcohols	Concentrated Group 1-A or 1-B wastes
Aldehydes	Group 2-A wastes
Halogenated hydrocarbons	
Nitrated hydrocarbons	
Unsaturated hydrocarbons	
Other reactive organic compounds and solvents	

Potential consequences: Fire, explosion, or violent reaction.

Table 7-13 (cont.)

Group 5-A	Group 5-B
Spent cyanide and sulfide solutions	
Chlorites	
Potential consequences: Generation of toxic hydrogen cyanide or hydrogen sulfide gas.	
Group 6-A	**Group 6-B**
Chlorates	Acetic Acid and other organic acids
Chlorine	Concentrated mineral acids
	Group 1-B wastes
Group 2-A wastes	
Chromic acid	
Group 4-A wastes	
Hyphoclorites	Other flammable and combustible wastes
Nitrates	
Nitric acid, fuming	
Perchlorates	
Permanganates	
Peroxides	
Other strong oxidizers	
Potential consequences: Fire, explosion, or violent reaction.	

***Hazardous Waste Management Law, Regulations, and Guidelines for the Handling of Hazardous Waste. California Department of Health, Sacramento, California, February, 1975.**

Table 7.14 Weight for Incompatibility of Chemicals

Incompatibility	Weight
No incompatible substances are present	0
Present, but do not pose a hazard	1
Present and may pose a future hazard	2
Present and posing an immediate hazard	3

Table 7.15 ERIS Air Route Worksheet

Reactivity and Incompatibility	______
	×6.67
Normalized Score	______

Table 7.16 Weights for Accessibility

Barrier	Weight
A 24-hour surveillance system (e.g., television monitoring or surveillance by guards or facility personnel) which continuously monitors and controls entry onto the facility **or** an artificial or natural barrier (e.g., a fence combined with a cliff), which completely surrounds the facility and a means to control entry at all times through the gates or other entrances to the facility (e.g., an attendant, television monitors, locked entrances, or controlled roadway access to facility)	0
Security guard, but no barrier	1
A barrier, but no separate means to control entry	2
Barriers do not completely surround the facility	3

Table 7.17 ERIS Direct Contact Worksheet

Accessibility	________
	×6.67
Normalized Score	________

Table 7.18 NFPA Ignitability Levels and Assigned Weights

NFPA level		Weight
4	Very flammable gases, very volatile flammable liquids, and materials that in the form of dusts or mists readily form explosive mixtures when dispersed in air	3
3	Liquids which can be ignited under all normal temperature conditions. Any material that ignites spontaneously at normal temperatures in air	
2	Liquids which must be moderately heated before ignition will occur and solids that readily give off flammable vapors	2
1	Materials that must be preheated before ignition will occur. Most combustible solids have a flammability rating of 1	1
0	Materials that will not burn	0

Table 7.19 Weight for Ignitability

Ignitability	Weight
Flashpoint >200°F or NFPA level 0	0
Flashpoint 140–200°F or NFPA level 1	1
Flashpoint 80–140°F or NFPA level 2	2
Flashpoint <80°F or NFPA levels 3 or 4	3

Table 7.20 ERIS Fire/Explosion Worksheet

Ignitability	________
Reactivity	________
Incompatibility	________
Total Raw Score	________
	×2.22
Normalized F/E Score	________

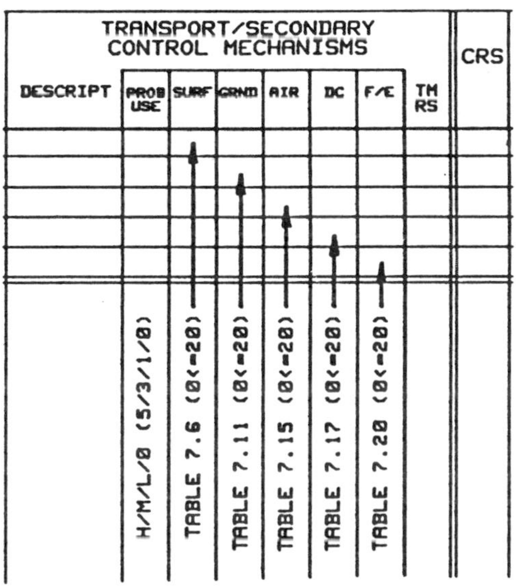

Figure 7.11 ERIS Transport Mechanism Inventory Worksheet Detail.

REFERENCES

169. *26 CFR 300 Appendix A — Uncontrolled Hazardous Waste Site Ranking System: A User's Manual.*

170. *Rainfall Frequency Atlas of the United States, Technical Paper No. 40;* U.S. Department of Commerce; U.S. Government Printing Office; Washington, DC; 1963.

171. Drawing and drawing explanation taken from *29 CFR 300 Appendix A — Uncontrolled Hazardous Waste Site Ranking System: A User's Manual.*

172. Drawing and explanation taken from *29 CFR 300 Appendix A.*

173. Author's note: In *29 CFR 300 Appendix A* the description under the figure reads ". . . then the distance to the nearest well would be 2 miles (assuming well No. 1 does not exist." This caption is incorrect due to the statement in *29 CFR 300 Appendix A "Distance to nearest well* is measured from the hazardous substance (not the facility boundary) to the nearest well that draws water from the aquifer of concern. . . . If a discontinuity in the aquifer occurs between the hazardous substance and all wells, give this factor a score of 0, *except where it can be shown that the contaminant is likely to migrate beyond the discontinuity."* (Author's emphasis.)

174. *Climatic Atlas of the United States;* U.S. Department of Commerce; National Climatic Center; Ashville, NC; 1979.

175. *Climatic Atlas of the United States;* ibid.

176. Derived from: Davis, S. N.; *Porosity and Permeability of Natural Materials in Flow-Through Porous Media;* R. J. M. DeWest, ed.; Academic Press; New York; 1969; and Freeze, R. A. and J. A. Cherry; *Groundwater;* Prentice-Hall, Inc.; New York; 1979.

8 ANALYSIS OF POTENTIAL TARGETS AND RISK SOURCES

INTRODUCTION

The analysis of potential targets for a given risk source involves two distinct steps: (1) the identification of the targets, and (2) the evaluation of the likely impact of the risk source on the target at the point of impact. The language of risk source target analysis may sound like a problem in ballistics, but unfortunately, the transport mechanisms for risk source distribution are not nearly as well defined.

Two general targets may be identified: (1) sensitive environments, and (2) humans. The analysis of the effects of a risk source on humans may be characterized as being in its early stages. Although much is known, a great deal more needs to be known. The analysis of risk source impacts on sensitive environments is in a lesser stage of development, particularly with respect to risk sources believed to have a global atmospheric effect. However, the negative default principle requires that if there is a reasonable basis for suspecting an adverse impact, precautions must be taken as if the impact had been confirmed.

This scenario of knowledge and lack of knowledge and of the need to take action in the absence of concrete knowledge is not unusual. Billions of dollars are spent each year to purchase insurance either directly or in the form of risk reduction actions based on the possibility of the occurrence of an event. The major difference between what are considered to be "insurable" risks and environmental risks is a lack of data on which to base expected outcome evaluations. We have only limited means for determining the monetary value to be assigned to environmental damage, to human health, or to life. Therefore this text has adopted primarily non-financial measurements of risk.

TARGET POPULATION

The HRS contains different methods for counting the number of human beings that may be targets depending on the transport mode. The Environmental Risk Inventory System counts individuals according to the HRS methods, but uses the number of individuals (in 1,000's) directly in addition to the HRS method of converting the number into a weight factor. The reason for the difference is to emphasize the importance of protecting human life/health.

Fire/Explosion

For fire and explosion the HRS requires that the population be counted within a 2-mile radius of the likely center of a fire/explosion according to one of the following methods.

1. The population at risk is to include residents and workers who would regularly be present within reach of the risk source. It does not include transients such as travelers passing through the area. A census of the population is considered the most reliable indicator.
2. If census data is not available, aerial photographs may be used with each dwelling within the assigned radius assumed to have 3.8 occupants.

Surface and Ground Water

In preparing to evaluate this factor, the analyst again has two methods for determining the population likely to be affected by a source of surface or groundwater contamination. The difference between these two methods and the methods provided for the Fire/Explosion mode are based on the idea that fire and explosion have highly localized impacts whereas the impact of the contamination of a source of water may reach far beyond (even hundreds of miles beyond) the source of contamination. For example, if the Colorado river system became contaminated, the major population centers of Arizona and Southern California would be impacted. In both of the methods, the issue will be the number of people who directly depend on the particular source for their water supply and the use to which that water supply is normally put.

The methods for determining population are:

1. The population at risk is to include residents and workers who would regularly use the source. It does not include transients such as travelers passing through the area. A census of the population is considered to be the most reliable indicator. Note that the number of people likely to be affected does

not depend on distance from the source of the water, only on the number of individuals who will use that source.

2. If census data is not available, the aerial photographs may be used based on the following assumptions:
   a. Each dwelling using the water source is to be assumed to have 3.8 occupants.
   b. If the source is used for irrigation of agricultural land, then for each acre of irrigated land, 1.5 people are to be considered affected.

Population Adjustments

The HRS, again, assumes that an uncontrolled release into the environment has already occurred, while environmental risk management must look to the prevention of an uncontrolled release. The Current Time and Time t_1 consider the population that may be effected by a release now. Times t_2 and t_3 consider releases that may occur at some distant time in the future when local population parameters may have changed (the area around the risk source became more densely populated for example).

1. Current Time — The number of individuals as counted according to the Hazard Ranking System rules. This is the number to be used in developing the initial ERIS inventory human risk scores.
2. Time t_1 — Specifically the number of individuals that would most likely be affected in the time span between a release and the time that emergency response measures can reasonably be expected to become effective in containing the release.
3. Time t_2 — The number of individuals that would probably be affected one year in the future.
4. Time t_3 — The number of individuals that would probably be affected if the release occurred at some more distant time in the future, say 10 years from now.

The idea is to gain some insight into the influence of an unexpected release and of the impact of the growth of the area within reach of the transport mechanisms over time.

HUMAN TARGET ASSESSMENT FACTORS

Surface Water

For surface water, distance is calculated from the farthest-on point of contamination to the nearest intake for the water supply. If the source water

Table 8.1 Weight for Population vs. Distance to Surface Water Intake or Ground Water Well

Population	Distance to intake or well > 3 miles	2–3 miles	1–2 miles	2000′–1 mile	0–2000 feet
0	0	0	0	0	0
1–100	0	4	0	8	10
101–1,000	0	8	12	16	20
1,001–3,000	0	12	18	24	30
3,001–10,000	0	16	24	32	35
>10,000	0	20	30	35	40

Table 8.2 Weight for Surface Water Use

Surface water use (fresh or salt water)	Assigned Value
Not currently used	0
Commercial or industrial	3
Irrigation, economically important resources (e.g., shellfish), commercial food preparation, or recreation (e.g., fishing, boating, swimming)	6
Drinking water	9

Table 8.3 ERIS Surface Water Route — Human Target Worksheet

Current Population (1,000's)	______
Population at t_1	______
Population at t_2	______
Population at t_3	______
Surface Water Use	______
Population/Distance	______
Raw Total Score	______
	×2.04
Normalized Score	______

supply is from a flowing body of water, the intake must be located within three miles of the farthest-on point. If the source is a still body of water (e.g., a lake), the intake must be located within one mile of the farthest-on point. These are rather arbitrary limits and the analyst should exercise judgment as to whether they are too conservative. It is the author's opinion that they should be treated as lower limits of distance, especially considering the discussions to follow on persistence and toxicity and the earlier discussions of transport mechanisms. The Surface Water Route ERIS Score is calculated in Table 8.3.

The question of the use to which the water can be put is important and the weight for this factor is given in Table 8.2.

Ground Water

Groundwater is normally used in a more localized area than are surface water sources. The population is the number of individuals served by the well. The analyst should note that municipalities often have well fields many miles from the legal boundaries of the city or town, so care in determining what populations are actually served by a given well should be exercised. The population served/distance from the risk source to the nearest well is given an HRS weighting, provided by Table 8.1.

Associated with the population to be served by a groundwater source is the use to which the water is put. This factor is provided with its own weight by Table 8.4. The ERIS Ground Water Route Score is calculated in Table 8.5.

Table 8.4 Weight for Ground Water Use

Ground Water Use	Weight
Unusable (e.g., extremely saline aquifer, extremely low yields, etc.)	0
Commercial, industrial or irrigation and another water source presently available, not used, but usable	3
Drinking water with municipal water from alternate unthreatened sources presently available (i.e., minimal hookup requirements); or commercial, industrial or irrigation with no other water source presently available	6
Drinking water; no municipal water from alternate unthreatened sources presently available	9

Table 8.5 ERIS Ground Water Route — Human Target Worksheet

Current Population (1,000's)	______
Population at t_1	______
Population at t_2	______
Population at t_3	______
Ground Water Use	______
Population/Distance	______
Raw Total Score	______
	× 2.043
Normalized Score	______

Table 8.6 Distance to Population From Hazardous Substance

Population	0–4 miles	0–1 mile	0–½ mile	0–¼ mile
0	0	0	0	0
1–100	9	12	15	18
101–1,000	12	15	18	21
1,001–3,000 ..	15	18	21	24
3,001–10,000	18	21	24	27
>10,000	21	24	27	30

Table 8.7 Weights for Land Use (Air Route)

Assigned Value =	0	1	2	3
Distance to Commercial-Industrial	> 1 mile	½–1 mile	¼–½ mile	< ¼ mile
Distance to National/State Parks, Forests, Wildlife Reserves, and Residential Areas	> 2 miles	1–2 miles	¼–1 mile	< ¼ mile
Distance to Agricultural Lands (in Production within 5 years)				
Ag Land	> 1 mile	½–1 mile	¼–½ mile	< ¼ milie
Prime Ag Land[177]	> 2 miles	1–2 miles	½–1 mile	< ½ mile
Distance to Historic/Landmark Sites (National Register of Historic Places and National Landmarks) ...	....	....	....	Within view of site or if site is subject to significant impacts.

Table 8.8 ERIS Air Route - Human Target Worksheet

Current Population (1,000's)	________
Population at t_1	________
Population at t_2	________
Population at t_3	________
Population/Distance	________
Land Use	________
Raw Total Score	________
	×3.03
Normalized Score	________

Air Route

The HRS requires that the analyst prepare an estimate of the population for four radial distances from the risk source using Table 8.6 to assign the appropriate (i.e., the highest identified) weight factor. Land use weights are developed in Table 8.7 and the ERIS Score is calculated in Table 8.8.

Direct Contact

For the direct contact mode, the target factor is the population within one mile of the risk source, given by Table 8.9. The ERIS Score is calculated in Table 8.10.

Table 8.9 Weight for Population Within One Mile Radius

Population	Weight
0	0
1–100	1
101–1,000	2
1,001–3,000	3
3,001–10,000	4
>10,000	5

Table 8.10 ERIS Direct Contact — Human Target Worksheet

Current Population (1,000's)	______
Population at t_1	______
Population at t_2	______
Population at t_3	______
Population within 1-mile	______
	×20.0
Normalized Score	______

Fire/Explosion

The Fire/Explosion mode considers the population within a 2-mile-radius and the land use within that radius as important target parameters. The land use weight factor is found in Table 8.7. Weights for population within two miles are obtained from Table 8.11 and the ERIS Score is calculated in Table 8.12.

Table 8.11 Weight for Population Within 2-Mile Radius

Population	Weight
0	0
1–100	1
101 1,000	2
1,001–3,000	3
3,001–10,000	4
>10,000	5

Table 8.12 ERIS Fire/Explosion — Human Target Worksheet

Current Population (1,000's)	______
Population at t_1	______
Population at t_2	______
Population at t_3	______
Population within 2 miles	______
Land Use (Table 8.7)	______
Total Raw Score	______
	× 12.50
Normalized Score	______

SENSITIVE ENVIRONMENT TARGET ANALYSIS

The HRS assumes that sensitive environments are the subject of environmental risks through four modes of transport: (1) direct contact with the risk source, (2) transportation of the risk through surface waters, (3) through airborne migration of the substance, and (4) transportation of the risk through fire/explosion. The ERIS Score is calculated in Table 8.14.

Surface Water Route

Sensitive environments, such as areas of endangered species habitat (plant or animal) or a fragile natural setting that could suffer an especially severe impact from pollution, are assigned weights for surface water migration of the risk source using Table 8.13.

Table 8.13 Weights for Sensitive Environment (Surface Water)

Weight =	0	1	2	3
Distance to Wetlands[178] (5-acre minimum)				
Coastal	> 2 miles	1–2 miles	½–1 mile	< ½ mile
Fresh Water	> 1 mile	¼–1 mile	100′–¼ mile	< 100 feet
Distance to Critical Habitat (of endangered species)[179]	> 1 mile	½–1 mile	¼–½ mile	< ¼ mile

Table 8.14 ERIS Surface Water Route — Sensitive Environment Worksheet

Sensitive Environment	______
	× 33.33
Normalized Weight	______

Table 8.15 ERIS Air Route — Sensitive Environment Worksheet

Land Use (Table 8.7)	________
	×33.33
Normalized Weight	________

Air Route

For airborne releases the type of the land use within range of the transportation mechanism must be evaluated according to Table 8.7. The ERIS Score is calculated in Table 8.15.

Direct Contact

The Direct Contact HRS Score is obtained from Table 8.16, and the corresponding ERIS Score from Table 8.17.

Fire/Explosion

Weights for the relationship between fire/explosion and wetlands or other critical habitats are provided by Table 8.18. Table 8.19 provides HRS weights for the relationship between fire/explosion and the number of buildings within a specified distance, and Table 8.20 provides an additional weighting factor for the distance to the nearest building. Table 8.21 provides for the calculation of the associated ERIS score.

Table 8.16 Weight for Distance to a Critical Habitat

Distance	Weight
$<$ 1 mile	0
$^{1}/_{2}$–1 mile	1
$^{1}/_{4}$–$^{1}/_{2}$ mile	2
$>$ $^{1}/_{4}$ mile	3

Table 8.17 ERIS Direct Contact — Sensitive Environment Worksheet

Distance to Habitat	________
	×33.33
Normalized Weight	________

Table 8.18 Weight for Sensitive Environments (Fire or Explosion)

Weight =	0	1	2	3
Distance to Wetlands[180]	> 100 feet	....	....	< 100 feet
Distance to Critical Habitat[181]	> 1/2 mile	1000′–1/2 mi	100–1000 feet	< 100 feet

Table 8.19 Weights for Buildings (Fire and Explosion)

Number of buildings	Weight
0	0
1–26	1
27–260	2
261–790	3
791–2600	4
>2600	5

Table 8.20 Weight for Distance to Nearest Building

Distance	Weight
> 1/2 mile	0
201 feet–1/2 mile	1
51 feet–200 feet	2
0 to–feet	3

Table 8.21 ERIS Fire/Explosion — Sensitive Environment Worksheet

Distance to Sensitive Env.	______
Number of Buildings	______
Distance to Nearest	______
	×9.09
Normalized Weight	______

RISK SOURCE ANALYSIS

Persistence and toxicity are probably the two most important topics of environmental risk management and deserve a great deal of study on the part of the analyst. This study can only be introduced in a book of this nature.

Persistence

Persistence is a measure of the period of time over which a given risk source remains a threat to a target. The life of a hazardous substance may be thought of in terms of whether or not the material will degrade to non-hazardous components over time, either through chemical reactions with the environment, or through the action of biological organisms on the material (biodegradation).

For hazardous substances, the chemical or biological half-life (for non-radioactive risks) is the normal measure and is defined as the period of time required for one half of the substance to degrade to non-hazardous substances under the existing environmental conditions. The half-life for many substances may extend for thousands of years, effectively forever for planning purposes. Exhibits 8.1 and 8.2 provide measures for persistence in terms of abstract weights that instead of representing the half-life directly, represent the relative strength of the risk sources on a scale of 0 to 3.

With respect to the HRS weighing factors for persistence, there is nothing magic about the weights assigned, they result from a type of Delphi approach. If strict use of the HRS is not required for a given analysis, other measures of persistence such as the direct use of the half-life (if available) can be substituted in the decision support tools to be introduced later in this book. Half-life values for most substances are based on laboratory measurements of the pure substance in most cases, and the actual half-life of the substance in field conditions may not directly relate to the laboratory values. Laboratory conditions simply cannot duplicate the complex situation existing in nature.

The persistence and toxicity of a risk source are generally so closely related that the HRS again uses the double-decision method to arrive at a final weight factor. This involves determining the weight to be assigned to the persistence or toxicity of the risk source as separate subjects, and then utilizing these individual weights in a matrix to arrive at an overall weighting factor for combined persistence and toxicity. Exhibits 8.1 and 8.2 provide basic information on some of the toxicity and persistence data associated with a number of common materials. The HRS weights for persistence are in Table 8.26.

Table 8.22 HRS Risk Factor for Persistence

Substance	Assigned Value
Easily biodegradable compounds	0
Straight chain hydrocarbons	1
Substituted and other ring compounds	2
Metals, polycyclic compounds and halogenated hydrocarbons	3

EXHIBIT 8.1 Persistence (Biodegradability) of Some Organic Compounds[182]

Weight = 3 Highly Persistent Compounds	Weight = 2 Persistent Compounds	Weight = 1 Somewhat Persistent Compounds	Weight = 0 Nonpersistent Compounds
aldrin	acenaphthylene	acetylene dichloride	acetaldehyde
benzopyrene	atrazine	behenic acid, methyl ester	acetic acid
benzothiazole	(diethyl) atrazine	benzene	acetone
benzothiophene	barbital	benzene sulfonic acid	acetophenone
benzyl butyl phythalate	borneol	butyl benzene	benzoic acid
bromochlorobenzene	bromobenzene	butyl bromide	di-isobutyl carbinol
bromoform butanal	camphor	e-caprolactam	docosane
bromophenyl phyntyl ether	chlorobenzene	carbon-disulfide	eicosane
chlordane	1,2-bis-chloroethoxy ethane	o-cresol	ethanol
chlorohydroxy benzephenone	b-chloroethyl methyl ether	decane	ethylamine
bis-chloroisoprophyl ether	chloromethyl ether	1,2-dichloroethane	hexadecane
m-chloronitrobenzene	chloromethyl ethyl ether	1,2-dimethoxy benzene	methanol
DDE	3-chloropyridine	1,3-dimethyl naphthalene	methyl benzoate
DDT	di-t-butyl-p-benzoquinone	1,4-dimethyl phenol	3-methyl butanol
dibromobenzene	dichloroethyl ether	dioctyl adipate	methyl ethyl ketone
dibutyl phthalate	dihyrocarvone	n-decane	2-methylpropanol
1,4-dichlorobenzene	dimethyl sulfoxide	ethyl benzene	octadecane
dichlorodifluoroethane	2,6-dinitrotoluene	2-ethyl-n-hexane	pentadecane
dieldrin	cis-2-ethyl-4-methyl 1,3-dioxolane	o-ethyltoluene	pentanol
diethyl phthalate	trans-2-ethyl-4-methyl-1,3-	isodecane	propanol
di(2-ethylhexyl) phthalate	dioxolane	isoprophyl benzene	propylamine
dihexyl phthalate	guaiacol	limonene	tetradecane
di-isobutyl phthalate	2-hydroxyadiponitrile	methyl ester of lignoceric acid	n-tridecane
dimethyl phthalate	isophorone	methane	n-undecane
4,6-dinitro-2-aminophenol	indene	2-methyl-5-ethyl pyridine	
dipropyl phythalate	isoborneol	methyl naphthalene	
endrin	isoprophenyl-r-isopropyl benzene	methyl palmitate	
heptachlor	2-methoxy biphenyl	methyl phenyl carbinol	
heptachlor epoxide	methyl biphenyl	methyl stearate	
1,2,3,4,5,7,7,-	methyl chloride	naphthalene	

heptachloronorbornene	methylindene	nonane
hexachlorobenzene	methylene chloride	octane
hexachloro-1,3-butadiene	nitroanisole	octyl chloride
hexachlorocyclohexane	nitrobenzene	pentane
hexachloroethane	1,1,2-trichloroethylene	phenyl benzoate
methyl benzothiazole	trimethyl-trioxo-hexahydrotriazine	phthalic anhydride
pentachlorobiphenyl	isomer	propylbenzene
pentachlorophenol		1-terpineol
1,1,3,3,-tetrachloroacetone		toluene
tetrachlorophenyl		vinyl benzene
thiomethylbenzothiazole		xylene
trichlorobenzene		
trichlorobiphenyl		
trichlorofluormethane		
2,4,6-trichlorophenol		
triphenyl phosphate		
bromodichloromethane		
bromoform		
carbon tetrachloride		
chloroform		
chlbromochloromethane		
dibromodichloroethane		
tetrachloroethane		
1,1,2-trichloroethane		

EXHIBIT 8.2 Waste Characteristic Values for Some Common Chemicals

Chemical/Compound	Reactivity[185]	Toxicity[183]	Persistence[184]	Ignitability[185]
Acetaldehyde	3	0	3	2
Acetic Acid	3	0	2	1
Acetone	2	0	3	0
Aldrin	3	3	1	0
Ammonia, Anhydrous	3	0	1	0
Aniline	3	1	2	0
Benzene	3	1	3	0
Carbon Tetrachloride	3	3	0	0
Chlordane	3	3	0[186]	0[186]
Chlorobenzene	2	2	3	0
Chloroform	3	3	0	0
Cresol-O	3	1	2	0
Cresol-M&P	3	1	1	0
Cyclohexane	2	2	3	0
Endrin	3	3	1	0
Ethyl Benzene	2	1	3	0
Formaldehyde	3	0	2	0
Formic Acid	3	0	2	0
Hydrochloric Acid	3	0	0	0
Isopropyl Ether	3	1	3	1
Lindane	3	3	1	0
Methane	1	1	3	0
Methyl Ethyl Ketone	2	0	3	0
Methyl Parathion (in Xylene Solution)	3	0[187]	3	2
Naphthalene	2	1	2	0
Nitric Acid	3	0	0	0
Parathion	3	0[187]	1	2
PCB	3	3	0[187]	0[187]
Petroleum, Kerosene (Fuel Oil No. 1)	3	1	2	0
Phenol	3	1	2	0
Sulfuric Acid	3	0	0	2
Toluene	2	1	3	0
Trichlorobenzene	2	3	1	0
αTrichloroethane	2	2	1	0
Xylene	2	1	3	0

The persistence of a substance will greatly influence the performance requirements of Phase I audits. It results in a need to carefully examine the history of a specific property to at least the time that the underlying land was in a virgin status, and sometimes prior to that point in time.

Toxicity

The basic concept of toxicology is that the size of the dose determines the effect. Dose is the ratio of the mass of the toxic substance administered to the body weight of the individual and the time over which that dose is

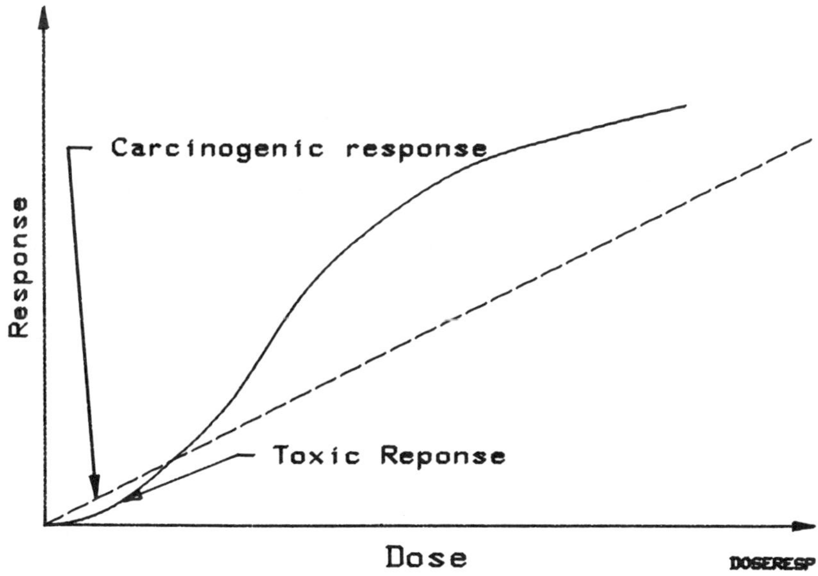

Figure 8.1 Generalized Dose-Response Relationship.

administered. This gives rise to the idea of a dose-response relationship, the generalized relationship being shown in Figure 8.1.

There are a series of technical issues that are associated with the dose-response relationship and only the most important will be discussed here. The reader is advised to consult additional sources for more information on this extremely important topic.

The difference between carcinogens and non-carcinogens results in a major factor of concern to the analyst. The curves shown in Figure 8.1 display two very different behaviors at the low-dosage level. The carcinogenic response is assumed to be a straight line passing through the zero-dose/zero-response point, while the non-carcinogenic line is curved and does not necessarily pass through the zero-dose/zero-response point. USEPA policy holds that for a carcinogen, there is no safe dose.[189] For non-carcinogens, there may be doses that show no adverse health effects and therefore safe dosage levels may exist. The bottom line is that any incremental dosage of a carcinogen above ambient background is officially too much.[190]

For non-carcinogens, the issue is considerably more complicated and the concept of exposure limits enters the picture. Environmental toxicology recognizes that some substances exhibit a threshold level, a dose below which no toxic or adverse effects may be detected. This has lead to the idea of establishing a set of allowable limits for exposure with the best known of the published standards for such limits being the Threshold Limit Values

(TLV[191]).[192] The TLV's are **not** legally defined standards except as they have been specifically adopted or referenced by a law or regulation.

A TLV is defined as being the level of airborne exposure concentrations at which adult healthy males working at 40 hours per week generally do not suffer adverse health effects. TLVs are published at two levels, an eight-hour Time Weighted Average concentration of exposure (the TWA) and the maximum 15-minute average concentration to which a worker may be exposed for a short term—the Short Term Exposure Limit (STEL). OSHA has adopted certain of the TLVs as Permissible Exposure Levels (PELs). The PELs are legally enforceable limits for industrial and certain other workers (AHERA adopts the PELs for asbestos for all work subject to AHERA, for example). The No Observed Adverse Effect Level (NOAEL) is used as a starting point for allowable exposure limit setting for humans in a number of situations where only animal study data is available. The level for humans will then be set at a fraction of the NOAEL determined by the animal studies, generally in the range of $1/1{,}000^{th}$ to $1/1{,}000{,}000^{th}$ of the NOAEL.

It is very important to note carefully the wording of the definition of the TLV, "the level of airborne exposure concentrations at which adult healthy males working at 40 hours per week generally do not suffer adverse health effects," (quoting from Griffin, see footnote 188). Age, sex, physical condition, body weight, and type of activity in the presence of the substance all can result in different reactions to the same concentration of the substance. In general, the aged, infants, women, and especially pregnant women, and those individuals whose activities require a high level of physical exertion can all experience adverse effects at lower concentrations.

Note also that the TLVs are established for a single substance. Synergism and antagonism between multiple substances can result in completely different effects. Synergism describes the enhancing effect of two different environmental contaminants upon the human body, such as asbestos combined with smoking tobacco where the result is an increase in the incidence of asbestos-related diseases by a reported factor of 50 over simple exposure to just asbestos. Antagonistic effects exists where one impact is lessened by having another substance present, and occur so rarely that it is not certain that they occur in the general environment.

In general, there are six types of toxic agents:[193]

1. Allergic agents — Cause itching, sneezing or rashes.
2. Asphyxiants — Displace oxygen.
3. Irritants — May cause pulmonary edema at very high concentrations, or unpleasant sensations throughout the body.
4. Necrotic agents — Directly cause cell death.
5. Carcinogens — Cause cancer.
6. Systemic poisons — Attack the entire body.

In addition to concerns about the specific reaction of the organism to exposure at the time of exposure, there should be concern with respect to effects that are only manifest a considerable period of time after exposure. Asbestos-related diseases have a latency period of 15 to 40 years, making it extremely difficult to identify the cause of a disease and to develop consistently applied daily behavior programs that will prevent unnecessary exposure.

Long latency periods are generally associated with carcinogens, although this is not always the case. In terms of the identification of environmental risks, the implications of long latency periods are with respect to the influence of exposures in the past manifesting themselves in the current or future periods. Closely related to long latency periods are the possibility that the organism will accumulate the substance until the retained level is sufficient to result in adverse effects. The TLVs attempt to deal with this issue, but data may not be available to account for this possibility adequately at the present time.

To deal with the complicated issue of toxicity, the HRS again provides a series of weighting factors that will, in this case, be used in conjunction with the persistence weighting factors, to arrive at an overall weight for persistence/toxicity. There are two major methods for the measurement of toxicity; the SAX Toxicity Rating scheme (Table 8.23), and the National Fire Protection Association (NFPA) rating scheme (Table 8.24). Table 8.25 is the HRS crossover table for the two methods.

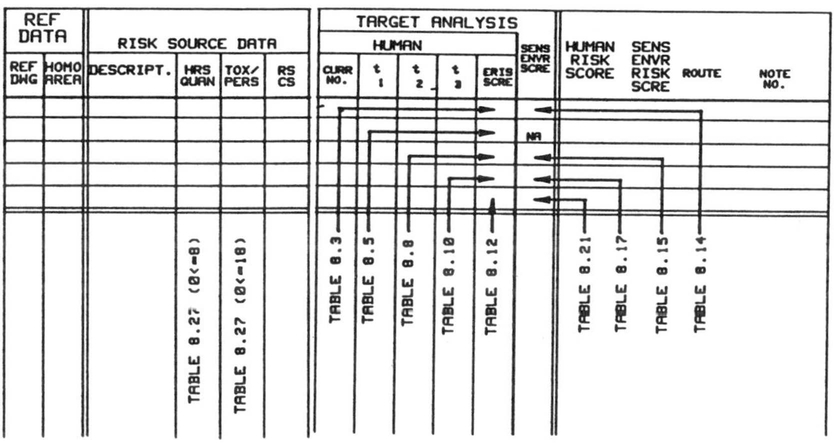

Figure 8-2 Target and Risk Source Sections of ERIS.

Table 8.23 SAX Toxicity Ratings

0 = No Toxicity[194] (None)[195]

This designation is given to materials which fall into one of the following categories:

(a) Materials that cause no harm under any conditions of normal use.

(b) Materials that produce toxic effects on humans only under the most unusual conditions or by overwhelming dosage.

1 = Slight Toxicity* (Low)**

(a) *Acute local.* Materials that on single exposures lasting seconds, minutes, or hours, cause only slight effects on the skin or mucous membranes regardless of the extent of the exposure.

(b) *Acute systemic.* Materials that can be absorbed into the body by inhalation, ingestion, or through the skin and that produce only slight effects following single exposures lasting seconds, minutes, or hours, or following ingestion of a single dose regardless of the quantity absorbed or the extent of the exposure.

(c) *Chronic local.* Materials that on continuous or repeated exposures extending over periods of days, months, or years cause only slight and usually reversible harm to the skin or mucous membranes. The extent of exposure may be great or small.

(d) *Chronic systemic.* Materials that can be absorbed into the body by inhalation, ingestion, or through the skin and that produce only slightly usually reversible effects extending over days, months, or years. The extent of the exposure may be great or small.

In general, those substances classified as having slight toxicity produce changes in the human body that are readily reversible and that will disappear following termination of exposure, either with or without medical treatment.

2 = Moderate Toxicity* (Mod)**

(a) *Acute local.* Materials that on single exposure lasting seconds, minutes, or hours cause moderate effects on the skin or mucous membranes. These effects may be the result of intense exposure for a matter of seconds or moderate exposure for a matter of hours.

(b) *Acute systemic.* Materials that can be absorbed into the body by inhalation, ingestion, or through the skin and produce moderate effects following single exposures lasting seconds, minutes, or hours, or following ingestion of a single dose.

(c) *Chronic local.* Materials that on continuous or repeated exposures extending over periods of days, months, or years cause moderate harm to the skin or mucous membranes.

(d) *Chronic systemic.* Materials that can be absorbed into the body by inhalation, ingestion, or through the skin and which produce moderate effects following continuous or repeated exposures over periods of days, months, or years.

Those substances classified as having moderate toxicity may produce irreversible as well as reversible changes in the human body. These changes are not of such severity as to threaten life or to produce serious physical impairment.

3 = Severe Toxicity* (High)**

(a) *Acute local.* Materials that on single exposure lasting seconds, minutes or hours cause injury to skin or mucous membranes of sufficient severity to threaten life or to cause permanent physical impairment or disfigurement.

Table 8-23 (cont.)

(b) *Acute systemic.* Materials that can be absorbed into the body by inhalation or ingestion or through the skin and that can cause injury of sufficient severity to threaten life following a single exposure lasting seconds, minutes, or hours, or following ingestion of a single dose.

(c) *Chronic local.* Materials that on continuous or repeated exposures extending over periods of days, months, or years can cause injury to skin or mucous membranes of sufficient severity to threaten life or cause permanent impairment or disfigurement, or irreversible change.

(d) *Chronic systemic.* Materials that can be absorbed into the body by inhalation or ingestion, or through the skin and that cause death or serious physical impairment following continuous or repeated exposures to small amounts extending over periods of days, months, or years.

Table 8.24 NFPA Toxicity Ratings[196]

0 Materials that on exposure under fire conditions would offer no health hazard beyond that of ordinary combustible material.

1 Materials only slightly hazardous to health. It may be desirable to wear self-contained breathing apparatus.

2 Materials hazardous to health, but areas may be entered freely with self-contained breathing apparatus.

3 Materials extremely hazardous to health, but areas may be entered with extreme care. Full protective clothing, including self-contained breathing apparatus, rubber gloves, boots and bands around legs, arms and waist should be provided. No skin surface should be exposed.

4 A few whiffs of the gas or vapor could cause death, or the gas, vapor, or liquid could be fatal on penetrating the fire fighters' normal full protective clothing which is designed for resistance to heat. For most chemicals having a Health 4 rating, the normal full protective clothing available to the average fire department will not provide adequate protection against skin cancer with these materials. Only special protective clothing designed to protect against the specific hazard should be worn.

Table 8.25 SAX/NFPA to Weight Factor Conversion Table

Toxicity	Weight
Sax or NFPA level 0	0
Sax or NFPA level 1	1
Sax or NFPA level 2	2
Sax 3 or NFPA 3 or 4	3

The interrelationship between persistence and toxicity is evaluated using the weights from Table 8.22 (for persistence) and Table 8.25 (for toxicity) as the indices for Table 8.26. This value is to be directly recorded on the ERIS Detailed Inventory Worksheet.

Table 8.26 Persistence vs. Toxicity Matrix

Weight for Toxicity	Weight for persistence 0	1	2	3
0	0	0	0	0
1	3	6	9	12
2	6	9	12	15
3	9	12	15	18

Table 8.27 Quantity to Weight Conversion Table

1 ton = 1 cubic yard = 4 drums
1 drum = 50 gallons

Tons in cubic yards	Number of drums	Weight
0	0	0
1–10	1–40	1
11–62	41–250	2
63–125	251–500	3
126–250	501–1,000	4
251–625	1,001–2,500	5
626–1,250	2,501–5,000	6
1,251–2,500	5,001–10,000	7
>2,500	>10,000	8

QUANTITY OF MATERIAL

A standardized unit for the quantity of the risk source present in a given situation (static), may be obtained by again using the HRS approach shown in Table 8.28. A measurement for a dynamic quantity as in a current production facility may be obtained using the same table, but for a specified period of production time approximating the amount of time that would elapse from the beginning of an unexpected release until the release could be stopped. The value from Table 8.27 will be directly recorded on the ERIS Detailed Inventory Worksheet.

The product of the Toxicity/Persistence score and the Quantity score will be normalized by dividing the actual score by 144 and multiplying by 100. As with the other sections of the inventory worksheet, the Risk Source section will have a maximum value of 100.

REFERENCES

177. Defined in the Code of Federal Regulations, *7 CFR 657.5*, 1981.
178. Wetland is defined by EPA in the Code of Federal Regulations *40 CFR Part 230, Appendix A*, 1980.
179. Endangered species are designed by the U.S. Fish and Wildlife Service.
180. Wetland is defined by EPA in the Code of Federal Regulations *40 CFR Part 230, Appendix A*, 1980.
181. Designated by the U.S. Fish and Wildlife Service.
182. JRB Associates, Inc.; *Methodology for Rating the Hazard Potential for Waste Disposal Sites;* May 5, 1980.
183. Sax, N. I.; *Dangerous Properties of Industrial Materials;* Van Nostrand Rheinhold Co.; New York, 4th ed.; 1975. The highest rating listed under each chemical is used.
184. JRB Associates, Inc.; *Methodology for Rating the Hazard Potential for Waste Disposal Sites;* May 5, 1980.
185. National Fire Protection Association; *National Fire Codes*, Vol. 13, No. 49; 1977.
186. Professional judgment based on information contained in U.S. Coast Guard CHRIS Hazardous Chemical Data, 1978.
187. Professional judgment based on existing literature.
188. Griffin, Rodger D.; *Principles of Hazardous Materials Management;* Lewis Publishers, Inc.; 121 South Main Street; Chelsea, MI; 1988 provides a reasonable introduction to both the topic of toxicology and some of the mathematical models used to deduce the target risk associated with a given substance at a given environmental concentration.
189. *Risk Assessment and Management: Framework for Decision Making;* USEPA Report 600/9-85-002 (1984), p. 11.
190. *Carcinogen Risk Assessment, Estimating Exposures, Mutagenicity Risk Assessment, Health Assessment of Suspect Development Toxicants, Health Risk Assessments of Chemical Mixtures;* USEPA Office of Research and Development, Risk Assessment Guidelines, Federal Register, September 24, 1986.
191. TLV is a registered trademark of the American Conference of Governmental Industrial Hygienists.
192. *Threshold Limit Values and Biological Exposure Indices for 1988–1989;* published annually by the American Conference of Governmental Industrial Hygienists; 6500 Glenway Ave., Bldg. D-7, Cincinnati, OH 45211-4438.
193. Griffin, ibid.
194. Sax, N. I.; *Dangerous Properties of Industrial Materials;* Van Nostrand Rheinhold Co.; New York, NY, 4th edition; 1975.
195. Sax, N. I.; *Dangerous Properties of Industrial Materials;* Van Nostrand Rheinhold Co.; New York, NY, 5th edition; 1979.
196. National Fire Protection Association; *National Fire Codes*, Vol. 13, No. 49, 1977.

9 ENVIRONMENTAL RISK — THE INVENTORY RISK "VALUE"

INTRODUCTION

To this point, we have been looking at the pieces of the environmental risk inventory, but have not yet stated how the risk "value" of the inventory will be calculated for purposes of management control. There are several issues that have been avoided that must now be explicitly dealt with in order to proceed.

The first issue, and possibly the most important, is to recognize that any risk value obtained by the methods described so far are not absolute values, values equivalent to, for example, the balance in your checking account. Obtaining an absolute risk value is a practical impossibility. There are neither the resources, the theoretical means or the methods for calculating an absolute risk value in all but the very simplest circumstances. What will result from the calculations is what might be termed a risk number, whose importance lies in the relationship of one risk number to another. It is the relative level of risk that is being sought in order to intelligently allocate scarce resources to meet the objective of the Environmental Risk Management Program (ERMP), *"Maximum risk reduction at minimum feasible cost."*

The first major step will be to calculate an intermediate value, called the Composite Risk Score (CRS) on the ERIS Detailed Inventory Worksheet. Keeping in mind that the risk system component scores have already been normalized (scaled to a maximum value of 100), this number will also be normalized to a maximum value of 100. The purpose is to place the various risk values for the risk sources in context with each other and to place emphasis on the relative consequences.

The second major issue is the concept of expected values. Risk calculations by definition deal with events that have only some probability of occurrence—there is no guarantee that the event will or will not occur. The Composite Risk Score and the Human and Sensitive Environment Risk Scores are a form of expected value of the occurrence of an event placed in the worst case context. The worst case context is defined as the maximum amount of human or environmental damage resulting *if the event occurs*. This worst case is tempered by the probability of the event occurring; the relative ability of the risk source to damage the target; the relative probability of a primary control mechanism failure; the relative probability of a given transport mechanism being used, and its efficiency to place the risk source and target in proximity.

The expected value of a compound event may be calculated based on the following formula.

$$\text{Expected Value} = [(p_1)*(p_2)*\ldots*(p_n)*v] \quad (9.1)$$

Where: p_i = Probability of event i occurring, n such possible events in the chain of events necessary to realize v.

v = Value of the event

For example, suppose that the probability of the use of surface water to transport a substance is rated as being high and the rated value of the surface water to transport that substance is 16. The expected value, given that we say that a high probability is equivalent to a numeric value of 0.80 (80% probability), would then be 0.80*16 = 12.8. This does not say that the event will or will not occur, simply that the expected value for the event is 12.8, whatever 12.8 may mean.

The importance of an expected value will lie in its relationship to the expected values for the other transport mechanisms. If the air route has an expected value of 15, then it is relatively of more importance for management action purposes than the surface water route. If the expected value for the ground water route is 9, then it is of less importance than both the air or surface water routes. Resources would be allocated to prevent a problem accordingly if this were the only consideration. Naturally, this is not the only consideration, there are still the primary control mechanisms, the risk source, and the targets to be considered.

If there are events that may be considered to be independent of each other, the expected value for each event may be summed to result in a total expected value for a combined event. In the case of the ERIS, the Risk Source, Primary Control Mechanism, and the Transport Mechanism are considered to be independent events and the composite scores for each will be summed to yield the raw Composite Risk Score. However, the target impacts

Table 9.1 G/F/P and AHERA Decision Tree Translation to ERIS Score

G/F/P Scale	AHERA Scale	ERIS Scale
Good	7+	1
Fair	4–6	3
Poor	1–3	5

are not independent of the Risk Source, Primary Control Mechanism, and Transport Mechanisms, and the Composite Risk Score will be multiplied against the target information to yield the final risk inventory "value."

Translation of G/F/P to a Numeric Value

The Good/Fair/Poor or AHERA (augmented decision tree) rating for the current condition of the physical control system, for example, can be translated into numeric values according to Table 9.1.

Note that these translations may appear to be "backwards", with the apparent "best" value receiving the smallest numeric value (e.g., Good = 1). This apparent reversal is due to the desire to express a relatively poorer condition as a higher composite score, so a poor rating should have a higher numeric value at this point.

Translation of H/M/L to a Numeric Value

The use of the High/Medium/Low rating was to express some idea of the probability of failure and the translation of the H/M/L to a numeric value may seem more in line with intuition than the G/F/P translation. The H/M/L translation could take place as shown in Table 9.2.

Table 9.2 H/M/L and Probability to ERIS Score Translation

H/M/L Score	"Normal" Probability	ERIS Score
H	0.83–1.00	5
M	0.17–0.82	3*
L	0.01–0.16	1
0	0.00	0

*Note that this range of normal probability values of 0.17 to 0.82 being translated into an ERIS score of 3 or an H/M/L score of M is not arbitrary. The probability distribution is assumed to be normal about the mean value of 0.50, and two-thirds of the probability values will therefore fall within one standard deviation of the mean value or in the range of 0.17 to 0.82. One sixth of the probability values will fall above this central range and one-sixth below (by the way, this is the mathematical basis for the "80/20 rule").

Note that for the earlier calculations the possibility of a zero value for a given risk system component is only a mathematical possibility that must be allowed for but that would not make much sense in the context of environmental risk analysis. For example, a risk source composite score of zero would imply that there is no risk source, or a primary control mechanism score of zero would imply a somehow "perfect" primary control mechanism for the risk source. In these cases, why include the possibility in the ERIS at all?

On the other hand, it is realistic to think in terms of a transport mechanism composite score of zero, implying that one or more of the transport mechanisms are simply not physically available to this particular risk source/primary control mechanism combination. For example, if the risk source is solid lead and it cannot be vaporized by any reasonable means under the circumstances, then the air route is essentially unavailable except to lead in the form of bullets or thrown bricks. Transport mechanisms may therefore have a zero probability of use and a zero composite score for a particular transport mechanism.

COMPOSITE RISK SCORE (CRS) CALCULATION

The calculation of the Composite Risk Score requires the following steps:

1. Sum the composite risk scores from the Risk Source, Primary Control Mechanism, and Transport Mechanisms. Each transport mechanism composite score is to be separately added to the score for the other two risk system components to arrive at five raw Composite Risk Scores for each Risk Source/Primary Control Mechanism/Transport Mechanism combination.
2. The maximum total possible raw CRS will be 300. To normalize to a scale of 0–100, divide the raw CRS score by 3.

These relationships can be displayed in equation form as shown in Table 9.3.

THE FIRST EXERCISE IN MANAGEMENT CONTROL

At this stage, a large amount of management control information is available as a basis for action. Without going any further, it will be possible to examine by inspection the CRS scores to obtain a very solid feel for the technical vulnerability to risk based on the relative values of the CRS scores and to begin to focus on those areas obviously in need of attention, the objective being to lower the CRS scores through appropriate action.

In addition, some indication of detailed actions to be taken can be

Table 9.3 Equation Relationships for Risk Scores

Human Risk Score$_i$	=	[Current Number*ERIS Human Target Score]*[CRS$_i$]	(9.2)
Where: CRS$_i$	=	[RSCS + PCMRS + TMRS$_i$]3	(9.3)
i	=	Surface, Ground, Air, Direct Contact or Fire/Explosion	
RSRS	=	[HRS Quantity * Toxicity/persistence] * 0.694	(9.4)
PCMRS	=	[Current Condition * HRS Score * Awareness * Training * Equipment * Discipline] * 0.000464	(9.5)
TMRS$_{surf}$	=	Probability of Use$_{surf}$ * Surface Water Score * 1.33	(9.6)
TMRS$_{grnd}$	=	Probability of Use$_{grnd}$ * Ground Water Score * 0.91	(9.7)
TMRS$_{air}$	=	Probability of Use$_{air}$ * Air Route Score * 6.67	(9.8)
TMRS$_{DC}$	=	Probability of Use$_{DC}$ * Direct Contact Score * 6.67	(9.9)
TMRS$_{F/E}$	=	Probability of Use$_{F/E}$ * Fire/Explosion Score * 2.22	(9.10)

obtained by examination of the risk system component scores; an examination that can impart some feel for the impact of an action. Take as an example the human component scores of the primary control mechanism. If one or all of these individual scores are high, relatively inexpensive training, awareness, equipment or discipline improvement actions might be taken and a qualitative analysis of the impact of those actions developed.

Suppose that training had a numeric rating of 5, indicating that the training programs were Poor in content or quality. An improvement in the training programs sufficient to lower this score to 3 (Fair), would have the following impact on the CRS score (see Table 9.4).

The result of an improvement in training, presuming that somehow improved training does not also improve awareness, results in a decrease in the CRS of 4.6%. An analysis of all of the CRS components along these lines

Table 9.4 Primary Control Mechanism Composite Score Elements

	Current	Projected
Physical Component		
Condition	3 (Fair)	3 (Fair)
ERIS Score	69	69
Human Component		
Training	5 (Poor)	3 (Fair)
Awareness	3 (Fair)	3 (Fair)
Equipment	3 (Fair)	3 (Fair)
Discipline	3 (Fair)	3 (Fair)
Raw Score	27,045	16,767
Normalized Score	12.96	7.78
Risk Source Composite Score	50.00	50.00
Transport Mech. (Surf. Water)	50.00	50.00
CRS Score	37.65	35.93

can yield pointers for management action that can have significant risk reduction impacts. This type of analysis will also provide excellent information with respect to designing the so-called Phase I, II and III audits to obtain the kind of data needed by management to properly analyze the situation and develop action plans.

TARGET RISK ANALYSIS

The development of a risk score for targets was the subject of the preceding chapter and little needs to be covered here except to again point out that the result will be a number of no intrinsic importance. The importance of the scores will lie in their relative relationship to other target risk scores and the underlying consistency with which the various scores are arrived at to complete this section of the analysis and inventory. Consistency is the key.

THE RISK "VALUE"

The "value" of the final human and sensitive environment risk scores will determine the relative importance of management action to reduce risk. It must again be pointed out that a deliberate choice was made by the author in developing the final scores; that human beings are far more important than gollywumps.

No matter the risk to a plant or animal species, given a choice, people must come first. To implement this argument, the target analysis actually counts people twice; once in terms of the actual numbers within reach of a risk source through a given transport mechanism (possibly weighted by proximity and directness of the individual's relationship to the risk source and/or control mechanisms), and again through the use of the HRS population/distance to target weight matrix. The result is that sensitive environments will be a distant second to people.

This may at first appear to be somewhat disturbing to certain sections of the environmental advocacy community, but two points must be made in this regard. First, there are simply not going to be enough resources to immediately deal with every environmental risk, implying that resources must be rationed on some basis and the basis chosen here is "people first." Second, in most circumstances, if the risk to people is reduced, the risk to sensitive environments will also be reduced.

The final risk values are calculated as follows:

- Human Risk Score

  Multiply the CRS score (for each transport mechanism independently) by the Current Number of potential human targets and by the ERIS risk

REF DATA		RISK SOURCE DATA				PRIMARY CONTROL MECHANISM ANALYSIS											TRANSPORT/SECONDARY CONTROL MECHANISMS								CRS	TARGET ANALYSIS						HUMAN RISK SCORE	SENS ENVR RISK SCRE	ROUTE	NOTE NO.
							PHYSICAL COMPONENT					HUMAN COMP.				PCM RS										HUMAN					SENS ENVR SCRE				
REF DWG	HOMO AREA	DESCRIPT.	HRS QUFN	TOX/FERS	RS CS	DESCRIPT	CURR COND	t 1	t 2	t 3	HRS SCRE	AWARNESS	TRAINING	EQUIPMNT	DISCIPLN		DESCRIPT	PROB USE	SURF	GRND	AIR	DC	F/E	TM RS		CURR NO.	t 1	t 2	t 3	ERIS SCRE					
15.1	001	BENZENE	5	18	62	CLAY	P	P	P	P	47	G	F	P	G	2	RIVER	L	8					8	24	8	10	14	20	90	Ø	17280	Ø	SURF	
																	WELLS	M		11				33	32	8	10	14	20	84	NA	21504	NOT APP	GRND	
																		Ø						Ø	Ø									AIR	
																	DIG.	L				Ø		Ø	Ø									DC	
																		Ø						Ø	Ø									F/E	
15.1	002	CREOSOT	7	12	58	CLAY	P	P	P	P	47	P	P	P	G	14	RIVER	L	12					12	28	8	10	14	20	90	Ø	20160	Ø	SURF	
																	WELLS	M		11				33	35	8	10	14	20	84	NA	23520	NOT APP	GRND	
																		Ø						Ø	Ø									AIR	
																	DIG.	L				Ø		Ø	Ø									DC	
																		Ø						Ø	Ø									F/E	
																																		SURF	
																															NA		NOT APP	GRND	
																																		AIR	
																																		DC	
																																		F/E	
																																		SURF	
																															NA		NOT APP	GRND	
																																		AIR	
																																		DC	
																																		F/E	
																																		SURF	
																															NA		NOT APP	GRND	
																																		AIR	
																																		DC	
																																		F/E	
																																		SURF	
																															NA		NOT APP	GRND	
																																		AIR	
																																		DC	
																																		F/E	
																																		SURF	
																															NA		NOT APP	GRND	
																																		AIR	
																																		DC	
																																		F/E	

ERINVTRY1

Figure 9.1 Example ERIS Detailed Inventory Worksheet.

score for human targets. The product is the Human Risk Score (for each transport mechanism) expressed in thousands (or if appropriate, millions) of Risk Units.

- Sensitive Environment Risk Score

  Multiply the CRS score (for each transport mechanism independently) by the ERIS Sensitive Environment Score. The product is the Sensitive Environment Risk Score expressed in thousands (or if appropriate, millions) of Risk Units. Note that for the Sensitive Environment Risk Score, the Ground Water Transportation Route is not applicable, an argument consistent with the Hazard Ranking System Approach.

The Environmental Risk scores will determine the priority of action based on their relative magnitude. The highest risk score for a particular Risk Source/Primary Control Mechanism will govern the relative importance of each in prioritizing for risk management purposes.

Figure 9.1 provides an example ERIS Detailed Inventory for two Risk Source/Primary Control Mechanism scenarios. For the purposes of allocating scarce resources, the cyanide/steel tank combination should be dealt with first because of the relative size of the human risk score compared to the asbestos/calcium carbonate matrix combination. Note in this example that some resources can probably be allocated to reducing the risk from the asbestos/matrix combination by simply increasing the level of training and awareness of the individuals involved, while the cyanide/steel tank combination will probably require both capital expenditures and a continuing overhead cost increase in order to properly deal with the situation.

10 COST MODELS AND COST ESTIMATION—BENEFIT MODELS AND BENEFIT ESTIMATION

INTRODUCTION

Given that we now have a way of consistently measuring the level of environmental risk, at least in uniform, if artificial, risk units, it then becomes possible to develop the data necessary for the analysis of the benefits and associated costs of possible risk reduction alternatives. This will also allow for the selection of an optimum feasible strategy for the reduction of risk given the existing constraints of technology, local conditions, and economics. In order to proceed, a model for costs and benefits within the context of the ERIS must be developed.

This book is not a treatise on cost accounting, but the basic ideas from cost accounting must be utilized and the reader should review those ideas as necessary. The purpose here is to develop a generalized model for costs and benefits, not a detailed model that will deal with every conceivable situation. The general model will have to be tailored to the specific situation encountered in practice.

GENERAL MODEL CONSIDERATIONS

There are several key concepts concerning the ownership of environmental risks that must be recognized and that are slightly different from other ownership considerations.

1. The responsibilities associated with the ownership of a risk source may only be ended in one of three general ways:
   a. Transfer of title to another party who has been fully informed of the existence of the risk source and is capable of fully evaluating the con-

sequences and possible consequences of the ownership of the risk source, and who has sufficient financial resources to deal with the consequences of ownership.

b. The complete and absolute destruction of the risk source in a manner that no risk source will exist after said destruction. Note that removal of a risk source from its original site and disposal of that risk source, even in a properly designated facility and by properly designated methods, will leave a residual risk to the owner or operator at the time of removal. This risk results from the possibility that at a later date the owner or operator at the time of removal may become a Potentially Responsible Party (PRP) in the disposal site.

c. That the risk source is not, in fact, a risk source. The simple fact of the matter is that some substances or conditions currently defined as being or as likely to be risk sources may be found not to be risk sources with the accumulation of scientific information over time.

2. That responsibility for a risk source does not end with the productive value of the risk source or the property associated with the risk source.[197] In the past it was possible to "walk away from" a property that no longer had productive value, but if that property contains a risk source this will no longer be possible.

3. Responsibility for a risk source and damages resulting from a failure to control the risk source may be personal to the owners or managers of the entity responsible. Many of the environmental laws and cases fix responsibility not only on the owner or operator as a corporate entity, but also personally on the owners and managers of the corporate entity.

The net result of these considerations will be several new elements in the cost models to account for the differences between owning a "normal" property and owning a property containing an environmental risk.

TIME AND PROBABILITY CONSIDERATIONS

Critical to an understanding of any cost model are several factors associated with the behavior of financial models in general.

The first factor is the behavior of costs with respect to time, which has three general types of impact on environmental risks. One of the impacts is the time value of money. In general, a dollar in hand today is worth more than a dollar in hand at some time in the future. To account for this idea, the Discounted Cash Flow concept is employed and results in an estimate of the present value of a future sum of money.

Probably the simplest way to think about present value is to consider its inverse, the amount of money that would have to be deposited in a savings account today, at a given interest rate, in order to have a specified sum of money available in that account at a later date. For example, to have $10,000 available in 10 years, the amount that must be deposited in a savings account today at 10% compound interest (compounded annually) will be $3,855. In fact, $3,855 is the present value of $10,000 received in 10 years at a 10% annual discount rate.

The present value is determined by the amount to be paid, the distance into the future at which the payment is to be made, and the discount rate applicable to the situation. The present value of a future payment may be determined by the formula:

$$PV = FA/(1 + k) \tag{10.1}$$

Where: PV = Present Value of the Future Amount
FA = Future Amount
k = The applicable discount rate per time period.
t = The number of time periods until the FA is received.

The difficulty in calculating the present value of a future payment lies in determining the appropriate discount rate to apply. In the simple example cited above the discount rate was equal to the interest rate that could be earned on the savings account, but this is not usually the correct discount rate for risk management purposes.

Many organizations have something known as an *internal hurdle rate* used to determine if a project meets the organization's investment criteria. In many cases the internal hurdle rate will be more appropriate to environmental risk decision making than, for example, the so-called prime rate. Even more appropriate may be the internal hurdle rate plus some increment representing the impact of negative leverage associated with environmental risks. This amounts to increasing the internal hurdle rate by an amount that represents a "self-insurance" premium, since outside insurance is not generally available. As with all of the other components of the ERIS and ERMP, consistency will be the key in the long run. Given a discount rate, however established, it must be used consistently for all analyses.

Although it may seem strange to discuss present values of risk reduction benefits measured in non-financial units, it is as appropriate to consider this issue as it is to consider the calculation of the present value of a cash stream. The difficulty will again be in the determination of an appropriate "discount" rate to use in the analysis. For example, the population at risk in the future indicated by the t_1, t_2, and t_3 population values might be discounted to a present value by an annual growth rate for the general population, emphasizing the effect of unusual local growth rates.

The second area of impact of time on the cost models is the relationship between response time to environmental accidents and the costs for dealing with such accidents. Although no precise relationship between response time and cost can be established, past experience indicates that increases in response time tend to result in disproportionate increases in costs. In fact, for major environmental incidents the costs appear to increase geometrically or exponentially with response time, although not without limit.

The third impact area of time is the often overlooked fact that environmental risk alternative costs may be very strongly influenced by the timing of other management decisions. As a simple example, consider the effects on cost if an environmental risk is to be removed from a property in a manner that will require the demolition and reconstruction of a part of the property. The costs will be substantially different if the action takes place twenty years prior to the planned demolition of the property as opposed to one or two years prior to the planned demolition. Environmental risk control actions cannot be absolutely governed by considerations of the remaining useful life of a property, but the selection of the particular response option to be implemented at a particular time can, and should be, influenced by such considerations, and may (will), in turn, influence the timing of obsolescence decisions.

In addition, the concept of expected value will be critical to the analysis of alternatives as a result of the probabilistic nature of risk. The probabilistic nature of predicting the future is fundamental to virtually any decision, but its explicit consideration in environmental risk management is necessary. The expected value of some future event or group of events was discussed in the preceding chapter, and will be revisited here in terms of the expected value of financial impacts and the financial component of the benefits to be derived from environmental risk management.

ENVIRONMENTAL RISK COST MODEL

The environmental risk cost model shown in outline form in Table 10.1—in common with all cost models for management decision making purposes—depends on the overall context of the entity using the model. As discussed in the previous section, it would be very improper and erroneous to divorce environmental risk decisions from the total plans, objectives, purposes, and resources of the organization itself. How this issue will be dealt with in a specific case will be highly dependent on the organization and its internal decision-making structure and behavior, but in general, the environmental risk model must fit into the overall planning and performance measurement structure of the organization. As mentioned in the preceding section, the decision to remove a risk source cannot be totally divorced from other asset/liability management decisions having apparently little or nothing to do with the environmental risk.

The model proposed here is composed of a series of major component parts, but these parts are not separable. In particular, the Long-term Control Costs will exist so long as any meaningful quantity of a given risk source exists. Therefore, although a part of the long-term costs may vary depending on the quantity of risk source or the number of risk sources, until or unless the risk sources are totally eliminated, at least some minimum level of long-term control costs will be experienced by the organization.

Cost Model Part I—Initial Costs

Certain costs must be experienced by the organization to simply know if there are risk sources that will require management. The costs for a Phase I Environmental Risk Assessment will have to be borne by virtually all organizations at some time, even if the results of the Phase I are negative; that is, no risk sources are identified. The additional costs for Phase II and Phase III ERA work, given that risk sources have been found, may run into the hundreds of thousands of dollars and are fundamental to the construction of the ERIS and the management of the risks through the ERMP.

The development and analysis of the data necessary to make informed and appropriate risk management decisions through the selection of risk reduction strategies must also be included in the category of initial costs; costs that may not be "recoverable" in the traditional financial sense. However, one product of the initial costs should be the development of the ERIS "beginning inventory" and the associated recordkeeping system that will allow the tracking of management performance over time. In addition, given that environmental risks have been discovered, the initial costs should also cover the development of a minimum notification and reporting program as required by the various governing regulations and the considerations of liability limitation and good public relations practice.

All of these initial costs are in the nature of basic investments in the development of the ERMP, investments whose value will only be recovered through the reduction of risks to human, environmental, and liability exposures. It is one of the unfortunate, but nevertheless true situations of environmental risks that the value of the investment cannot be predicted until the majority of the investment has been made. The investment may be a "sunk" cost, unrecoverable by normal standards.

Cost Model Part II—Long-Term Control Costs (O&M Program Costs)

The costs for the development and implementation of a long term control program fall into the three classic categories of all costs: (1) fixed costs, (2)

Table 10.1 Environmental Risk Management Cost Model

I. Initial Costs

- Phase I, II and III ERA costs
- Strategy selection costs
- Recordkeeping system setup and beginning inventory
- Basic notification and reporting costs

II. Long-Term Control Costs (O&M Program Costs)

- *Initial Fixed Costs (minimum configuration)*
  - Training and retraining of minimum number of personnel
  - Equipment and equipment maintenance
  - Awareness (awareness program setup costs and minimum notification)
  - Discipline (may include negotiations with labor unions, attorney fees, etc.)
  - Recordkeeping (personnel, control activities, and reporting)
  - Monitoring (surveillance, analytical testing)
  - Basic control program performance auditing
- *Variable Costs (dependent on facility activity level above the minimum configuration)*
  - Training and retraining
  - Equipment and equipment maintenance
  - Awareness (periodic notification)
  - Recordkeeping (personnel, control activities, and reporting)
  - Monitoring (surveillance, analytical testing)
  - Control supplies
  - Variable cost of program performance auditing

III. Emergency Response Costs

- *Fixed Costs*
  - Standby emergency response capabilities
    - Equipment
    - Personnel training and drills
    - Supplies (including replacement of dated supplies)
  - Accruals for anticipated emergency response costs in lieu of insurance
  - Emergency prevention engineering
- *Variable*
  - Actual response costs (personnel, expended supplies and equipment, damages, etc.)

IV. Planned Response Action Costs (other than Long-Term Control and Removal Options)

- *Miscellaneous*
  - Design and engineering costs
- *Direct Costs*
  - Direct cost of action
  - Work monitoring
  - Risk of emergency costs
- *Elective Costs*
  - Pre-work demolition
  - Restoration
  - Lost revenues
  - Relocation costs (temporary)
    - Opportunistic basis
    - Dedicated basis
    - Equipment and services relocation (out and in)
  - Tenant relocation (out and in)

V. Removal and Disposal Costs

Miscellaneous

- Design and engineering costs

Direct Costs

- Direct cost of action
- Work monitoring
- Transportation accident risk costs
- Disposal site risk costs
- Functional replacement of risk source costs

Elective Costs

- Pre-work demolition
- Restoration
- Lost revenues
- Relocation costs (temporary)
  - Opportunistic basis
  - Dedicated basis
  - Equipment and services relocation (out and in)
  - Tenant relocation (out and in)

Cost Offset

- Reduction in O&M costs attributable to removed materials (Note: Leaves a residual risk of PRP status at disposal site)

VI. Removal and Destruction Costs

Miscellaneous

- Design and engineering costs

Direct Costs

- Direct cost of action
- Work monitoring
- Transportation accident risk costs
- Destruction costs
- Functional replacement of risk source costs

Elective Costs

- Pre-work demolition
- Restoration
- Lost revenues
- Relocation costs (temporary)
  - Opportunistic basis
  - Dedicated basis
  - Equipment and services relocation (out and in)
  - Tenant relocation (out and in)

Cost Offset

- Reduction in O&M costs attributable to removed materials (Note: Complete destruction of risk source assumed)

---

variable costs, and (3) semi-variable costs or costs having both fixed and variable characteristics. Because of the long-term nature of these costs the evaluation must generally take into account the time value of money through the Discounted Cash Flow concept. Decisions on long-term courses of action should generally be made on the total amount of the present value of the stream of expenditures.

Fixed Cost Portion of Long-term Control Costs

With regard to fixed costs, the analyst must often take care to insure that the specialized concerns noted earlier concerning the characteristics of ownership of environmental risks are addressed. This will probably mean that there will be two basic fixed cost categories for long-term control, fixed costs purely attributable to the fact ownership without regard to productive activities, and fixed costs attributable to some minimum level of productive activities. The reason for this distinction is to clearly draw management attention to the fact that the closure of a facility from a productive point of view will not end all of the costs associated with that facility—a fact that may alter the closure decision substantially.

A property may have a negative or zero value in the marketplace, but have a positive value to the current owner if it is maintained in productive use. Also, in certain cases, the closure and particularly the abandonment of a facility may trigger additional, and potentially very large, liability exposures.

The fixed costs associated with the implementation of a long-term control program for a productive facility can be substantial. These may involve: (a) the costs of training a minimum number of individuals and their periodic retraining; (b) their proper equipping and the maintenance costs for the equipment; (c) the costs of operating a minimum awareness program for all employees and tenants; (d) the development of disciplinary systems and the legal ramifications of those systems on union contracts and various laws; (e) basic costs for monitoring the primary control mechanisms and the transport and secondary control mechanisms including analytical tests and surveillance; and (f) the costs for periodic management control performance audits.

The costs of maintaining the complex recordkeeping systems for compliance with the various laws and with the management reporting systems necessary to evaluate and control risks can be substantial and will normally fall into the category of a semi-variable cost as opposed to being a true fixed cost. The variability of the recordkeeping costs will most likely come from the number of individuals involved for a given level of productive activity with the fixed cost component being defined by the number and type of risk sources/primary control mechanisms to be dealt with in the particular situation.

Variable Portion of Long-term Control Costs

The variability of most of the long-term control costs will be associated with the level of productive activity (the number of units of product produced, the number of new tenants in a building each year, the number of employees in the facility, etc.). Variable costs are assumed to be directly proportional to the amount of activity with which they are associated *over some range* of activity.

It is important to recognize that the closer an activity approaches the maximum level of activity supported by the facility, the more likely it will be that the long-term control costs will increase disproportionately to the increased level of activity. There is nothing unusual about this characteristic of environmental control costs, virtually all of the supporting systems for a given activity will experience the same behavior. The closer to design capacity a given system is, the more likely it will be to experience failure and problems, resulting in disproportionately higher operating costs. The explicit recognition of this fact is critical to risk management for obvious reasons, and particularly for the next major model component, emergency response costs.

The list of variable cost items for long-term control is much the same as the list for fixed costs with two exceptions; the costs for expendable supplies and the variable portion of the program performance auditing costs.

Cost Model Part III—Emergency Response Costs

The development of budgets, which in effect is what is being done here, always serves two purposes. The obvious purpose is to allocate resources, but behind that lies the organizational discipline of simply evaluating all of the alternatives prior to allocation. The development of the actual allocations for emergency response will be less important than the discipline of thinking through the possible emergency situations and developing scenarios for dealing with those possible emergencies. The reader is reminded of the fundamental nature of the failure of primary control mechanisms discussed in an earlier chapter in dealing with these issues.

Fixed Portion of Emergency Response Costs

The bulk of emergency response costs will tend to range between fixed and semi-variable with a large fixed component because of the necessity of having available specific assets to deal with possible emergencies; oil booms, skimmers, neutralizing agents, and the like. A portion of these assets will tie up resources in items that have relatively long life spans and the techniques of depreciation analysis will be necessary to assign a budgetary number to the asset values.

The analyst is cautioned here to totally ignore the issues of depreciation as they relate to taxes. The depreciation allowed for an asset for tax purposes will have very little or no bearing on the useful life of an environmental risk emergency response asset if for no other reason than the rapid development of technology in this field. The negative consequences of holding on to obsolete technology, given that an emergency occurs, will normally far outweigh any possible tax benefits.

The size of the fixed investment will usually be related to the available transport mechanism and target characteristics. The costs for equipment, personnel training and drills, and supplies (including the replacement of supplies that deteriorate with time) will remain relatively fixed. In addition, the organization should probably accrue funds in an emergency response budget over time because of the lack of insurance coverage to cover environmental emergencies as a form of self insurance.

A major component of emergency response costs should be an allocation to pay for the engineering required to reduce the likelihood of an emergency or the magnitude of the emergency event if one should occur. Deliberate and specific engineering attention devoted to the subject of loss prevention has long been recognized as a beneficial investment of funds in classic emergency cases that do not contain the negative leverage potential of environmental risks and are therefore even more appropriate here.

Variable Portion of Emergency Response Costs

Some emergency response costs do vary with the level of activity, but great care must be exercised in determining the productive activity to which the variability will be tied. For example, if oil spills from tankers entering or leaving a harbor or port are of concern, an increase in the number of tankers entering and leaving may necessitate an increase in standby capacity to deal with spills. However, the real activity variable may not be the number of tankers, but the capacity of the tankers; larger tankers offering the potential for larger spills. The actual activity variable to which the emergency response costs should be tied must be very carefully evaluated for appropriateness to prevent under- or over-burdening the environmental risk budget resulting in an improper allocation of resources and thereby increased risks.

Cost Model Part IV—Planned Response Action Costs

Planned response actions cannot generally be divided into fixed and variable components; they are normally of a project cost nature with miscellaneous, direct and elective cost components. The timing of planned response actions will dictate the present value impact of the project costs for decision-making purposes.

Planned response actions in this model section normally constitute the repair, encapsulation, enclosure and isolation response options. Removal is treated separately because of its special effects on such items as emergency response costs and long-term control costs.

Miscellaneous Portion of Planned Response Action Costs

The miscellaneous portion of planned response action costs will normally consist of the response action design and engineering costs. While these costs are normally small relative to the total size of the response action, the development of soundly designed and engineered response actions is critical to both the success and the total cost of carrying out the response action. The response may be designed far in advance of the actual implementation of the project in order to allow for the accrual of resources with which to carry out the project.

Direct Portion of Planned Response Action Costs

The direct cost of carrying out the response action for a given situation must be carefully separated from associated, but elective costs in order to provide sound decision-making data. The direct cost of removing a risk source, improving or replacing a primary control mechanism, modifying the availability of transport mechanisms or the efficiency of secondary control mechanisms are all part of the direct costs. In addition, the costs for independent monitoring and/or supervision of the actual response action are an appropriate part of the direct costs along with the costs of the temporary relocation of equipment and utilities that may be required because of the technical specifications of the work to be undertaken.

Elective Portion of Planned Response Action Costs

The cost of elective activities should be kept separate from and not confused with the direct costs because the elective costs may be avoided on an objective basis. For example, the organization may elect to conduct planned response actions in advance of the normal obsolescence cycle of a facility, essentially triggering a rehabilitation of the facility in advance of normal. The increased cost for demolishing a portion of a structure prior to the end of its useful life imposes an additional cost burden that is not a proper portion of the response action cost if the long-term control program would be able to maintain the risk source in a low-risk configuration.

The objective of distinguishing between direct and elective costs is to identify the incremental cost components for management decision making.

Decisions are made on the basis of increment costs versus incremental benefits, not on the basis of total costs. If the organization should choose to accept additional incremental costs because of lost productive capacity by accelerating the remediation activity schedule, it may be a reasonable decision. The direct costs of the remediation activity will not change, but the total cost will as a result of the acceptance of the elective costs.

Elective costs are normally found in such categories as pre-work demolition costs, restoration costs, lost revenues, and the like. They are normally in the form of the costs of shorter-than-normal useful lives for the asset being demolished and restored, or the earlier-than-normal acceptance of lost revenue or the acceptance of lost revenue when the loss could be avoided. A convenient example of this situation can be found in asbestos abatement work. Provided that a long-term control program is properly designed and implemented and will provide a reasonable assurance of a low level of risk, two alternative strategies for the timing of removal are available. These have been denoted in the literature as ''opportunistic'' and ''dedicated.''[198]

As defined, the opportunistic basis argues that as sufficient space in the building becomes available through tenant vacancy to provide an economic removal environment, the opportunity to remove is taken. On this basis, there is only a small or no loss of revenue, no tenant relocation costs, and possibly no demolition/restoration costs (assuming that the space would be renovated prior to a new tenant occupancy) attributable to the removal action.

The dedicated basis assumes that the owner deliberately decides to remove on a schedule that disregards or takes little notice of naturally occurring tenant vacancy and results in increased costs for lost revenue, relocation of tenants, possibly premature demolition and replacement of demising walls and other costs. These additional costs are elective with respect to the actual cost of removing the asbestos-containing material, but must be considered in developing an appropriate course of action.

Cost Model Part V—Removal and Disposal Costs

Removal and disposal of the risk source is different from other response actions in that it will normally provide an offset to some portion of other costs, particularly in the area of the long-term control costs. It will also generate a new cost estimation area that may be difficult or impossible to quantify in the form of a type of insurance premium against the possibility of the disposal site becoming a Superfund site and the owner or operator at the time of removal becoming a PRP in the site.

Miscellaneous Portion of Removal and Disposal Costs

The miscellaneous portion of removal and disposal costs include the same cost elements as the other planned response action categories, the costs to design and engineer the work to be accomplished. In addition, the owner/operator may choose to reduce the risks of becoming involved as a PRP by properly disposing of the material in a site owned by the owner/operator and controlled by them into the future. This option will involve a substantial additional investment for engineering design, governmental approval, acquisition, and site administration, but it is an option that has been elected by a number of organizations to date.

Direct Portion of Removal and Disposal Costs

Again, this portion is very similar to the other planned response actions, but with the addition of three major cost elements. The cost to transport the removed materials to the disposal site and the cost of disposal can represent substantial amounts of money that should be evaluated in looking at this option. In addition, the cost to provide a functional replacement for the risk source must be considered and included if appropriate.

Elective Portion of Removal and Disposal Costs

The elective portion of removal and disposal costs are the same as for the other planned response actions.

Cost Offsets Resulting from Removal and Disposal

The removal and disposal of the risk source, assuming that it is not replaced by another risk source that will come under the long-term control program, should result in some incremental reduction in at least the variable portion of the long-term control costs. In addition, there may be reductions in insurance premiums, financing costs, and possibly an improvement in facility operating efficiencies that result from a more efficient risk source replacement material.

Cost Model Part VI—Removal and Destruction Costs

The word "destruction" is used advisedly and primarily to denote the

Figure 10.1 Example Cost Analysis Worksheet

IDENTIFICATION: ____________________

DATE PREPARED: _____/_____/_____

PREPARER: ____________________

DISCOUNT RATE: _____._____%

INITIAL COSTS		
Phase I ERA	$_______	
Phase I ECA	_______	
Phase II ERA	_______	
Phase III ERA	_______	
Strategy Development and Selections	_______	
Initial Notification		_______
Initial Regulatory Reports	_______	
Other Costs	_______	
TOTAL INITIAL COSTS		$_______

OPERATIONS & MAINTENANCE		
Fixed Costs	**Initial**	**Annual**
Base (Property not active)		
Control Systems Maintenance	$_______	$_______
Monitoring, Testing and Analysis	_______	_______
Recordkeeping and Reporting	_______	_______
Minimum Active Facility Costs		
Training (# Trained _____)	_______	_______
Equipment	_______	_______
Awareness Program	_______	_______
Discipline Program	_______	_______
Recordkeeping Costs	_______	_______
MINIMUM COST TOTALS	$_______	$_______

Incremental Costs	Initial	Annual	Activity Variable Description
Training	$______	$______	______
Awareness	______	______	______
Equipment	______	______	______
Expendable Supplies	______	______	______
Discipline Program	______	______	______
Recordkeeping Costs	______	______	______
TOTAL INCREMENT COSTS	$______	$______	
TOTAL PRESENT VALUE OF O&M PROGRAM		$______	
Estimated Risk Reduction Benefit	______	______ (Units)	

PLANNED RESPONSE ACTIVITIES

Activity: ______ Date: ____/____/____		**Avoidable**
Design and Engineering Costs	$______	
Premature Obsolescence Costs	______	Y N
Temporary Tenant Relocation Costs	______	Y N
Temporary Equipment Relocation Costs	______	Y N
Lost Income	______	Y N
Direct Activity Costs	______	
Monitoring and Construction Super.	______	
Functional Replacement Costs	______	
Disposal Costs	______	
Disposal Risk Costs		$______ [Annual]
Appurtenance Replacement Costs	______	
O&M Program Cost Reduction		($______) [Annual]
Expected Lifetime ______yrs		
		Avoidable: $______
TOTAL PRESENT VALUE		**Unavoidable: $______**
Estimated Risk Reduction Benefit	______	______ **(Units)**

fact that the fate of the risk source after removal falls into two general categories; permanent reduction of risk to a zero level (destruction), and permanent/semi-permanent reduction to a small but non-zero risk level (disposal). Destruction may be achieved by a number of methods, some of which may not actually involve the physical/chemical reduction to non-hazardous components of the risk source.

For example, suppose that asbestos could be buried in a mine shaft thousands of feet underground. Because of the inaccessibility of the asbestos in such a location, this would be considered as destruction even though the asbestos fibers were still physically and chemically intact. Destruction could also involve heating the asbestos fibers to a sufficiently high temperature that they melt and change to a non-hazardous form. However, if the fibers are simply buried in the local municipal dump, this would be defined as disposal, not destruction.

The differences between disposal and destruction result in a change of the costs attributable to the long-term risks associated with the two methods. The residual risks associated with a disposed risk source are non-zero and so are the costs associated with the residual risks even if they cannot be adequately estimated. The residual risks associated with destruction are virtually zero and consequently so are the cost items. Figure 10.1 provides a summary worksheet outline for cost estimation.

BENEFIT MODEL AND BENEFIT ESTIMATION

The primary model for the estimation of benefits will be the procedures already used to develop the ERIS. To some extent the idea of estimating benefits by "plugging" the reasonably expected changes resulting from a risk reduction alternative into the ERIS and examining the result was shown in the discussion in Chapter 9 using the CRS component of the ERIS.

In general, the risk reduction benefits are the incremental changes in the human and sensitive environment risk scores resulting from a change in one or more of the various components. The Human Risk Score equations from Table 9.4 are repeated in Table 10.2 for convenience.

Working through these equations as appropriate will yield a projected risk score and the difference between the projected and the original risk score will be the incremental risk reduction (or increase) benefit. Other benefits may also accrue of a financial nature resulting from decreases in operating costs for long-term control programs, and other areas.

Table 10.2 Human Risk Score Model Equations

Human Risk Score$_i$	=	[Current Number*ERIS Human Target Score]*[CRS$_i$]	(10.2)
Where: CRS$_i$	=	[RSCS + PCMRS + TMRS$_i$]3	(10.3)
i	=	Surface, Ground, Air, Direct Contact or Fire/ Explosion	
RSRS	=	[HRS Quantity * Toxicity/persistence] * 0.694	(10.4)
PCMRS	=	[Current Condition * HRS Score * Awareness * Training * Equipment * Discipline] * 0.000464	(10.5)
TMRS$_{surf}$	=	Probability of Use$_{surf}$ * Surface Water Score * 1.33	(10.6)
TMRS$_{grnd}$	=	Probability of Use$_{grnd}$ * Ground Water Score * 0.91	(10.7)
TMRS$_{air}$	=	Probability of Use$_{air}$ * Air Route Score * 6.67	(10.8)
TMRS$_{DC}$	=	Probability of Use$_{DC}$ * Direct Contact Score * 6.67	(10.9)
TMRS$_{F/E}$	=	Probability of Use$_{F/E}$ * Fire/Explosion Score * 2.22	(10.10)

Other Risk Reduction Benefits

Other financial benefits beyond the reductions in operating costs are properly a separate topic[199] and will only be mentioned here. Among the benefits are the reduced exposure to civil litigation, fines, and regulatory reporting costs. In addition, there may be property value enhancements or restoration of property values to an unencumbered level. These benefits can be very substantial if also very difficult to quantify.

Part of the quantification process will involve the treatment of the ERIS components as an indication of the probability of the occurrence of an event, which in a crude but useful fashion can be accomplished by converting the CRS values to estimates of the probability of the occurrence of an event and the Human and Sensitive Environment Risk Scores to estimates of the magnitude of impact. Given that these concepts are properly implemented, it is possible to arrive at a risk rate of return calculation based on the financial model known as the Capital Asset Pricing Model. The risk rate of return can then be used to develop a better estimate of the discount rate for the discounted cash flow calculations mentioned earlier.

Note however, that all of this apparent sophistication of approach may be only that—apparent. Much of the real and daily work of risk management can be accomplished with the tools already provided here and given the rather gross approximations that have been made to arrive at the various measures of relative risk, a more sophisticated approach may not be justified.

REF DATA		INITIAL RISK DATA				OPERATIONS & MAINTENANCE				ISOLATION					ENCAPSULATION					ENCLOSURE					REMOVAL & DISPOSAL					REMOVAL & DESTRUCT			
							BENEFITS				BENEFITS					BENEFITS					BENEFITS					BENEFITS					BENEFITS		
REF DWG	HOMO AREA	DESCRIPT.	HUMAN RISK SCORE	SENS. ENVIR. RISK SCORE	ROUTE	ANN COST	NEW HUMAN R.S.	NEW SENS. R.S.	COMMENTS	COST	NEW HUMAN R.S.	NEW SENS. R.S.	NEW O&M COST	EFF. DATE	COST	NEW HUMAN R.S.	NEW SENS. R.S.	NEW O&M COST	EFF. DATE	COST	NEW HUMAN R.S.	NEW SENS. R.S.	NEW O&M COST	EFF. DATE	COST	NEW HUMAN R.S.	NEW SENS. R.S.	NEW O&M COST	EFF. DATE	COST	NEW HUMAN R.S.	NEW SENS. R.S.	EFF. DATE
					SURF																												
				NOT APP	GRND																												
					AIR																												
					DC																												
					F/E																												
					SURF																												
				NOT APP	GRND																												
					AIR																												
					DC																												
					F/E																												
					SURF																												
				NOT APP	GRND																												
					AIR																												
					DC																												
					F/E																												
					SURF																												
				NOT APP	GRND																												
					AIR																												
					DC																												
					F/E																												
					SURF																												
				NOT APP	GRND																												
					AIR																												
					DC																												
					F/E																												
					SURF																												
				NOT APP	GRND																												
					AIR																												
					DC																												
					F/E																												
					SURF																												
				NOT APP	GRND																												
					AIR																												
					DC																												
					F/E																												

Figure 10.2 ERIS Cost & Benefits Worksheet.

ERIS COST AND BENEFITS WORKSHEET

Figure 10.2 shows at least one way of recording a summary of the various risk source/primary control mechanism situations and the analysis results for the various alternatives for dealing with them.

On this worksheet the initial risk data comes directly from the ERIS Detailed Inventory Worksheet and provides the reference basis for the evaluation of risk reduction options. For each available option that is *technologically appropriate* the projected human and sensitive environment risk scores (New Human R.S. and New Sens. R. S.), the projected new long-term control program operating costs (New O&M Cost) and the effective date for the implementation of the option are to be provided as appropriate. It is important to provide the total estimated values for each of the options available as later decision-making procedures will utilize this information to develop a feasible risk reduction strategy.

The management decision-making process, disregarding some of the more sophisticated financial issues noted above, will be the subject of a later chapter.

REFERENCES

197. For example, under Ohio law, in part, (H) ''Owner'' means: (2) in the instance of an underground storage tank system in use before November 8, 1984, that was no longer in use on that date, the person who owned the underground storage tank system immediately before the discontinuation of its use'' Ohio Revised Code, Section 3737.87 as amended.
198. Ramsland, Maxwell O., Jr., MAI, CRE; *Appraisers; Creating a Valuation Model;* Asbestos Issues; April, 1990.
199. Wilson, A. R.; *Environmental Risk: Evaluation and Finance;* forthcoming from Lewis Publishers, Inc. in 1991.

11 ENVIRONMENTAL RISK MANAGEMENT DECISION MAKING

INTRODUCTION

The objective of this book has been to develop a systematic basis for environmental risk management decision making. To accomplish this objective, three goals had to be satisfied.

1. The development of a systematic method for quantifying environmental risks and the costs and benefits associated with those risks.
2. The development of management decision support tools that would be required to implement the ERMP strategic objective, *"Maximum feasible reduction of environmental risk."*
3. The development of a set of guidelines for Phase I, II and III Environmental Risk Assessments that will provide the information needed to identify and manage environmental risks.

The first nine chapters have concentrated on the first goal; the development of the ERIS. The preceding chapter and this one concentrate on the second goal; decision tools to *support* environmental risk management decisions. The remaining chapters will concentrate on the third goal; the definition of Phase I, II and III ERAs.

The emphasis was deliberately placed on the adjective "support." The tools discussed here only assist in the making of decisions. They are quantitative tools but by their very nature the bulk of a management decision will be based on qualitative concerns that cannot be a direct part of a quantitative decision support tool.

The point has been made throughout that the tools developed here do not, at any time, require the assignment of a monetary value to human health, life, or environmental damage. This fact will require the use of newer quantitative decision support tools based on the concepts of multi-objective linear programming. Multi-objective programming has been available for several decades but relatively unused except in the fields of oil production and airline scheduling.[200,201,202] Multi-objective programming explicitly does not require comensurability of measurements allowing the optimal feasible solution of a problem to be developed from constraints having many units of measurement.

The qualitative portion that will dominate the final decisions of management can be illustrated by simply pointing out that a decision must be made as to the total amount of resources that the organization can afford to devote to risk management. There is no black and white answer to this issue. The quantitative tools can, and should, assist in making this decision but they cannot determine the outcome in the final analysis as will be illustrated in this chapter.

QUANTITATIVE DECISION SUPPORT

There are several reasons for utilizing a quantitative and systematic decision support approach to environmental risk management. First is that simple approaches will soon be shown to result in misleading if not counterproductive conclusions and risk reduction strategies. Second, if an approach, at least in its quantitative portions, is in fact systematic and quantitative, the results should be reproducible. In particular, two analysts should be able to arrive at approximately the same risk reduction strategy; a result that should allow for comparative analysis of risk management performance and contribute to the overall creditability of the ERMP. Third, there are a number of areas for legitimate disagreement on how or how much environmental risk reduction is needed or appropriate, but there should be an objective basis for the evaluation of the alternatives—an objective basis that can only be provided by a quantitative and systematic approach.

This does not mean that the full mathematical and computer sophistication of multi-objective programming must be employed in every case to arrive at an objectively supportable risk reduction strategy. The use of multi-objective programming to analyze risk reduction strategies has allowed for the development of a relatively simple set of decision rules that—when used with a calculator or spread-sheet program—will yield the same risk reduction strategy as the "optimum feasible" strategy identified with the more sophisticated tools 90% of the time. The relatively straightforward case of asbestos in a building governed under the AHERA rules and regulations will be used to illustrate both a simplified version of the ERIS and the application of these decision rules.

MANAGEMENT CONTROL UNITS

To proceed further, it will be necessary to define the unit of control—the management control unit. In general, the management control unit is defined both by practical considerations and by the risk system. In particular, a management control unit is an identifiable physical entity, such as a room, building or volume of space that contains a set of risk source/primary control mechanisms and generally the same targets, meaning that the transport mechanisms will have approximately the same "reach."

In the case of asbestos in a building, a convenient management control unit might be a room. The room will contain a set of homogeneous areas of asbestos-containing materials. If the asbestos were to escape the primary control mechanisms, generally an identifiable population of the building will be at risk; a population defined by the air movement patterns associated with the room. The room is then a logical choice for designation as a management control unit.

OBJECTIVE ASBESTOS PROJECT SELECTION

An additional reason for beginning with the asbestos example is that under AHERA, the Building Inspection and Management Plan is supposed to result in a full asbestos risk management program. This is the desired end point of all risk management programs, but in a simpler, single substance and confined (to the building) context.

The methods presented here have the primary benefit of being reproducible. Parts of the decision process rely on subjective decisions, notably decisions concerning the primary control mechanism condition, transport mechanism efficiency, and the weighting factor to be applied to different categories of individuals with respect to risk of exposure. Given the possibility of different subjective conclusions being reached by different observers, the process outlined here will insure that independent evaluations of the same situation will arrive at essentially the same selection and priority ordering of risk reduction projects.

This section concentrates on the problem of asbestos in buildings as governed by the federal AHERA rules and regulations. The methods developed should be equally applicable to any carcinogenic single-substance problem, but extensions to non-carcinogenic or multiple-substance problems have not been tested using this simplified method and should be approached with caution.

Response Actions

The available response actions under AHERA are defined as follows.

1. Operations and Maintenance (O&M)—or a formalized system for continual observation and maintenance of the primary control mechanism for the substance over long periods of time. The objective of O&M is specifically to maintain the primary control mechanism in a satisfactory condition such that the substance will not become a direct threat to human health and/or the environment.
2. Repair—or an action designed to restore the primary control mechanisms to functional condition.
3. Isolation—or an action designed to isolate the substance from available transport mechanisms, human beings and the general environment. Entry permitted only to properly trained and equipped individuals.
4. Encapsulation—or an action designed to isolate the substance from available transport mechanisms through the application of an impermeable membrane.
5. Enclosure—or an action designed to isolate the substance from available transport mechanisms or potential primary control mechanism damage through the construction of a structurally sound and impermeable barrier.
6. Removal and disposal—or an action designed to safely remove the substance from its present location and either reduce it to non-hazardous components or to place it in a location and condition that will reduce the risk to human health and/or the environment.

From these response actions, for any given situation, a balance must be chosen that will provide the maximum amount of risk reduction within the limits of the technology, resources, and political constraints applicable. The investigator must appreciate that the political constraints are an extremely important part of the manager's decision, but are generally beyond the scope of the investigator's knowledge or expertise.

CONDITION ASSESSMENT

Consider the data shown in Figure 11.1, a homogeneous area sketch. On this sketch you will note that the boundaries of the homogeneous area are clearly defined as are the boundaries of the salients within the homogeneous area, the locations of analytical samples taken from it, and the primary physical features of the space such as doors to other areas and stairwells.

This information and pertinent other data are summarized for this management control unit, the 2nd Floor Hallway, in Figures 11.2 and 11.2a, the Room Cross Reference and Assessment. In essence, the Room Cross Reference and Assessment forms an inventory of the hazardous substance control mechanism condition, transport mechanism status, and occupant activity in-

REPORT TITLE: JEFFERSON ELEMENTARY REPORT PAGE NO.: ____

HOMOGENEOUS AREA SKETCH

DATE: ____ ____, ______

DOCUMENT NUMBER	Building Number and Name	
	105	JEFFERSON ELEMENTARY
	Sample Area/Lot Number and Name	
	200C	2nd Fl. Hallway Ceiling
	Drawing/Sketch Number and Name	
	200C	2nd Fl. Hallway Ceiling

PROVIDE THE SIZE OF THE AREA (SQ. FT., LN. FT. OR # OF ITEMS

3,350 SQ. FT.

DESCRIBE HOMOGENEOUS MATERIAL

SCALE: 1" = 35′

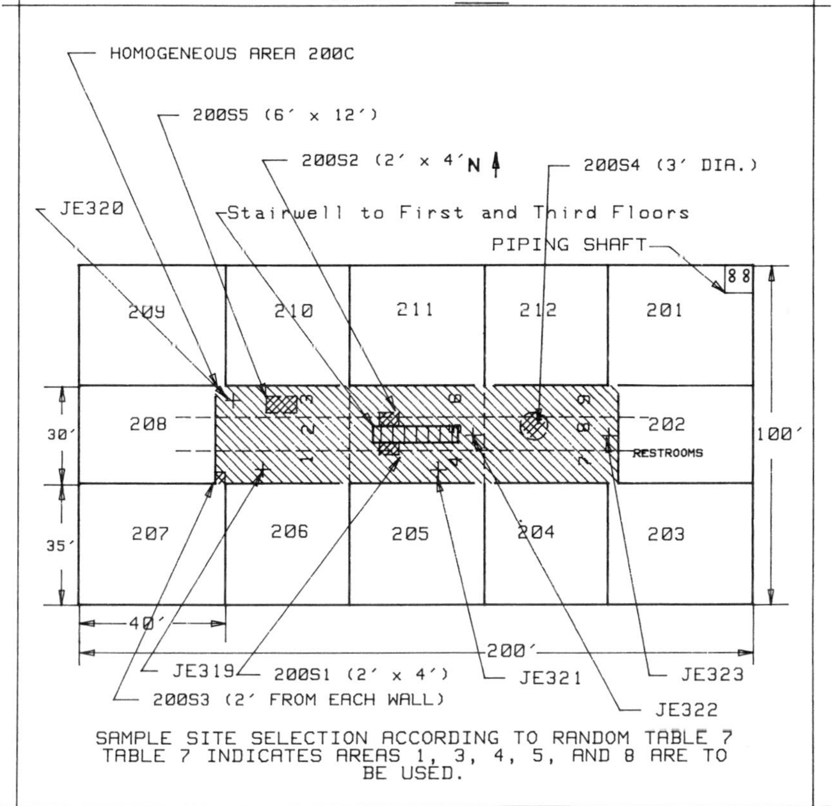

FORM 1

Figure 11.1 Homogeneous Area Sketch.

REPORT TITLE: JEFFERSON ELEMENTARY REPORT PAGE NO. 31

ROOM CROSS-REFERENCE AND ASSESSMENT

PAGE 1 OF 2

PAGE 8 OF 15

Building Number and Name: JEFFERSON ELEM.

Building Location: 105 N. JEFFERSON, ANYTOWN, USA

DATE: FEB. 26, 1989

ROOM NUMBER/NAME: 2nd FLOOR HALLWAY

HOMOGENEOUS AREA OR SALIENT ID	HOMOGENEOUS AREA/SALIENT DESCRIPTION	ACBM			CONDITION				
		ASSUMED ?	CONFIRM ?	FRIA. ?	MATER. TYPE	DAMAGE COND.	POT. DAMAGE EST.	EQUIVALENT AREA	C.C. NO.
200C	2nd FLOOR HALLWAY CEILING		Y	N	S	ND	PD	NON-FRIABLE ACM	8
SALIENT 1	PATCH AROUND LIGHT	Y		N	S	ND	PD	•	8
SALIENT 2	DAMAGE AT STAIRWELL, RIGHT SIDE		Y	N	S	D	PSD	•	8
SALIENT 3	DISCOLORED AREA		Y	N	S	D	PD	•	8
SALIENT 4	DAMAGE AT STAIRWELL, LEFT SIDE		Y	N	S	D	PSD	•	8
SALIENT 5	PATCH IN CORNER	Y		N	S	ND	PD	•	8
200D	FLOOR TILE	Y		N	M	ND	PD	•	8

MANAGEMENT PLANNER'S ROOM HAZARD ASSESSMENT

INSPECTOR'S ROOM CONDITION ASSESSMENT:

MANAGEMENT PLANNER'S SIGNATURE CERT. NO. & STATE

FORM9A

BUILDING INSPECTOR'S SIGNATURE CERT. NO. & ST.

Figure 11.2 Room Cross-reference and Assessment — Page 1.

REPORT TITLE: **JEFFERSON ELEMENTARY** REPORT PAGE NO. **32**

ROOM CROSS-REFERENCE AND ASSESSMENT
GENERAL ROOM DATA
PAGE 2 OF 2

ROOM USEAGE DESCRIPTION
HALLWAY ACCESSING 2nd FLOOR CLASSROOMS AND 1st AND
3rd FLOORS

PERSON-RISK HOURS OF OCCUPANCY PER YEAR

	ACTIVE	PASSIVE	TOTAL PERSON-RISK HOURS/YEAR	
AVG. NO. OF PEOPLE/HR	5.25 a1	300 a2		329.9
AVG. NO. OF HOURS/DAY	2 b1	6 b2	(a1 x b1 x c1) x 5 = .5	(1,000)
AVG.NO. OF DAYS/YR	200 c1	183 c2	(a2 x b2 x c2) =329.4	

DAMAGE FACTORS

PHYSICAL DAMAGE		WATER DAMAGE		DETERIORATION	
SIGNIFICANT	[ ]	EXTENSIVE	[ ]	HEAVY	[ ]
DAMAGED	[X]	MODERATE	[ ]	MODERATE	[ ]
NO DAMAGE	[ ]	SLIGHT	[X]	SLIGHT	[X]
		NONE	[ ]	NONE	[ ]

DISTURBANCE FACTORS

PROXIMITY TO REPAIR ITEMS		ACCESSIBILITY		ADJACENT ROOMS	
LESS THAN 1 FT.	[X]	WITHIN REACH	[ ]	GYMNASIUM	[ ]
1 TO 5 FT.	[ ]	BARELY REACHABLE	[X]	MUSIC ROOM	[ ]
MORE THAN 5 FT.	[ ]	NOT REACHABLE	[ ]	MECHANICAL ROOM	[ ]
				OTHER	[ ]

AIR FLOW FACTORS

BARRIERS		VENTILATION		AIR MOVEMENT	
PERMANENT ENCLOSURE	[ ]	YES	[X]	HIGH	[ ]
ENCLOSED	[ ]	NO	[ ]	MODERATE	[X]
ENCAPSULATED	[ ]	INTAKE	[X]	LOW	[ ]
NONE	[X]	EXHAUST	[X]	NOTE IF: AIR PLENUM	[X]

FORM 9A

Figure 11.2a Room Cross-reference and Assessment — Page 2.

formation for this risk management control unit. Not shown are the other homogeneous area sketches applicable to this management control unit or the detailed information on such supporting documents as the Sample Logs, Laboratory Analysis Reports, photographs or Photograph Logs, or the Salient Logs.[203]

Decision Tree

The decision tree is shown in Figure 11.3 and it summarizes the Condition Assessment decision process. Note that the first page of the Room Cross Reference provides the answers to each decision element for the control mechanism and the second page for the transport mechanism.

First Decision Element—Is it Asbestos Containing (an ACBM)?

The first element concerning a decision about a suspected asbestos-containing material is whether or not the material is, in fact, asbestos containing. As with all materials suspected of being or containing a hazardous substance, the legal environment requires the application of the negative default principle.

On this basis the question to answer is: "Do we have an asbestos-containing material?" and the only means available for answering NO are either that there were no grounds for suspecting the material in the first place or that analytical results are available that prove that the material does not contain asbestos. If neither condition is met, continued use of the decision tree will be required. If the material is not asbestos containing a Condition Category 9 may be assigned, primarily for the sake of completeness.

Second Decision Element—Is it Friable?

Friability of an asbestos-containing material is defined in the AHERA rules as a material that may be crushed or reduced to powder by hand pressure. If the asbestos-containing material can be crushed or reduced by hand pressure, then continuation down the decision tree is necessary because it is felt that such material is easily capable of releasing asbestos fibers into the air.

If the material is not friable, then it is placed in Condition Category 8—non-friable asbestos-containing material. Note that under the right conditions all materials can be made to release fibers and placing the material in Condition Category 8 does not relieve the owner of responsibility for that possibility. In fact, the rules specifically state that an O&M program is required whenever a building contains friable material, or "material about to become friable" and Category 8 material must therefore be treated with the respect it deserves. To this end, the standard decision tree has been supplemented to allow the grading of non-friable materials. If the supplemental decision tree had been used on the example 2nd Floor Hallway, the salients near the stairwell would have received a lower Condition Category rating, and would therefore have received attention earlier than they might otherwise. This is wholly consistent with the "Negative Default" principle.

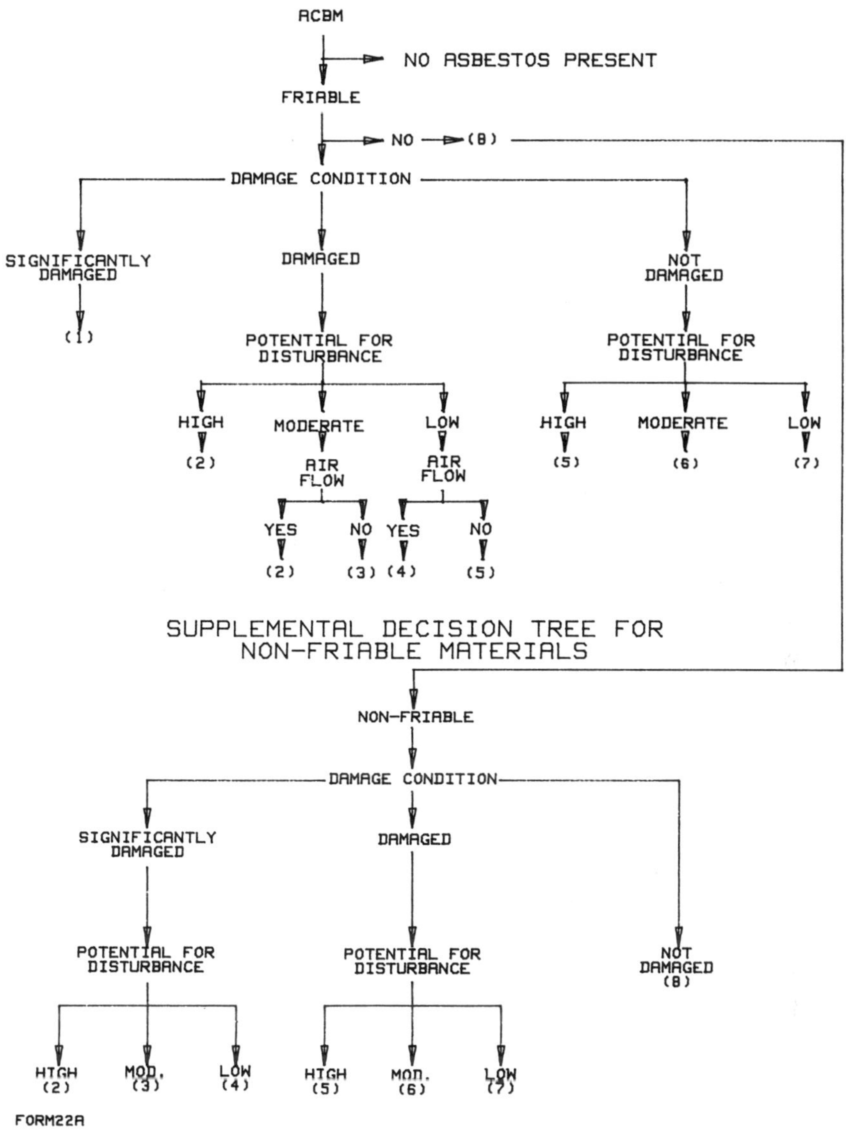

Figure 11.3 Condition Assessment Decision Tree.

Third Decision Element—What is the Current Condition of the Control Mechanism (Damage Condition)?

This decision element requires that the material be placed into one of three subgroups relative to the existence or absence of damage to the primary control mechanism—Not Damaged, Damaged, or Significantly Damaged. Note that if the material is graded as Significantly Damaged an imperative response is invoked; immediate isolation of the area is required to possibly be followed by some other action to render the area accessible again. In deciding on project priority later it will be seen that the categorization of a material into Condition Category 1 will strongly influence what will be done and in what order.

To this point, a Significantly Damaged material has met the following conditions, in order (or following the branches of the decision tree if you prefer): (1) of being either a suspected or confirmed Asbestos-Containing Building Material (ACBM) that is (2) friable and (3) Significantly Damaged and is therefore in Condition Category 1 that requires immediate isolation of the area. **Note that the area to be isolated is the area within reach of the available transport mechanisms and that the isolated area may therefore extend to many times the size of the actual damaged area.**

If the material is judged to be Damaged or Not Damaged, other factors must be considered by moving further down the Decision Tree.

Fourth Decision Element—What is the Potential for Damage to the Control Mechanism (Potential for Damage)?

The potential for damage to the primary control mechanism may be thought of as originating from two types of sources; physical damage and disturbance. Within the physical damage type the primary concerns are with physical damage caused by impact, water damage, and deterioration with time. Within the disturbance factor, consideration may be given to the proximity of the material to repair items, the accessibility of the material to occupants, and the possibility that adjacent rooms may contain sources of vibration or disturbance such as music rooms, fan or boiler rooms, or elevators.

The common element is that even if the primary control mechanism is currently undamaged the possibility for future damage always exists. Given the presence of the Disturbance Factors (See Figure 11.2a), the probability of damage occurring must be judged to be relatively high, moderate or low. This judgment will define which branch of the tree to take next. For damaged material with a high probability of potential for disturbance a Condition Category 2 is appropriate. For undamaged material, the potential for disturbance for high, moderate, or low ratings will yield Condition Categories of 6, 7, or 8 respectively.

Fifth Decision Element—Is a Transport Mechanism Available?

In the case of asbestos the transport mechanism of primary concern is air movement because of the potential for the respiration of asbestos fibers. This is a simple Yes-No determination. If the transport mechanism is available to the material in the event of a primary control mechanism failure, the "Yes" branch is taken. Otherwise the "No" branch is used.

Condition Categories

Each of the Condition Categories defines a set of response actions that are available with respect to the risk management of the material.

Condition Category 1

A material judged to be in Condition Category 1 must be immediately isolated after which any of the other response actions may be applied to return the isolated area to an occupiable status. As mentioned earlier, the designation of any homogeneous area or salient as a Condition Category 1 requires the isolation of all areas within reach of the transport mechanism's ability to move the asbestos fibers from the Category 1 material site. Note that all of the Condition Categories apply to a judgment with respect to the primary control mechanism condition and the availability of a transport mechanism, the influence of risk to human health and/or the environment has not yet been evaluated.

Condition Category 2

A material in Condition Category 2—while not requiring immediate isolation of the transport mechanism area—is of a very high level of concern and should be dealt with as soon as possible by any of the applicable response actions. Note again that if any of the substance remains after a response action, an O&M program must be in effect.

Condition Category 3

A material in Condition Category 3 requires response actions on a slightly lower priority basis than a Condition Category 2 material, but is otherwise essentially the same.

Condition Category 4

A material in Condition Category 4 requires response actions on a slightly lower priority basis than a Condition Category 3 material because it either is less damaged or has a lower potential for damage, but is otherwise essentially the same.

Condition Category 5

A material in Condition Category 5 requires response actions on a slightly lower priority basis than a Condition Category 4 material due to a lack of a readily available transport mechanism (air flow in this case) or the material is not damaged, but has a high potential for disturbance.

Condition Category 6

Operations and Maintenance program implementation coupled with actions designed to reduce the potential for disturbance are indicated.

Condition Category 7

Operations and Maintenance program implementation.

Condition Category 8

It is strongly suggested that Condition Category 8 (non-friable, asbestos-containing or suspected of being asbestos-containing) materials be further evaluated using the Supplemental Decision Tree and at least placed under the O&M program with the primary objective of preventing the materials from becoming friable.

Condition Category 9

This category exists primarily for the sake of completeness, but having been documented during the inspection process the inclusion in the records for possible future use seems appropriate.

RISK/HAZARD ASSESSMENT

Risk/Hazard Assessment, or the inclusion of the human/environmental damage factor in the evaluation process is both the most important element of the process, the "bottom line" if you will, and the part of the process most often misunderstood and subject to controversy. One of the major areas for controversy comes in the attempt to assign a value to human health or environmental damage in order to perform a so-called "cost-benefit" analysis.

The position here is that no assignment of dollar value is desirable or necessary and the method presented does not require any such assignment. Instead the prime objective to be served is a reduction in the level of risk to human health and/or the environment given the resources available to accomplish such risk reduction, and further that the amount of resources to be made available is subject to alteration as a result of the risk reduction imperative.

Let us begin by disposing of the first question of risk analysis, "Risk from what source?" The Condition Assessment process, and particularly the assignment of a ranked order of material condition relative to the material's probability of releasing asbestos fibers into the air answers this question. The source of the risk and at least a qualitative assessment of how great the risk might be from a given source are known in terms of the Condition Category number assigned through the decision tree.

The second question in risk analysis is "Risk to whom or what?" With respect to asbestos, there is some concern for risk to the general environment but this is typically not the major concern. Concern lies primarily with the possibility of damage to human health and the welfare of individuals within the structure.

The third question is "How may an individual be at risk?" Clearly, if an individual cannot be exposed to asbestos fibers that individual cannot be at risk, and conversely, if the individual is exposed to high levels of asbestos fibers without protection, then a risk exists. The answer to this question is that the individual must be in a position to be exposed to asbestos fibers.

If unlimited resources were available and no other considerations entered into the picture, the process could stop here. The solution to the problem is to remove the source of the risk. Unfortunately, this simplistic solution is not possible for a number of reasons. In addition to the low level of competent performance of removal work, consideration should be given to the possibility that the replacement materials available may not perform safety functions as well as the existing ACBMs and/or may represent the same kind of risks in the future as asbestos does now.

Economic considerations also enter the picture and AHERA requires their explicit consideration in the development of remediation alternatives. The funds to immediately remove the ACBMs may not be available and major commitments of funds may disrupt the primary mission of the organization.

It can be shown that, in general, the immediate removal of materials will have the greatest negative impact on the financial position of the organization of all of the available response actions in addition to probably increasing, rather than decreasing, risk.

The point of this discussion is that the simplistic solution of immediate removal of all of the material is simply neither prudent or generally possible, whether from a health risk or an economic point of view. The implication is that some alternative risk reduction strategy must be employed to minimize risk given the resource limitations, or constraints, that exist in the real world. The remaining risk reduction question is then "How can the maximum risk reduction be achieved given the constraints placed on the process?"

To answer this question, three issues must be explored. First is a need for a more refined method for answering the earlier question of "How may an individual be at risk?" The second issue will be the determination of the amount of the necessary resources that can be devoted to risk reduction. The third issue will be how to apply the available resources in a manner that will achieve the maximum of risk reduction. The second and third issues are the concern of the next section, Project Selection and Prioritization.

To refine the risk evaluation method, it must first be recognized that the possibility of exposure to asbestos fibers is the basic element of risk. There are a large number of refinement methods available but a simplistic approach to exposure measurement based on its reliability, ease of use, and reproducibility was chosen. Specifically, it is assumed that the population of a building may be divided into two major groups; those who are essentially passive in the presence of the asbestos-containing material and those who are active in its presence.

The individual is defined as being passive if, in the course of normal activities, he/she is not likely to physically disturb the control mechanism and conversely, an active individual is one who may or will disturb the control mechanism of the materials. On this basis, a student or teacher would not be likely to disturb the control mechanisms of asbestos-containing materials and would therefore be in the passive group, but facilities maintenance personnel may be required to do so and would therefore be in the active group.

The obvious means of measuring the exposure to risk will be the number of hours a given individual is in the presence of a hazardous substance, possibly weighting the number of hours based on the active and passive groupings. The number of thousands of person-hours spent within reach of the transport mechanisms from a given risk source is the basic unit of exposure to risk measurement. This factor is shown on the "Room Cross-reference and Assessment" form in Figure 11.2a.

The weighting factor applied to differentiate between active and passive groups was arbitrarily chosen to be 5. There is no scientific basis for this weighting factor at this time except to say that available evidence indicates that a maintenance person may be exposed to a level of fibers between 3 and 10 times the normal background level. This method then provides a single number, the weighted number of person-risk hours per year for a given risk management control unit for use in evaluating the exposure potential. Note that the passive category for the 2nd Floor Hallway has a very high number of person-risk hours due to its function as an air plenum connecting three floors, resulting in the potential of exposure to the total school population during regular occupancy hours.

PROJECT SELECTION AND PRIORITIZATION

At this point all of the information necessary to make intelligent decisions regarding risk management is available except the amount of resources that may be applied to perform the work. It has been recommended that the organization consider establishing a series of budgets[204] to allocate resources generally as follows:

Operations & Maintenance (O&M) Budget

This is defined as a budget to cover the daily activities of the operations and maintenance program such as premium pay or contractor services for the maintenance of the primary control mechanisms for the hazardous substances. In the case of asbestos, indications are that such a program may increase the normal operating costs for a facility by 5% to 15%. Note that, as a result of the OSHA Hazard Communications Standards and the ''Community Right-to-Know'' provisions of SARA, some type of O&M program will probably always be necessary, with or without the presence of ACBMs. This factor will also require budgeting under the other four categories shown here. For this reason, the O&M Program's costs are considered to be a fixed component of overhead.

Training Budget

If individuals in the employ of the owner are to perform tasks directly associated with the control mechanism, some level of training, and retraining at regular intervals will be required. An explicit budget to cover these training costs should be prepared. This is also an overhead cost.

Personal Protection Budget

The individuals who will be active in the presence of the substances defined earlier will require personal protective equipment such as masks, suits, gloves, boots, and maintenance supplies for masks, etc. An explicit budget to cover these costs should be prepared and is a part of overhead.

Emergency Response Action Budget

The only law to which no exceptions have yet been found is Murphy's. Substantial resource savings and human health/environmental risk reduction benefits can be obtained through the discipline of developing a budget to cover emergency requirements. The primary benefits are not so much in having the funds available as they are in having answered the questions of "What can go wrong?" and "How do we deal with what might go wrong?" This is a financial reserve budget and not a part of the reduction strategy selection process.

Planned Response Action Budget

Given that the condition and exposure to risk data has been gathered and properly evaluated, certain response actions will be indicated and should be planned for including the allocation of resources to implement the plan. It is with this budget that we are concerned in terms of its development and the selection and prioritization of the projects to be implemented given the available resources.

The Planned Response Action Budget may be developed in three steps:

1. An initial approximate assignment of resources to deal with the projects;
2. The application of the Project Selection and Prioritization process given those resources, and
3. The adjustment of the amount of resources to be made available given the projects selected and their prioritization to account for any major issues uncovered during the selection and prioritization process.

This procedure will insure that the appropriate amount of resources are made available to deal with the major risk control projects through the exercise of management judgment as opposed to the application of a purely mathematical process that cannot possibly include a number of non-quantifiable decision parameters. Despite the sophistication of any mathematics-based decision process, good judgment must always be exercised to produce the best possible final results.

The "Response Action Prioritization Table" (Figure 11.4) will initially be a listing of all of the possible projects that might be undertaken to reduce risk; a "menu". For purposes of a manual evaluation of this list, it is strongly recommended that the projects be listed in Condition Category order as this will allow the elimination of a number of projects by simple inspection. That is, by looking at the list of projects it will be obvious, given the maximum amount of resources available, that a number of projects with high Condition Category numbers will not be accomplished within the budget period as the resources will simply not stretch that far. Something must be done with the Condition Category 1 projects even if that means isolating the area semi-permanently.

Note that this table provides the following information:

1. The management control unit (e.g., room) identification and a brief description of the problem(s) associated with that room along with the Condition Category numbers associated with each project in Condition Category number order. In saying that projects should be listed in Condition Category order note that risk reduction theory will assign to the management control unit the worst Condition Category number for all material conditions present in that room. An example of this type of listing is shown in Figure 11.4.
2. The current number of person-risk hours per year for personnel within reach of the transport mechanisms associated with the material in that room. This information all comes directly from the "Room Cross Reference and Assessment" document.
3. For each project there will be at least one response action. In general, there will be two or more—the basic two being isolation and removal—it always being possible to either isolate the room and/or to remove all of the substance from the room. All of the projects should have a cost estimate for the applicable response actions.
4. For the room (or management control unit), an overall set of response action possibilities should be shown separately from the individual projects. Total room renovation cost information should be developed on the basis of the cost for dealing with all of the problems in the room at one time. Note that, for the total room situation, the room will assume the Condition Category number of the worst Condition Category material in the room; this policy being consistent with the negative default principle.

To illustrate the suggested process and its impact, look carefully at Figure 11.4 and assume that a policy decision has been made to remove all Condition Category 1 materials. This will have the following result (Table 11.1)

This table is simply the summary of effects of removing all of the Condition Category 1 materials with the exception that since it only costs \$1,000 more to completely clean the boiler room, that option was used. The Expended Funds Efficiency is simply the Total Cost divided by the number of thousands of hours of risk reduction accomplished.

ROOM OR FUNCT. AREA	DESC-RIPTION	COND CATEG	INITIAL EXPOSURE (1,000's HRS/YR)	ESTIMATED RESPONSE ACTION COST AND EXPOSURE EFFECT										NEW EXPOS. (1,000's HRS/YR)	PRIORITY
				ISOLATE (90%)	NEW EXPOS.	REPAIR (50%)	NEW EXPOS.	ENCAP. (60%)	NEW EXPOS.	ENCLOS. (70%)	NEW EXPOS.	REMOVE (100%)	NEW EXPOS.		
201	SAL 201-6	1	20.			500	10	3,000	8			2,000	0		
	HA-6	4	20.									7,000	0		
	HA-9	8	20.									6,000	0		
	ROOM	1	20.									12,000	0		
TUNN 2	HA-11	1	.1	1,000	.01	25,000	.05					80,000	0		
B. R.	PIPE L.	1	3.			10,000	1.5					60,000	0		
	B. COVER	3	3.			2,000	1.5					12,000	0		
	ROOM		3.			9,000	1.5					61,000	0		
2nd FL	HA-5	3	100.	12,000	10	3,000	50	24,000	40	55,000	30	110000	0		
1st FL	HA-14	4	211.	17,000	21.1	5,000	105.5	60,000	84.4			105000	0		
AUD.	HA-21	4	50.	5,000	5	6,000	25	20,000	20	40,000	15	75,000	0		
COST/EXPOSURE TOTALS															

FORM 301

Figure 11.4 Example Initial Response Action Prioritization Table.

Table 11.1 Cost and Reduction Effects Remove All Condition Category 1 Materials

Area	Description	Action	Cost	Risk Reduction
201	SAL 201-6	REMOVE	$ 2,000	20.000
Tunnel 2	HA-11	REMOVE	$ 80,000	0.100
Boiler Rm	ROOM	REMOVE	$ 61,000	3.000
			$143,000	23.100

Expended Funds Efficiency = $6,190.48/1,000 hours or risk reduction.

Thus far it has been implicitly assumed that all response actions are equally effective in reducing risk; an assumption which is arguable at the very least. Such an assumption has some justification in the short-term in that a repair action properly performed is as effective as a removal action properly performed for as long as the repair action binds the asbestos fibers in place. Two other major issues in determining the effectiveness of a given response action are the length of time over which the action will be effective, and the problem of a potential for catastrophic and unanticipated failure.

Under the ERMP's operating conditions, particularly as implemented in AHERA's semi-annual surveillance and tri-annual re-inspection requirements, the probability of a catastrophic and unanticipated failure should be reduced to a relatively low level. So long as the material remains, the possibility of such a failure must also remain and to account for this the amount of risk reduction benefit applicable to a given alternative will be adjusted according to Table 11.2.

The issue of the influence of time on the various response actions could be accounted for if the information necessary to determine how long a particular action would serve its intended purpose were available. This information is not currently available but can be approximated. If the manager wishes to apply this additional level of sophistication to the method outlined here, it may be introduced through the concepts of the present value of a stream of cash flows required to repeat a response action over time. This will be done later as a second refinement step in this analysis using the Lifetime Cost Adjustment Factors in Table 11.2. These factors are based on the as-

Table 11.2 Risk Reduction Benefit Adjustment Factors

Response Action	Risk Adjustment Factor	Lifetime (20 year) Cost Adjustment Factor at 12%
Repair	50%	2.1758 + Removal @ 0.1037
Encapsulation	60%	1.3220 + Removal @ 0.1037
Enclosure	70%	1.0000 + Removal @ 0.1037
Isolation	90%	1.0000 + Removal @ 0.1037
Removal	100%	1.0000

Table 11.3 First Iteration Lowest Expenditure Strategy

Area	Description	Action	Cost	Risk Reduction
201	SAL 201-6	REPAIR	$ 500	10.000
Tunnel 2	HA-11	ISOLATE	$ 1,000	0.090
Boiler Rm	ROOM	REPAIR	$ 9,000	1.500
2nd FL	HA-5	REPAIR	$ 3,000	50.000
1st FL	HA-14	REPAIR	$ 5,000	105.500
Aud.	HA-21	ISOLATE	$ 5,000	45.000
			$ 23,500	212.090

Expended Funds Efficiency = $110.80/1,000 hours of risk reduction.

sumption that the Mean Time Between Failures (MTBF) for the response actions are: Repair, 5 years; Encapsulation, 10 years; Enclosure, 20 years; and Isolation, 20 years.

At this point the simplifying assumption is that over the budget period of one year, all actions are equally effective.

From this point a repetitive process of trying to find a successively better and better set of projects that will best utilize the available resources to maximize the amount of risk reduction will be used. To begin this process, a starting point must be chosen and to do so, we adopt the strategy defined as follows:

1. Choose to do individual projects that have a Condition Category 1 designation by the least expensive applicable method.
2. If resources remain after paying for these projects, choose Condition Category 2, 3, etc. projects on the least expensive applicable method basis until all resources have been expended.
3. Add up the total number of person-risk hours associated with each of the selected projects.

Projects on the list will be replaced with other projects, continually attempting to increase the total number of person-risk hours saved until we can no longer increase the savings within the budget. In doing this remember that something *must* be done with the Condition Category 1 materials. Note that if a Condition Category 1 material is dealt with in a room containing other possible projects, that room assumes the next lowest Condition Category number within that room (see Room 201 in the example where, after doing the Condition Category 1 project, the room will drop to a Condition Category 4). Also note that whenever the cost of dealing with the projects in a given room approximates the cost of a total room alternative, the total room alternative should be considered the project to be used as the candidate in the next iteration. The results of this step are shown in Table 11.3.

Note that based on this first step, the number of hours of risk reduction have increased from 23,100 to 212,090 while reducing the amount of funds expended from $143,000 to $23,500. These two facts clearly show up in the calculated number, Expended Funds Efficiency, which went from $6,190.48 to $110.80 per 1,000 hours of risk reduction.

This presents the concept that at least one way to help in deciding on which projects to execute and on whether or not the best solution has been found, is to look at the Expended Funds Efficiency for each possible project. This is done in Figure 11.5.

A careful examination of this table will show that this time, the most efficient possible use of funds for this particular situation has been found because there are no other candidates that may be selected that will further decrease the Expended Funds Efficiency. No other alternative has an individual Expended Funds Efficiency that is lower than the selected alternative.

This does not mean that the building owner, giving due consideration to a number of unquantifiable factors such as public relations, program credibility, effective utilization of space, and the like, will necessarily agree with this particular solution. In fact for a school, these considerations may lead to the following actual decision (Table 11.4).

In this solution the building owner has decided to expend more funds to try to achieve results deemed desirable over and above the most efficient solution. In fact, note the influence of repairing the Auditorium instead of isolating it; the number of risk hours saved by repair is 20,000 less than isolation, but the owner has decided that this is acceptable from some point of view. The differentiation between the risk reduction capabilities of the various response actions may be more apparent than real due to the number of approximations used and the difference may not be significant.

Table 11.4 Final Determination by Building Owner

Area	Description	Action	Cost	Risk Reduction
201	SAL 201-6	REPAIR	$ 500	10.000
Tunnel 2	HA-11	ISOLATE	$ 1,000	0.090
Boiler Rm	ROOM	REPAIR	$ 9,000	1.500
2nd FL	HA-5	ENCAPSUL	$ 24,000	60.000
1st FL	HA-14	REMOVE	$105,000	211.000
Aud.	HA-21	REPAIR	$ 6,000	25.000
			$147,500	307.590

Expended Funds Efficiency = $482.80/1,000 hours of risk reduction.

ROOM OR FUNCT. AREA	DESC- RIPTION	COND CATEG	INITIAL EXPOSURE (1,000's HRS/YR)	ESTIMATED RESPONSE ACTION COST AND EXPOSURE EFFECT										NEW EXPOS. (1,000's HRS/YR)	PRIORITY
				ISOLATE (90%)	NEW EXPOS.	REPAIR (50%)	NEW EXPOS.	ENCAP. (60%)	NEW EXPOS.	ENCLOS. (70%)	NEW EXPOS.	REMOVE (100%)	NEW EXPOS.		
201	SAL 201-6	1	20.			50		250				100			
	HA-6	4	20.									350			
	HA-9	8	20.									300			
	ROOM	1	20.									600			
TUNN 2	HA-11	1	.1	1111		50000						800000			
B. R.	PIPE L.	1	3.			6667						20000			
	B. COVER	3	3.			1333						4000			
	ROOM		3.			6000						20333			
2nd FL	HA-5	3	100.	133		60		400		786		1100			
1st FL	HA-14	4	211.	89		47		473				498			
AUD.	HA-21	4	50.	111		240		667		1143		1500			
COST/EXPOSURE TOTALS															

FORM301

Figure 11.5 FUNDS EFFICIENCY (Dollars/1,000 Person-risk Hours Reduction).

REFINEMENT: THE INCLUSION OF DISCOUNTED CASH FLOW

The inclusion of discounted cash flow has the effect of increasing the initial cost for a given response action as shown in Figure 11.6 that uses the information from Table 11.2 to adjust the costs of response actions for their repetition over time and the eventual need to remove at the end of the time horizon, assumed to be 20 years in the future.

The inclusion of discounted cash flow does not change the selection of projects based on the strategy of lowest dollar outlay. The equivalent to Table 11.3 is then Table 11.5.

This is a more efficient solution than the original "Remove" solution (which does not change because the discounted value of a dollar spent now is unchanged). However, review of the Funds Efficiency analysis of all of the alternatives using the present value refinements reveals some interesting information (see Figure 11.7).

The first change to note is with respect to SAL 201-6 in room 201. The Repair action, while the lowest initial dollar cost choice, is not as efficient as the Remove choice (a Funds Efficiency of $130/1,000 versus $100/1,000) indicating that a more efficient use of funds would be to remove the salient as opposed to repairing it. Using this same reasoning, the most efficient of all strategies is shown in Table 11.6.

This is the most efficient solution when efficiency is measured by the Discounted Expended Funds Efficiency. Note that the First Iteration yielded a Discounted Expended Funds Efficiency of $417.93/1,000 hours versus this solution's $292.29/1,000 hours.

The owner may not be inclined to accept this particular solution for very valid reasons—most of the building is unusable. Many other considerations should now be evaluated by the owner to arrive at a final selection of response actions and projects.

Table 11.5 First Iteration: Lowest Expenditure Strategy Using Discounted Cash Flow

Area	Description	Action	Cost	Discounted Cost	Risk Reduction
201	SAL 201-6	REPAIR	$ 500	$ 1,295	10.000
Tnl 2	HA-11	ISOLATE	$ 1,000	$ 9,296	0.090
B.R.	ROOM	REPAIR	$ 9,000	$ 25,908	1.500
2nd FL	HA-5	REPAIR	$ 3,000	$ 17,934	50.000
1st FL	HA-14	REPAIR	$ 5,000	$ 21,768	105.500
Aud.	HA-21	ISOLATE	$ 5,000	$ 12,777	45.000
			$ 23,500	$ 88,978	212.090

Expended Funds Efficiency = $110.80/1,000 hours of risk reduction.
Discounted Expended Funds Efficiency = $417.93/1,000 hours of risk reduction.

ROOM OR FUNCT. AREA	DESC-RIPTION	COND CATEG	INITIAL EXPOSURE (1,000's HRS/YR)	ESTIMATED RESPONSE ACTION COST AND EXPOSURE EFFECT										NEW EXPOS. (1,000's HRS/YR)	PRIORITY
				ISOLATE (90%)	NEW EXPOS.	REPAIR (50%)	NEW EXPOS.	ENCAP. (60%)	NEW EXPOS.	ENCLOS. (70%)	NEW EXPOS.	REMOVE (100%)	NEW EXPOS.		
201	SAL 201-6	1	20.			1295	10	4173	8			2,000	0		
	HA-6	4	20.									7,000	0		
	HA-9	8	20.									6,000	0		
	ROOM	1	20.									12,000	0		
TUNN 2	HA-11	1	.1	9296	.01	62691	.05					80,000	0		
B. R.	PIPE L.	1	3.			27980	1.5					60,000	0		
	B. COVER	3	3.			5596	1.5					12,000	0		
	ROOM		3.			25907	1.5					61,000	0		
2nd FL	HA-5	3	100.	23407	10	17934	50	43135	40	66407	30	110000	0		
1st FL	HA-14	4	211.	27889	21.1	21767	105.5	90209	84.4			105000	0		
AUD.	HA-21	4	50.	12778	5	20832	25	34218	20	47778	15	75000	0		
COST/EXPOSURE TOTALS															

FORM381

Figure 11.6 First Level Response Action Prioritization Table Adjusted for Discounted Cash Flow.

ROOM OR FUNCT. AREA	DESC-RIPTION	COND CATEG	INITIAL EXPOSURE (1,000's HRS/YR)	ESTIMATED RESPONSE ACTION COST AND EXPOSURE EFFECT										NEW EXPOS. (1,000's HRS/YR)	PRIORITY
				ISOLATE (90%)	NEW EXPOS.	REPAIR (50%)	NEW EXPOS.	ENCAP. (60%)	NEW EXPOS.	ENCLOS. (70%)	NEW EXPOS.	REMOVE (100%)	NEW EXPOS.		
201	SAL 201-6	1	20.			130		348				100			
	HA-6	4	20.									350			
	HA-9	8	20.									300			
	ROOM	1	20.									600			
TUNN 2	HA-11	1	.1	9296		1253820						800000			
B. R.	PIPE L.	1	3.			18653						20000			
	B. COVER	3	3.			3731						4000			
	ROOM		3.			17271						20333			
2nd FL	HA-5	3	100.	260		359		719		949		1100			
1st FL	HA-14	4	211.	147		206		713				498			
AUD.	HA-21	4	50.	284		833		1141		1365		1500			
COST/EXPOSURE TOTALS															

FORM301

Figure 11.7 FUNDS EFFICIENCY (Discounted Dollars/1,000 Person-risk Hours Reduction).

Table 11.6 Second Iteration: Most Efficient Solution

Area	Description	Action	Cost	Discounted Cost	Risk Reduction
201	SAL 201-6	REMOVE	$ 2,000	$ 2,000	20.000
Tnl 2	HA-11	ISOLATE	$ 1,000	$ 9,296	0.090
B.R.	ROOM	REPAIR	$ 9,000	$ 25,908	1.500
2nd FL	HA-5	ISOLATE	$ 12,000	$ 23,407	90.000
1st FL	HA-14	ISOLATE	$ 17,000	$ 27,889	189.900
Aud.	HA-21	ISOLATE	$ 5,000	$ 12,777	45.000
			$ 46,000	$101,277	346.490

Expended Funds Efficiency = $132.76/1,000 hours or risk reduction.
Discounted Expended Funds Efficiency = $292.29/1,000 hours of risk reduction.

Looking again at Figure 11.7, the fact that changing from an Isolation to a Repair response action in the Auditorium will alter the efficiency of the solution of that project from $284/1,000 hours to $833/1,000 hours. Note also that this step implies a degradation of the number of person-risk hours from 45,000 to 25,000. Remembering that this degradation is more apparent than real for some time after the action is initially taken, this may be acceptable in the short run.

For the 1st and 2nd Floors, Repair is significantly more efficient than other available alternatives and would therefore probably be the owner's most efficient choice. Note however, that if the owner chose to do more than repair and still keep the space available for general occupancy, the efficiency numbers in Figure 11.7 imply that Encapsulation is the next alternative for the 2nd Floor, but Removal is the next alternative for the 1st Floor.

If Removal for the Auditorium and Repair for the 1st and 2nd Floors were the owner's choice, then the results of these decisions are shown in Table 11.7.

The pure objective of maximum risk reduction for expended funds could have been achieved by simply following the Second Iteration solution. This would have resulted in a total expenditure of $46,000 in the first year and a

Table 11.7 Owner's Final Solution

Area	Description	Action	Cost	Discounted Cost	Risk Reduction
201	SAL 201-6	REMOVE	$ 2,000	$ 2,000	20.000
Tnl 2	HA-11	ISOLATE	$ 1,000	$ 9,296	0.090
B.R.	ROOM	REPAIR	$ 9,000	$ 25,908	1.500
2nd FL	HA-5	REPAIR	$ 3,000	$ 17,934	50.000
1st FL	HA-14	REPAIR	$ 5,000	$ 21,768	105.500
Aud.	HA-21	REMOVE	$ 75,000	$ 75,000	50.000
			$ 95,000	$151,906	227.090

Expended Funds Efficiency = $418.34/1,000 hours or risk reduction.
Discounted Expended Funds Efficiency = $668.92/1,000 hours of risk reduction.

Table 11.8 Comparison of Results

	Total Cost	Discount Cost	Risk Red.	Funds Eff.	Disc. Funds Eff.
Remove All C.C. 1 (Table 11.1)	$143,000	N.C.	23	$6,190	N.C.
Lowest Expended Funds Strategy (Table 11.3)	$ 23,500	N.C.	212	$ 111	N.C.
Owner's Final Based on Initial Expenditures (Table 11.4)	$147,500	N.C.	308	$ 483	N.C.
Lowest Expended Funds Using DCR (Table 11.5)	$ 23,500	$ 88,978	212	$ 111	$ 418
Most Efficient Solution Using DCF (Table 11.6)	$ 46,000	$101,277	346	$ 133	$ 292
Owner's Final Choice Using DCF (Table 11.7)	$ 95,000	$151,906	227	$ 418	$ 669

risk reduction of 346,490 person-risk hours. In other words, 153% of the owner's final solution risk reduction could have been accomplished for approximately 48% of the first year cost or 67% of the discounted cost of the final solution, but the facility would have been essentially unusable. The point is that the mathematically "best" solution is not the "best" overall solution because of other considerations. Table 11.8 provides a summary of the six alternatives discussed.

REASONS FOR USING A SUPPORTABLE APPROACH

There are several major reasons for using an approach similar to the one outlined above. First is that for facilities coming under AHERA, the individual charged with responsibility for developing the facility's Management Plan—the Management Planner—there is an absolute requirement to document the Planner's initial recommendations, the owner's final decisions, and the reasons for the selection of the final decisions. The above approach provides a rational and supportable basis for both the initial recommendations and for the owner's final choices.

Second is the simple fact that this approach can form a basis for an informed choice on the part of the owner. The owner can intelligently trade off non-objective requirements for objectively arrived at benefits and recognize what the tradeoff involves. Better risk management decisions should result.

Third is a pure professional performance argument. This approach allows the professional to provide to the client a sound analytical basis for the conclusions reached as a result of the analyst's work. It is not a matter of pure subjective judgment, although the subjective judgment element will be a component, but rather a reasoned combination of subjective judgment and objective decision-support analysis.

NOTES ON THE PROCESS

The mathematical basis for this procedure is known as Multi-objective Linear Programming (MOLP) and, if the assumptions, objectives, and constraints are properly formulated, the MOLP process is *guaranteed* to provide the maximum risk reduction for the available resources given the *objective* constraints placed on the problem. This manual version does not carry this guarantee and is obviously tedious to the point where for a large number of risk reduction project possibilities, it will be necessary to use a computer to obtain a solution in a reasonable time. In fact with more that 50 possible projects, the problem cannot reasonably be solved by hand.

We noted earlier that a first step is to inspect the list with a view toward rejecting those projects that obviously cannot be achieved within the available resources—a strategy only implicitly shown in this example. Even in complex situations, this simplification is generally possible and will result in a final list of projects that—while not guaranteed to be the best possible list—will be very close to such a list. In any event, any improvement in the use of resources to reduce risk is a worthwhile effort and when the improvement can be as significant as the one shown here, it will be especially worthwhile.

CAUTIONS ABOUT THE PROCESS

Note that if, instead of a boiler room, an air-handling room with exactly the same costs and condition assessment had been present, the picture would be radically altered. The change is due to the fact that the air handler provides the best of all possible transport mechanisms and any problem in the air-handling room would expose all of the individuals in the building to a high level of risk. Great care must be exercised in calculating the person-risk hours per year. Person-risk Hours refer to the *total population exposed to a problem and who are accessible to the transport mechanisms for that problem.*

IMPROVEMENT POSSIBILITIES

Any method, and particularly one that deals with approximations such as were used here for the measurement of exposure to risk, mean time between failure for a given response action, or the influencing factors on risk, is subject to improvement. A number of such improvements can be made and some of them are mentioned here.

Improvement in the Measurement of Multiple Project Risk Assessment

This model suggests that, when a given room has multiple projects, all of the projects have a number of risk hours associated with them that are the same for all projects and the total room. This is obviously not the case and an improvement can be made if the inspector is required to record the amount of material in the room in common units of measure such as square feet. Then the amount of total risk exposure for the room could be allocated proportionately between the various homogeneous areas and salients based on their fractional portion of the total of all areas within the room. Such an improvement, while appearing very reasonable, may not be worth the effort expended to arrive at the result, but should be given serious consideration.

Improvement in the Measurement of Risk Reduction Associated with a Given Response Action

In a full-blown computer-based mathematical solution, a significant improvement can be obtained by basing the reduction of risk for a given response action on the change in Condition Category that would actually be associated with that response action rather than the relatively arbitrary weighting factors for the risk reduction efficiency of an alternative used here. Consider the fact that any response action taken other than removal will act to increase the Condition Category for a given situation, but not necessarily to the highest possible value.

For example, the repair of a significantly damaged area of surfacing material will change the Condition Category from 1 to some higher value, but the cause of the damage, and therefore the potential for damage reoccurrence later, will probably still exist. In this case the Condition Category may move from 1 to 4 or 5, but probably not to 7 and since the material will still be friable, it cannot generally move to a Condition Category of 8.

MULTIPLE RISK MANAGEMENT

The case of managing multiple risks is considerably more complex in practice, but not necessarily in theory. The ERIS data in the detailed inventory and the cost and benefits analysis worksheets is an expanded version of the system used in the asbestos case. Because of the number of variables that must be dealt with, the only realistic means to evaluate alternatives and develop a strategy will be to use a computer and probably a specialized piece of software, but the concepts are basically the same and the decision rules can be summarized as follows (Table 11-9).

Table 11.9 Generalized Environmental Risk Decision Rules

1. Define the management control units applicable to the situation in such a manner that all risk source/primary control mechanism combinations are assigned to a management control unit.
2. The initial risk level assignment for a management control unit will be based on the risk source/primary control mechanism and transport mechanism having the largest magnitude human risk or sensitive environment risk score.
3. Order the management control units in descending order from highest risk score.
4. Calculate the efficiencies for each risk reduction option based on the net difference between the existing risk score and the alternative's risk score divided by the present value of the net resources expended.
5. Determine the maximum possible risk reduction available by choosing the most efficient risk reduction option for each management control unit. The solution search is to begin with the management control unit having the highest risk score and continuing until the initial resources have been expended. Note that in some cases, an efficient risk reduction strategy within a management control unit may involve executing an option for dealing with all or a major portion of the risk source/primary control mechanisms within the management control unit.
6. Having determined the most efficient quantitative strategy,[205] non-quantitative considerations should be evaluated. The quantitative risk reduction impact for each non-quantitative issue should be evaluated to determine if the overall risk will increase or decrease, and if the change is appropriate to the non-quantitative issue's benefits.
7. The final strategy should be documented including the benefits and costs, as the management plan to be executed and against which performance will be measured when the risk system undergoes its next periodic audit.

Ideally, the analysis of multiple risks will also involve the analysis of those risks as projected into the future. Four time periods have been identified for use in the ERIS approach, the Current Time, and times t_1, t_2, and t_3. If these time periods are uniform throughout the analysis, the results of the human and sensitive environment risk scores pertaining to each time period can be analyzed to form a type of present value of risk. This concept is dealt with by implication in the AHERA example through the probable future condition assessment idea, but is not carried to its logical conclusion by the simultaneous requirement to develop population exposure data into the future as a part of the risk analysis process.

The importance of an evaluation of the future can be judged by a number of current problems. The "Love Canal" issue would not have been as important if the area surrounding the canal had not been developed for residential use—a situation that possibly might have been foreseen prior to using the Love Canal site. Although the Hooker Chemical Company attempted to prevent the residential development of the area surrounding the site, their attempts were unsuccessful for a number of reasons, not the least of which was the natural pressure on the part of developers to profit from the local growth patterns given a lack of appreciation for the risks involved.

Using the same site as an example, it is likely that it should not have been even considered as a hazardous waste burial site in the first place given the influence of time on the transportation of the materials through the local geologic structures and the local growth patterns. Again, a deliberate attempt to look into the future and analyze the likely impacts of time on the risks may have caused a different course of action to be implemented.

A third area not explicitly dealt with in the ERIS (although to some extent implicitly considered in the persistence and toxicity supporting data) is the possibility of delayed effects from exposure. It is becoming more obvious with increased scientific knowledge that humans, biological organisms, and the global ecosphere are all vulnerable to delayed effects.

Asbestos is one of the best readily available examples of the issues that must be dealt with in this respect. If asbestos were to cause the individual exposed to immediately suffer an adverse health effect, the problem of enforcing safety and protective measures on the workforce exposed to asbestos would not be a significant problem. However, with the effects delayed for periods of up to 40 years, two problems result:

1. It becomes difficult to convince individuals that there is a substantial future risk without the use of safety and protective measures.
2. It becomes difficult to convince management of the impact of asbestos and particularly the impacts resulting from lack of proof that proper actions were taken today when the possibility of the impact is so long delayed.

It is unfortunate, but true, that American business and government tend to make decisions based on short time horizons. Environmental risks tend to be long-term issues that cannot be effectively dealt with in the short term. An investment in a carefully designed and operated recordkeeping system today to reduce the possibility of a problem 20 to 30 years in the future is difficult to justify, particularly since the costs of the future problem when discounted to today are difficult to quantify or even evaluate. It is for this reason that the expected value is extremely important to risk management decision making. The expected value of an accumulation of events, even if that expected value may not be experienced for a period of decades, may provide nearly overwhelming justification whereas the expected value of a single event may not, given the discounting of that cost to a present value.

The choice of the discount rate is also very important to the justification of the amount of resources for a project. As explained earlier, the higher the discount rate, the lower the present value of a future sum. The present value of one dollar to be received or spent 20 years in the future at a 10% discount rate is $0.15, at a 20% discount rate it is $0.03. The point is that a poor choice of a discount rate and a failure to carefully evaluate the expected value of a series of events can combine to cause a valid and worthwhile risk reduction project to be dropped.

Because of the complexity of a multiple risk analysis and the fact that little can be learned from an example over the asbestos example already provided, the very tedious calculation steps for a multiple risk analysis will not be performed here.

REFERENCES

200. Ignizio, J. P.; *Multi-objective Mathematical Programming via the MULTIPLEX Model and Algorithm;* European Journal of Operational Research, v22 (1985); pp 338–346.
201. Ignizio, J. P.; *Linear Programming in Single- & Multiple-Objective Systems;* Prentice-Hall International Series in Industrial and Systems Engineering; Prentice-Hall, Inc.; Englewood Cliffs, NJ; 1982.
202. Wilson, A. R.; *The Development of a Multi-objective Linear Programming Analysis System;* unpublished thesis in partial fulfillment of the requirements for the degree of Master of Business Administration; Bowling Green State University, Bowling Green, OH; 1987.
203. See the *Model Management Plan and Inspection Report, Sample Management Plan and Inspection Report:* Hazardous Materials Institute, Inc.; 1550 Old Henderson Road; Suite N-232; Columbus, OH 43220.
204. Further details on these budgets and the basic elements of each may be found in the *Model Management Plan and Inspection Report.*
205. NOTE: This will be the most efficient strategy given the limitations on the method. The mathematically optimum feasible strategy may be different from the strategy determined by this method. One of the results of the decision rules set out here and one of the differences between the most efficient strategy and the optimum feasible strategy, is that there may be many most efficient strategies. There will be only one optimum feasible strategy.

PART II

ENVIRONMENTAL RISK INVESTIGATION

DATA SOURCES AND INVESTIGATION PROCEDURES FOR PHASE I, II AND III ENVIRONMENTAL RISK ASSESSMENTS

12 PHASE I ENVIRONMENTAL RISK ASSESSMENTS PART I — GENERAL CONTEXT OF PHASE I ERAs

INTRODUCTION

To this point the concern has been with defining the information required to make sound environmental risk management decisions. The issue now turns to developing the information itself through environmental risk and compliance assessments. It may seem quite logical to first define the information needed and then to develop the program for obtaining the information, but this has not been the case due to the *ad hoc* nature of the development of environmental rules and regulations.

The phrase "Phase I Environmental Audit" has been badly abused. There is not now, nor has there been in the past, a clear and concise definition of what constitutes a Phase I Environmental Audit. For this reason it is necessary to clearly distinguish between a Phase I Environmental Audit, and the two phrases that have been and will be used throughout this book—Phase I Environmental Risk Assessments (Phase I ERA), and Phase I Environmental Compliance Audits (Phase I ECA).

As stated in an earlier chapter, a Phase I Environmental Risk Assessment is defined by the question it will attempt to answer: *For the subject property, is there a reasonable basis to suspect the presence of an environmental risk?* This and the following chapter will attempt to provide a comprehensive set of guidelines for answering this question.

A Phase I Environmental Compliance Audit is contained within a Phase I ERA and must not be confused with it. Specifically, a Phase I Environmental

Compliance Audit will seek to assure the owner/operator of property that all environmental laws pertaining to the property have been complied with and will identify any deviations.

An example may serve to illustrate the difference. Consider the Underground Storage Tank rules outlined earlier. Full compliance with the federal UST rules will assure the owner/operator that the specific concerns of the published rules have been complied with as they pertain to the USTs covered by the rules. Note however that a very large number of USTs *are not* covered by the rules. Specifically, tanks taken out of service prior to November of 1974, farm tanks, certain process tanks, domestic heating oil tanks and some others are not covered and would not therefore be the proper subject of a Phase I Environmental Compliance Audit. *All* USTs would be the subject of a Phase I ERA, however, and the difference can be very substantial.

In an audit of the Phase I Environmental Audits for a major commercial lender, 73% failed to recognize the highly likely existence of substantial environmental risks. In one of these cases the discovery of four 5,000-gallon USTs very nearly caused severe injury to the demolition crews and resulted in the expenditure of an additional $70,000 in demolition costs on a 4,300-square-foot site. The gas station that previously occupied the site was listed in city phone directories and commercial directories available in the reference room of the local public library, on four fire insurance maps of the area, in the records of the local fire and real estate tax department, and a picture of the station in its active phase hangs on the wall of the city hall. The Phase I Environmental Audit made no mention of the possibility of the tank's existence, nor would a Phase I Environmental Compliance Audit have necessarily dealt with the issue, at least under the stated federal rules.

Guidance for Phase I ERAs is supplied to some extent, by a document published by the USEPA that will be referred to hereafter as the *de minimus* guidance.[206] In addition, the guidance offered by a number of other regulatory bodies,[207] current practice, case law, and the literature have been used to construct the checklists and discussions that follow. There can be no guarantee that the information is complete or that all of the items for a specific situation will have been listed, or that all of the items listed are necessary for a specific situation.

THE GEOGRAPHIC CONTEXT

The risk system concept is critical to the conduct of a Phase I ERA, particularly with respect to defining the aerial extent of investigation. The subject property for the investigation must always be examined in the context of its geographic relationship to other properties and particularly to other properties that may be, are, or might have been a source of risk either in the direct sense of the physical presence of a risk source on the subject property, or in the indirect sense of having a possible impact on the value of the subject property without a physical presence.

The transport mechanisms are the key to the geographical context of the subject property. Simple rules of thumb such as the search of all properties within some specified distance can be very misleading unless they are adjusted for local transport mechanisms. In general, the farther away a possible source of environmental risk is from a subject property, the less the degree of concern. However, if a strong transport mechanism exists in the local environment, this statement must be modified. For example, if the subject property is located within an area subject to frequent air inversions with a number of sources of air pollutants or if the subject property is located on a steep slope with potential sources upgradient or targets downgradient, or alongside a fast moving stream, then the area of concern must be adjusted accordingly.

The first step in preparing for a Phase I ERA must be to identify the transport mechanisms and from them the area of concern that must be investigated. Based on the ability of the transport mechanisms to relocate risk sources, the degree of investigation detail can be determined. Obviously, the closer a potential risk source is to the subject property or the closer the target to the subject property, the higher the level of concern.

Note that the concern with respect to the subject property is bi-directional. That is, it is not just the possibility of risk sources moving onto the subject property that should be investigated, but risk sources moving off the subject property. Risk sources that move onto the subject property may result in a zero value for the subject property. Risk sources moving off the subject property may impart a negative value to the subject property.

Surrounding properties for purposes of environmental risk analysis are explicitly defined as being those within reach of the available transport mechanisms relative to the subject property and the potential risk sources.

HISTORICAL CONTEXT

The historical context of a Phase I ERA investigation is so important and so often overlooked that it is the specific topic of the following chapter. Put simply, the Phase I ERA must trace the subject and surrounding properties back to at least virgin land status. Note that virgin land means the true natural state of the land prior to any meaningful human activities, activities especially including modern agricultural techniques.

There is very little in the way of meaningful data to go on, but there are strong indications that existing formal public lists of environmental risks cover less than 20% of all environmental risks. The existing formal public lists of environmental risks will be referred to as the codified records. The uncodified records such as telephone books, commercial directories, aerial photographs, industrial censuses and the like, must be diligently examined in order to raise the probability of environmental risk identification to even the region of an 80% confidence level.

Included within the historical context must be considerations of the technology available at the time a risk source came into existence and the environmental fate of historically deposited risk sources. Process technology tends to change rapidly and the processes familiar to us today may be significantly different compared to those common just a few decades in the past. For example, today most auto body shops utilize plastics to repair body damage, but as recently as the 1970s the repair material was solid lead. Even a small shop handling only six to eight cars per week could have generated several tons of lead waste that would now be in the ground surrounding the shop site.

The environmental fate of risk sources is only just beginning to be understood, both in terms of the decomposition of the risk source and its current location given the available transport mechanisms. Risk sources that have been previously thought to decompose to non-hazardous materials or to disperse widely to concentration levels that would not be of concern have now been shown to be otherwise.

THE PHASE I ERA

Most Phase I ERAs today result from and are driven by, financial considerations. In particular, they tend to be triggered by an imminent financial transaction; a sale, purchase, or refinancing activity. Unfortunately, such activities are highly time dependent and the time frame for most will not allow for a comprehensive Phase I ERA. In general, for any property, a comprehensive Phase I ERA will take between 60 and 120 days to complete and cost in the range of several tens of thousands of dollars. However, considering the stakes, Phase I ERAs are a necessity, not a luxury and until Community Environmental Risk Inventories (CERI) (CERIs will be discussed in the next chapter) become generally available, the only alternatives will be either for the property owner to execute a Phase I ERA in advance of an anticipated transaction, or for the closing of the transaction to be delayed, or for the buyer, seller and financing entity to accept higher than necessary levels of risk.

The following checklists and guidance for Phase I ERA performance are based on the transaction scenario and presume that a highly competent team of experts will execute the Phase I ERA. There is absolutely no possibility whatever of a single individual possessing the level of knowledge, competence, and expertise required to execute the following checklists. Experienced appraisers, engineers, hydrogeologists, industrial historians and researchers, industrial hygienists, biochemists, attorneys, etc., all have their legitimate and varied roles to play as members of the Phase I ERA team.

Traditionally the first step in property valuation involves an appraiser, and the appraiser has much to contribute toward an efficient Phase I ERA as the following checklist will show. However, the appraiser must adjust the normal appraisal routine to account for the fact that the information uncovered will be needed by other specialists. In particular, the appraiser or person performing the appraisal checklist (Table 12.1) must carefully document the work done with a view toward having other specialists use the information for additional work. The alternative will be the duplication of effort on the part of several expensive professionals in order to cover all of the necessary issues.

In addition, the importance of an experienced appraiser's local knowledge cannot be overlooked. Most of the environmental risk evaluation specialists will not have a working knowledge of the local development history or geography that is almost inherent in the local appraiser's background. Note however that in this discussion it is the highly trained professional appraiser, such as a member of the American Institute of Real Estate Appraisers or the Society of Real Estate Appraisers[208] (now jointly known as The Appraisal Institute) that are being referred to, not the local real estate agent who claims to do appraisals without significant formal training.

Table 12.1 is based on the recommendations for appraisal inspections provided by the guidance and training documentation of the American Institute of Real Estate Appraisers as modified by the author. The point of this list is that for a Phase I ERA, most of the items must be accomplished by an expert and if properly accomplished, will form a necessary and highly useful base of information for the balance of the Phase I ERA and later Phase II and III work.

Site Drawings

Original site drawings are one of the primary documentary resources for determining if there is a reasonable basis to suspect the existence of an environmental risk.

One of the first steps should be to examine the existing maps, plot plans, and general building outline drawings to generate a set of drawings on which to record the results of the investigation. The documentation drawings should follow the same general rules, definitions, and specifications as set out earlier with one possible exception—it will often be difficult or inappropriate to reduce all plot drawings and maps to an 8.5 × 11 format. This set of documentary drawings will be a part of the initial site inventory.

Table 12.1 "Appraiser's" Initial Site Checklist[209,210]

(An appraiser or other competent general property inspector will normally investigate a basic set of subject property items in the normal course of the performance of their function. If the following list of items has been evaluated as appropriate to the specific site, a substantial portion of the preliminary data for a phase I ERA will have been accomplished. The Phase I ERA checklist set out later assumes that the following has been accomplished and the documentation is readily available in an applicable format.)

I. **Land or Site Description**
   A. Legal Description of the Land (whichever is applicable)
      1. Metes and Bounds
      2. Rectangular Survey
      3. Lot and Block
   B. Title and Record Data
      1. Ownership information
         a. Chain of title
         b. Current and past uses
      2. Zoning and land use information (current and historical)
      3. Assessment and tax information
   C. General chracteristics of the site
      1. Size and shape (prepare a drawing to scale)
      2. Corner influence
      3. Plottage
      4. Excess land description
      5. Topography
         a. Surface soil description
         b. Subsoil description
         c. Grade
         d. Natural drainage
      6. Utilities
      7. Site improvements
      8. Location
         a. Time-distance relationships
         b. Site access
   D. Special characteristics of site
      1. Buildings
         a. Number, location, size
         b. Type of construction
         c. Age
         d. General Condition
         e. Flooring materials
         f. Floor drains; to where?
         g. Spill evidence; cleaned up?
      2. Physical features
         a. Parking areas (paved or under roof)
         b. Roads
         c. Power lines
         d. Public buildings
         e. Dwellings
         f. Structures/improvements
         g. Right-of-ways
      3. Adjacent land use
         a. Surface water
         b. Roads/utilities
         c. Housing

Table 12-1 (cont.)

- d. Vacant lands
- e. Vegetation types

4. Waste evidence
   - a. Drums, barrels, containers
   - b. Waste materials
   - c. Construction/demolition debris
   - d. Discolored soils
   - e. Odors
   - f. Leachate seeps
   - g. Discolored surface water
   - h. "Unnatural" soils
5. Chemical/fuel/storage area
   - a. Type of construction
   - b. Age
   - c. General condition
   - d. Security and access
   - e. Spill control systems
   - f. Spill evidence
   - g. Manholes/catch basins
   - h. Drains, Fill pipes
     - (1) Where are they?
     - (2) What are they for?
     - (3) What do they look like? Unusual appearance or odors?
6. Water features
   - a. Wells
   - b. Springs
   - c. Seeps
   - d. Swamps/wetlands
   - e. Ponds
   - f. Streams
   - g. Direction of runoff
   - h. Direction of runon
   - i. Surface erosion
   - j. Evidence of flooding
7. Geologic features
   - a. Topography and slope
   - b. Soil chracteristics
   - c. Rock outcrops
   - d. Sink holes
   - e. Excavations
   - f. Spoil piles
   - g. Mining activity
   - h. Quarries or pits
   - i. Diversion ditches
   - j. Soil stockpiles
8. Process tanks/wastewater tanks
   - a. Size and materials of construction
   - b. Purpose, contents
   - c. Inside/outside?
   - d. Above grade/below grade
   - e. Lined/unlined
   - f. Contents

(continued)

Table 12-1 (cont.)

- g. General condition
- h. Leaks
- i. Connecting piping integrity
- j. Chemical feed pumps
- k. System condition
- l. Spill evidence

9. Pits/ponds/lagoons
   - a. Size, location
   - b. Materials/construction
   - c. Purpose/contents
   - d. Above/below grade
   - e. Lined/unlined
   - f. General condition
   - g. Freeboard
   - h. Leaks
   - i. Fill/drain pipes
10. Disposal areas
    - a. Size
    - b. Location
    - c. Age
    - d. Contents
    - e. General condition
    - f. Debris
    - g. Drums
    - h. Sludge/residue/rubble
    - i. Discoloration
    - j. Odors
    - k. Monitoring wells
    - l. Cover material
    - m. Vegetation
    - n. Equipment condition
    - o. Surface contour
    - p. Erosion
    - q. Leachate (analytical results)
11. Chemical transfer points
    - a. Where?
    - b. What's involved?
    - c. Inside/outside
    - d. Paved/unpaved (when?)
    - e. Spill evidence

II. Building Description [General][211] (Include dates of changes, modifications, additions, etc. and mark drawings to show locations of same.)

A. Exterior description

1. Substructure
   - a. Footings
   - b. Slabs
   - c. Piles
   - d. Columns
   - e. Piers
   - f. Beams
   - g. Foundation walls

Table 12-1 (cont.)

- 2. Superstructure
  - a. Framing
  - b. Insulation
  - c. Ventilation
  - d. Exterior doors
  - e. Exterior walls
  - f. Windows and screens
  - g. Façade
  - h. Roof and drain systems
  - i. Chimneys, stacks, vents
  - j. Special features

B. Interior description
- 1. Interior walls
  - a. Doors
- 2. Division of space
  - a. Storage areas
  - b. Stairs, ramps, elevators, hoists
- 3. Interior supports
  - a. Beams and columns
  - b. Flooring system (subflooring)
  - c. Ceilings
- 4. Painting, decorating, and finishing
  - a. Basements
  - b. Floor coverings
  - c. Molding and baseboards
  - d. Fireplaces
- 5. Protection against decay and insect damage
- 6. Miscellaneous and special features

C. Equipment and mechanical systems
- a. Plumbing system
  - a. Piping
  - b. Fixture
- 2. Systems that use energy
  - a. Hot water system
  - b. Heating system
    - (1) Warm or hot air
    - (2) Hot water
    - (3) Steam
    - (4) Electric
  - c. Heating fuels
    - (1) Coal
    - (2) Fuel oil
    - (3) Natural gas
    - (4) Electricity
  - d. Air-conditioning and ventilation systems
  - e. Electrical systems
- 3. Miscellaneous equipment
  - a. Fire protection
  - b. Elevators, escalators, speed ramps
  - c. Signals, alarms, call systems
  - d. Unloading facilities
  - e. Attached equipment

(continued)

Table 12-1 (cont.)

III. Building Description (Commercial)

- A. Building function
  - 1. Office
    - a. Garden office
    - b. Low-rise
    - c. Mid-rise
    - d. High-rise
  - 2. Retail
    - a. Neighborhood center
    - b. Community center
    - c. Discount outlet
    - d. Regional mall
    - e. Super regional mall
    - f. Specialty shopping
      - (1) Small store/convenience
      - (2) Department store
      - (3) Large discount store
      - (4) Drycleaner
      - (5) Etc. (describe)
  - 3. Light industrial
    - a. Fabrication and assembly
    - b. Distribution facility
    - c. Food production/service
    - d. Office warehouse
    - e. Office showroom
    - f. High tech
  - 4. Industrial
    - a. Manufacturing plant
    - b. Factory
    - c. Warehouse
    - d. Cold storage
  - 5. Special use
    - a. Motel/hotel
  - 6. Mixed use
    - a. Office and retail
    - b. Hotel, retail and office
    - c. Retail and residential
  - 7. Transportation
    - a. Motor freight terminal
    - b. Airport
    - c. Aircraft hangar
    - d. Rail freight terminal
    - e. Parking structure
    - f. Automobile service station
  - 8. Medical
    - a. Hospital
    - b. Clinic
    - c. Medical office
    - d. Laboratory
  - 9. Recreational
    - a. Bowling alley
    - b. Country club

Table 12-1 (cont.)

- c. Nightclub
- d. Theatre
- e. Etc. (describe)

B. Building type
  1. Concrete frame
    - a. Cast-in-place
    - b. Pre-cast
      - (1) Factory units
      - (2) Tilt-up
  2. Steel frame
    - a. Single-story
    - b. Multi-story
    - c. Pre-engineered
  3. Masonry
    - a. Unit block construction
    - b. Brick construction
  4. Other
    - a. Air-supported structure
    - b. Space frame

C. Select construction features
  1. Foundation
    - a. Piers, pilings, caissons
    - b. Other substructure
      - (1) Grade beams
      - (2) Spread footings
      - (3) Continuous foundation
    - c. Building slabs
      - (1) Turndown
      - (2) Slab-on-grade
      - (3) Post-tensioned
    - d. Waterproofing
    - e. Mechanical and electrical planning
  2. Structure
    - a. Anchor bolts
    - b. Bearing plates
    - c. Columns
    - d. Beams and girders
    - e. Joists
    - f. Floor system
      - (1) Beam and slab
      - (2) Flat slab
      - (3) Flat plate
      - (4) One-way joist slab
      - (5) Waffle (two-way) joist slab
      - (6) Poured concrete deck
    - g. Walls
      - (1) Shear wall
      - (2) Exterior load bearing wall
      - (3) Curtain wall
    - h. Roof
      - (1) Built-up roofing
      - (2) Sheet metal roofing

(continued)

Table 12-1 (cont.)

- i. Interior finish-out
  - (1) Floors
  - (2) Walls
    - (i) Stud framed
    - (ii) Sheetrock/plaster skin
    - (iii) Paint/fabric/vinyl covering
    - (iv) Fiberglass insulation
    - (v) Case walls
  - (3) Ceiling
    - (i) Suspended grid/panels
    - (ii) Sheetrock/textured paint

The original drawings, maps, aerial photographs, and construction specifications *(hereafter, all of these and similar documents will simply be referred to as the "drawings" unless otherwise noted)* should be examined from the point of view of determining if there is reason to suspect the presence of an environmental risk.

Particular care should be exercised in examining these drawings with respect to subtle clues implying the possible existence of a problem. A note on construction drawings that a fireproofing material is to be applied to structural members should raise the obvious question of whether or not asbestos was used. An area marked as "vehicle fueling" should raise the issues of underground storage tank sites, likely sites of overflow spills of petroleum products and the like. Sumps, ponds, filled areas, production area drainage or sewer piping, transformer stations and other items often shown on drawings should all generate an increased level of concern given the time period of activities. All suspect areas should be documented on an initial set of the documentation drawings for further investigation.

PHASE I ERA AND ENVIRONMENTAL COMPLIANCE AUDIT (ECA) CHECKLIST OVERVIEWS

The "Appraiser's" Initial Site Checklist (Table 12.1) provides a beginning point for the balance of the Phase I ERA and ECA work. In addition to the resources consulted by the appraiser, a series of other resources must also be consulted. Many of these resources may be unfamiliar to even experienced environmental professionals as they are historic in nature and will be extensively discussed in the next chapter.

Table 12.2 Partial Documentary Resources List

Air Quality Authority	Zoning Department	Water Quality Authority
Federal and state OSHA	Citation and Fine History	Commercial Directories
Gazetteers	Fire Department	Fire Insurance Maps
National Priority List	Local Health Departments	Mining and Natural Resource Development Maps
Planning department	Permit and License Authorities	Sanitation Agency
State Health Department	Registration & Notification Agencies	Aerial Photograph Archives
U.S. Department of Agriculture	State Priority List	USGS Topgraphic and Special Map Series
Underground Storage Tank Lists	U.S. EPA	

Public Records Search

Information obtained in the review of public records should be documented in the Phase I ERA or ECA report. This information may be voluminous and discretion is needed to distinguish the relevant from the nonrelevant. Information about site usage and dates of usage should be reported. The format for reporting should be easy to read and understand. As appropriate, sections of the report should document records found in the resources listed in Table 12.2 and the historic data resources in the following chapter.

Site Visit

There is no substitute for a site visit by skilled and experienced personnel, particularly appraisers, engineers, hydrogeologists and others with particular knowledge to contribute. Even highly experienced individuals may miss important clues as to the presence of environmental risks unless they have taken the opportunity to carefully examine the site and its physical surroundings. The observations should be recorded as a part of the Phase I ERA or ECA report as appropriate and should consider the following observational issues.

- Visual Observation:
  - Areas of potential contamination
  - Areas of visible contamination
  - Boundaries
  - Description of adjacent properties
  - Description of area within reasonable reach of all transport mechanisms
  - Description of building materials
  - Description of chemical activities
  - Description of ground management practices
  - Description of site operations

  - Description of waste management practices
  - Geology
  - Hydrology
  - Proximity of surface water
  - Topography

- Interviews:
  - Site line personnel and managers
  - Neighbors and local parties
  - Senior real estate agents
  - Purchasers and previous land owners and operators
  - Local historians

Environmental Agency Correspondence/Environmental Applications and Reports

Most of the records of environmental agencies with respect to applications for discharge permits, hearings on sites being assessed and the like are available and a matter of public record or can be obtained through Freedom of Information Act (FOIA) actions. As noted earlier however, availability is a relative term with respect to the time available to complete a particular assignment and although the request for information may have been made in a timely manner, the response may not be forthcoming within the time frame of the investigation. The fact that information was requested but not obtained prior to the completion of the report should be the subject of an important limiting condition statement.

Insurance Records

Insurance records for a site and especially the subject property can be an important source of information, particularly with respect to property damage or general liability claims. The casualty loss of important quantities of raw materials, production equipment, or finished products or damage to real property resulting from fires or natural disasters should prompt questions concerning the release of hazardous or toxic substances or the generation of such substances. For example, a fire in a steel mill electrical substation or rolling mill hydraulic power unit may have resulted in the contamination of a large area of building and land surfaces by dioxins (a product of the combustion of PCP-containing dielectric or hydraulic fluid).

A record of the denial of insurance as a result of past casualty losses should alert the investigator to the nature and quality of past operating practices. If hazardous substances are involved, there would be little to suggest that a history of high casualty losses would translate into good hazardous materials management practices.

Corporate Management Records

Corporate management records such as memos, minutes of meetings, capital expenditure planning documents, and policy statements can provide important background information to the Phase I ERA with respect to the attitude and plans of the organization for dealing with environmental risks. For obvious reasons much of this material may not be available to individuals who do not come under the attorney work-product doctrine, but to the extent that such records exist, they should be examined by competent personnel.

Third Party Agreements Involving Hazardous Materials

Out-of-sight/Out-of-mind is more than a trite folk adage. Unfortunately, it simply does not work for the disposition of hazardous materials. Also unfortunately, past practices have assumed that waste materials are of no consequence to the organization except as a production expense and record-keeping for the disposition of waste materials prior to 1970 may be almost or entirely non-existent. Even during the period from 1970 to the present the records of many organizations, particularly small organizations, may be totally inadequate. These facts should not preclude a good-faith effort on the part of the investigator to attempt to trace at least the major quantities of disposed materials in an to attempt to determine their current location and condition.

It is almost impossible to transfer the title to a hazardous waste material. Title to such material will still be vested in the generator in almost all circumstances. A careful investigation of any existing third party agreement concerning the disposition of hazardous materials should be undertaken with particular attention being given to the quality of the third-party operations.

The basic ideas of Phase I ERA and ECA investigations can be obtained from a narrative listing of questions in some of the areas of concern.

Adjacent Properties

Were there previous land uses (or current uses) which might adversely affect the subject property?

Are there registered underground storage tanks up gradient of the subject site?

Are there nearby landfills, dump sites, lagoons, waste ponds or settling basins which might affect the property?

Are there any large quantities of stored materials or chemicals on the adjacent sites which might cause substantial property damage in the event of earthquake, fire, high wind or other disasters?

Have there been any substantial citations, enforcement actions or mandated cleanup actions on these sites?

Has the surface or groundwater up gradient from the site been contaminated?

Is the subject property downwind from known air polluters, odor sources or large open chemical storage sites?

Air Emissions

What are the air emissions from each source of air pollution on the subject property? What are the types and fuels burned in each piece of equipment?

What air pollution control equipment is used for each source of emission? Are these permitted? Are permits active and transferrable?

Are air emissions monitored? What is the monitoring frequency and what are the results?

Does the report provide copies of permits, licenses and registrations? Are they current and transferable?

Do area-wide air quality control standards restrict development from the owner's or purchaser's point of view?

Asbestos

Has a survey and testing been completed to locate asbestos on the site?

Does the survey contain types, amounts, locations and conditions?

Were air samples taken, and if so, were any concentrations exceeding permissible levels?

Does the report identify the asbestos problems by a priority ranking system?

Were asbestos removal projects completed in the past? If so, the description should include dates, amounts, contractors and landfill information.

Pesticides, Insecticides, Herbicides and Fertilizers

Have pesticides, insecticides, herbicides or chemical fertilizers been used on the site?

What were the dates and volumes? What were the products and who were the manufacturers?

What are the current residue concentrations? What are the consequences of these concentrations?

Were there chemical rinses, washes, mixes or other operations on the site which were likely to generate concentrations in soil or water?

Polychlorinated Biphenyls (PCBs)

Are there currently PCB-filled or PCB-contaminated transformers, capacitors or other equipment at the site?

What are the quantities of PCBs and are there any PCB leaks?

Have any of the transformers, capacitors, etc., been drained and cleaned of PCBs? If so, who did the work and where were the PCBs incinerated? What are the current PCB concentrations in the cleaned units?

Do any of the transformers belong to the utility, municipality, or other party?

Was any PCB-filled equipment removed from the site? If so, when and by whom?

Are there any PCB contaminations at the site? In structure, in soil, in liquids?

Radon

Was a radon survey done on the building and on the property?

Was the survey a "spot check" or a more definitive characterization?

Does the report provide a comparison of the radon data to the EPA's acceptable levels?

Soil Condition

Have the signs of soil contamination been investigated?

- Discolored soils
- Areas with unusual odors
- Earthen drainage systems
- Ditches and trenches
- Sunken areas
- Distressed vegetation areas
- Excavated areas
- Fill locations

- Illegal dumping areas
- Dry stream or lake areas
- Empty settling basins
- Dry wells
- Dust-control oiled areas
- Areas around tanks and other containments
- Locations of former buildings, tanks and process equipment
- Waste collection areas
- Near fuel dispensing locations
- Where culverts, berms and natural topography direct runoff

Are there any past reported spills or other releases of hazardous materials on the property?

If spills occurred, what cleanup measures were taken? Who did the cleanup and where was the waste taken?

Storage Tanks, Drums, Clarifiers, Sumps and Connecting Pipe Systems (The word "tanks" shall apply as appropriate to all piping.)

Are there any above- or below-ground storage tanks, drums, clarifiers, sumps or other storage vessels on the site?

Does the report provide a complete inventory including size, location, age, contents and tank material?

Are there any tank leaks? Were pressure and integrity tests performed? What were the results?

Have all permits and registrations been filed and are they current?

Are there cathodic protection systems in place? Are there any leak detection or other leak protection systems in place?

Have any tanks been removed? If so, were soils tested? What were the results?

Are there any records of leaking tanks, repairs to tanks or spills from tanks?

Waste Disposal Practices

What are the types of hazardous liquids and solid wastes generated now or in the past at the site?

How are these liquid and solid hazardous wastes disposed?

Are off-site disposal locations identified? Is there a description of the treatment, storage and disposal of these wastes?

What is the EPA identification number? Are copies included of waste disposal permits, licenses and registrations?

What are the volumes of each waste sent in the past? What is the current monthly volume of each waste?

Are copies of manifests, disposal data and reports provided? (If practical.)

Are solvents and degreasers used at the facility? If so, provide trade and scientific names, uses and monthly volume estimates.

Has the site been used for disposal of hazardous liquids or solids?

Are the locations, depths, quantities and types of wastes on the property defined?

Are properly permitted waste disposal sites available with sufficient capacity? What risks may be associated with their use?

Water Issues

Are surface and groundwater conditions sufficiently communicated to allow projection of on-site and off-site waterborne containment migration?

Are there any wetland or aquatic issues to be considered? Such as, lakes, rivers, streams, marshes, estuaries, coastlines, etc.

What are the sources of wastewater discharges to surface and groundwater, septic systems, holding ponds, and settling basins?

Are there any wastewater discharges to sewer systems or storm control systems? Are there evaporation or storage ponds on the property?

Are there any wastewater treatment facilities on the site?

What are the results of surface and groundwater sampling on the site?

How do these tests compare with historic or publicity available data on surrounding sites?

Are there restrictions on disposal that may restrict site development?

PHASE I ERA CHECKLIST

The following checklist, Table 12.3, is a compendium checklist from a number of sources and may, or may not be adequate for a specific situation. It assumes that relevant data from the "Appraiser's Checklist" (Table 12.1) and Chapter 13 are available and have been analyzed.

COMPLIANCE HISTORY

The environmental compliance history of the site should be carefully researched, but over-reliance on this source of information may not be jus-

Table 12.3 Phase I Environmental Checklist

I. Records Review

- A. Real Estate Records
  - 1. Data from "Appraisers" Initial Site Checklist
  - 2. City directories and atlases
    - a. Commercial gazetteers
  - 3. Fire insurance maps
    - a. Property-specific
    - b. City-wide
  - 4. Photographs of subject and adjacent properties
    - a. Ground level
    - b. Aerial
- B. US Environmental Protection Agency
  - 1. Superfund National Priorities List (NPL)
  - 2. Comprehensive Emergency Response, Compensation, and Liability Information System (CERCLIS) database
  - 3. CERCLA release notifications
    - a. Follow-up reports
  - 4. Emergency Remedial Response Information Center System (ERRICS) database
  - 5. Toxic Substances Release Inventory
    - a. Release inventory reports (Form B)
  - 6. Toxic Substances Control Act (TSCA) inspection and compliance records
  - 7. Lists of violating facilities
    - a. "Significant" violators
  - 8. National Pollution Discharge Elimination System (NPDES) permits
  - 9. Petroleum oil spill reports (under Section 311 of Clean Water Act)
  - 10. Underground storage tank records
    - a. Tank notification
  - 11. Hazardous Waste Data Management System database
    - a. Generators and transporters
    - b. Treatment, storage, and disposal facilities
- C. US Army Corps of Engineers
  - a. Wetlands dredge and fill permits (now under USEPA)
- D. US Geological Survey
  - a. Cartographic records
    - a. Topographic series (7.5 minute)
    - b. Hydrological series
    - c. Miscellaneous special series
- E. Security and Exchange Commission potential liability filings
- F. State Emergency Planning Commissions (SPECs)
  - 1. Local Emergency Planning Committees (LEPCs)
    - a. List of hazardous substances on site
    - b. Annual reports
- G. State environmental protection agencies responsible for:
  - 1. Air quality
  - 2. Water quality
  - 3. Coastal zone management
  - 4. Natural resources
  - 5. Solid waste management
  - 6. Hazardous waste management
  - 7. Fish and wildlife

Table 12-3 (cont.)

II. Site Inspection

- A. Subject property information
  - 1. Identification
    - a. Facility name
    - b. Address
    - c. County
    - d. Owner
    - e. Facility manager/operator
      - (1) Name
      - (2) Telephone number
      - (3) Years of service
    - f. Facility environmental manager(s)
      - (1) Name
      - (2) Telephone number
      - (3) Years of service
    - g. Applicable SIC codes
    - h. Facility products and services
  - 2. Site history
    - a. Year(s) constructed
    - b. Year first occupied
    - c. Year acquired by present owner/operator
    - d. Previous owners/operators
      - (1) Name
      - (2) Current address
      - (3) Period of ownership/operation
      - (4) Site use
    - e. Improvements by current owner
      - (1) Building construction
      - (2) Topographic alterations
        - (i) Fill operations
      - (3) Etc. (describe)
  - 3. General site information
    - a. Acreage
    - b. Source of water
      - (1) Condition of on-site wells
    - c. Gross building area (GBA)
    - d. Size, condition and use of unimproved areas
    - e. Vegetation cover
      - (1) Discoloration
    - f. Floodplain considerations
    - g. On-site and adjacent surface water
    - h. Soil boring history
    - i. Soil contamination test history
    - j. Groundwater
      - (1) Depth
      - (2) Flow direction
      - (3) Use
      - (4) Quality
    - k. Groundwater quality test history
    - l. Nature of previous federal, state or local environmental studies of subject property
    - m. Adjacent land usage

(continued)

Table 12-3 (cont.)

- n. Nature of previous federal, state or local environmental studies of adjacent properties
4. Operational considerations
   - a. Industrial processes
   - b. Raw materials
     - (1) Material
     - (2) Quantity
     - (3) Use
     - (4) Location
   - c. Raw material transport
     - (1) Method of delivery
     - (2) Storage media
   - d. Solvents
     - (1) Type
     - (2) Quantity
     - (3) Use
     - (4) Location
   - e. Use of the following substances/materials:
     - (1) Arsenic
     - (2) Asbestos
     - (3) Cyanide
     - (4) Heavy metals
     - (5) PCB-containing transformers/capacitors
     - (6) PCB-containing oils/fluids
     - (7) Cutting oils
     - (8) Hydraulic oils
     - (9) Acids
     - (10) Caustics
     - (11) Reactive materials
     - (12) Fluorides
     - (13) Radium or strontium
     - (14) CFCs as process materials
     - (15) CFCs as part of mechanical or building systems
     - (16) Other as listed by federal, state or local agencies as hazardous materials or substances
   - f. Sources of process and non-process water
   - g. Electrical power sources
5. Material management
   - a. Above-ground drum storage
     - (1) Material stored
     - (2) Location
     - (3) Number of drums
     - (4) Drum storage surface
     - (5) Preventive release features
     - (6) Condition of drums
     - (7) Frequency of inspection
     - (8) Spill/leakage history
   - b. In-ground and underground storage
     - (1) Material stored
     - (2) Location
     - (3) Storage media (e.g., tanks, drums, in-ground pits, ponds, lagoons)
     - (4) Leak detection/monitoring methods
     - (5) Tank integrity test results
     - (6) Soil/water test results
     - (7) Tank excavation/removal history

Table 12-3 (cont.)

- (8) Spill/leak history
- c. Above-ground tank storage
  - (1) Material stored
  - (2) Location
  - (3) Tank size
  - (4) Secondary containment features
  - (5) Spill/leak history
  - (6) Etc.
- 6. Air emissions
  - a. Emission activities
    - (1) Fuel combustion
    - (2) Waste combustion
    - (3) Process venting
    - (4) Material handling
  - b. Emission control equipment
    - (1) Process control
    - (2) Wet scrubber
    - (3) Mechanical collector
    - (4) Capture/collection/dispoal
    - (5) Other
- 7. Roof inspection
  - a. Visible emissions
  - b. Stains/discoloration
- 8. Wastewater discharges
  - a. To sewer
  - b. To land/water supply
  - c. Stormwater removal
- 9. Solid and hazardous waste inventory
  - a. Hazardous waste inventory
    - (1) Waste type
    - (2) Waste source
    - (3) Annual quantity
    - (4) On-site storage
    - (5) Disposal container
    - (6) Hauler(s)
    - (7) EPA landfill ID number
  - b. Waste disposal
    - (1) On-site
    - (2) Adjacent properties
- 10. Odor and noise
  - a. Complaint history
  - b. Facility response history

III. Personal Interviews

- A. Interview subjects
  - 1. Facility employees
    - a. Environmental managers
    - b. Workers in contact with hazardous materials
  - 2. Adjacent facility employees
  - 3. Adjacent residents/neighbors
- B. Methodology
  - 1. Questionnaires
  - 2. Handwritten interview notes
  - 3. Audiotape or videotape recording
    - a. Release forms

tified. Compliance histories often do not extend far enough into the past to reveal some of the major potential sources of problems, particularly those relating to early production and raw materials/waste materials storage and disposal activities. In addition, compliance reviews are often highly limited in scope being restricted to the specific requirements of the rules and regulations at the time and will fail to reveal concerns of a more recent origin that are not under specific regulation.

Environmental Audits/Prior Inspection Reports

Early compliance reviews may or may not have been performed to the currently recognized state of the art. There are two major areas of weakness in almost all compliance or Phase I, II or III investigations. First is the caliber of documentation. Only rarely can reliance be placed on a statement, implied or explicit, that no problem exists in a specific area without substantial additional information as to how that conclusion was reached. For example, taking a single sample of an extensive area of suspect risk generally will not be acceptable evidence that no risk exists if the finding is negative.

The second area is a general failure to document precisely what was done, and by implication, what was left undone. The negative default principle requires that virtually all areas will fall into one of four definitive categories of risk:

1. Areas in which there is no reason to suspect the presence of a risk.
2. Areas in which the presence of a risk is suspected but which have not been specifically examined.
3. Areas of suspected risk that have been specifically examined and in which, based on sound analytical proof, no risk can be found.
4. Areas of confirmed or assumed risk and the nature of the risks.

In general, if an examination of a past environmental audit shows that a risk exists and there is some evidence that this conclusion was reached in a reasonable manner, then acceptance of the finding is reasonable. If a suspect area has been declared to not contain a risk, the acceptance of the finding will depend entirely on the quality of the prior work and may not be acceptable. In this case, a Phase I ERA may have to conclude that reasonable grounds for suspicion still exist and must be confirmed or denied by additional Phase II or Phase III work. The most dangerous situation is encountered when nothing is said about an area leaving the investigator to wonder. The general conclusion in such circumstances should be to treat the area as if it had never been inspected.

Subject Property Data

To the extent that the above data can be obtained on the subject and surrounding area as a whole, the probability of missing an important risk source will be substantially reduced. However, most of the detailed data will be basically unobtainable except on the subject property. This fact should be carefully noted in the Phase I ERA report as a limiting condition to that report.

The following information should be available for the subject property through the current owner if it exists at all. In examining the following data requirements, keep in mind that the products produced on the site and functions or services performed on the site will be of particular concern, whether current or past.

Generic List of Questions Regarding Products, Services and Functions

1. What products are/were produced on the site (site: subject and surrounding property as defined by the reach of available transport mechanisms)?
   a. Did or do any of them involve hazardous materials as a raw material, end product, by-product, or as a part of the process?
   b. Verify the inventory of materials against the inventory accounting records for current operations.
   c. Do any of the materials on site require MSDSs? If so, are the MSDSs available?
   d. Are any of the materials regulated under RCRA, TSCA or Title III of SARA? If so, have the corresponding reports and notices been provided?
2. What services are performed on the site?
   a. Do any of these services involve the use of hazardous materials?
   b. What materials?
3. What functions are performed on the site?
   a. Do any of these functions involve the use or handling of hazardous materials or result in hazardous materials?
   b. What materials?

Obviously, to answer these questions an inventory of the materials on the site will be extremely useful and should preferably be compiled from audited accounting records, operating manuals, and service operations manuals. The preparation of a formal inventory of hazardous materials or operations that may use or produce hazardous materials is necessary to both a compliance audit and to a Phase I ERA. In addition to an inventory of materials, the drawings noted above should provide an inventory of structures verifiable against the books and records of the organization. Also from those books and records, an inventory of equipment should be prepared, but care must be exercised to insure that obsolete equipment, or equipment depreciated to a zero book value, is listed. The economic value of equipment, buildings, and materials is not of concern; their existence in the past or the present is the point of concern. An outline Phase I ERA checklist can be found in Table 12.4.

FREQUENTLY ENCOUNTERED PROBLEMS AT SITES

There are a series of problems frequently encountered at sites that can compromise the validity of the Phase I ERA or ECA findings. Some of these are listed in Table 12.5 below, but one of the most important is not. Severe restrictions can arise from the need to maintain confidentiality or from the desire of site-related personnel to hide information necessary to the investigation.

The confidentiality issue properly belongs to the legal counsel for resolution, but the issue of inaccurate or erroneous information supplied to the investigator is of particular importance. The investigator should carefully document all sources of data in his/her field log as it was described earlier, including the time, date and name of the individual supplying the information.

Table 12.4 Phase I Environmental Compliance Assessment Checklist

Compliance History

A. Review of past environmental audits and assessments
   1. Internal staff studies
   2. External studies
   3. Environmental impact statements

B. If the studies or audits indicate any unresolved issues, check internal records for further action.

C. Prior inspection reports
   1. UST inspections
   2. OSHA inspections
   3. Asbestos inspections
   4. Violation notices

D. Environmental agency correspondence
   1. Environmental notices received
   2. Agency inspection reports
   3. Violation notices

E. Review of environmental applications and reports
   1. Applications
      a. *Clean Air Act* violating facilities list
      b. PSD applicant list
      c. NSR applicant list
      d. *Clean Water Act* violating facilities list
      e. National Pollution Discharge Elimination System (NPDES), POTW, NSR and PSD permit applications lists
      f. Dredge and fill permits
      g. CERCLA National Priorities List (NPL)

Table 12-4 (cont.)

- h. CERCLA Release Notification list
- i. CERCLA abandoned Toxic Substance Disposal Facilities (TSDF) list
- j. Emergence and Remedial Response Information List
- k. List of generators, transporters, and TSDFs registered under RCRA
- l. UST listing from state agency responsible
- m. Monitoring information (DMR, regional air and water, RCRA manifests and biennial shipment reports)
- n. SARA Title III Right-to-know reports
- o. Hazardous substance release report
- p. Permit exceedence reports
- q. Permit denial findings and records
  (Is an adversarial relationship evident? Have the concerns expressed in the finding been resolved?)

2. Are there current or threatened environmental actions?
3. Public Health Service Annual Toxic Substance Disposal Report (ATSDR):
   - a. Biennial report
   - b. List of health assessments

F. Insurance records
1. Applications, policies and policy denials; casualty and liability
2. Claims
3. Notices of liability
4. Inventory of past policies showing coverage, carrier and dates of coverage
5. Is there evidence of environmental problems and/or liabilities?

G. Corporate management records: minutes of meetings
1. Regular management meetings
2. Board of Directors meetings
3. Stockholders meetings

H. Third party agreements involving hazardous substances
1. Removal, transportation and disposal contracts
2. Removal, transportation and disposal contractors
   - a. Are contractors and contractor's personnel fully qualified and permitted for the function performed?
   - b. Have their qualifications and permits been independently verified?
   - c. What is their record of performance? Citations, violations, on-site and equipment condition (very good, good, bad). Obtain Better Business Bureau and credit reports.

Current compliance activities

A. Compliance program information
1. Organization
   - a. Who is the individual responsible?
     - (1) Training/background qualifications
     - (2) Organizational definition of position (reporting lines, explicit authority and responsibility, etc.)
     - (3) Who reports to this individual?
     - (4) Determine the unstated authority and responsibility for this individual
   - b. Formal program organization
     - (1) Are there current procedure manuals?
       - (i) Emergency Action Plan
       - (ii) Oil Spill and Hazardous Substances Contingency Plan

(continued)

Table 12-4 (cont.)

(ii) Other emergency planning as indicated by the hazardous materials inventory for this site
(iv) Electrical transformer labeling, storage and disposal plans
(2) Types and adequacy of the general work practices for dealing with hazardous substances?
(3) Are there formal training and retraining programs?
(4) Inventory of equipment provided and required by the types of hazardous materials
(i) Equipment type
(ii) Equipment condition
(iii) Equipment availability relative to the likely location of hazardous materials problems
(5) What is the status of the OSHA Hazard Communications Program?
(6) Where are the MSDS files?
(7) Are the MSDS files complete?
(8) Have the required notices been posted?
(9) Is there evidence that the required training has been provided and that the training program is adequate and current?
(10) Are materials properly labeled?

2. Permits, reports and records
a. Inventory and location of monitoring reports
(1) Mandatory reports required by law or formal agreement
(2) Informal reports as part of the organization's recordkeeping or the program manager's control program
(3) Audit reports for completeness, continuity and adequacy
b. Inventory and location of permits
(1) Audit permits for validation dates and continuity
(2) Audit permit records for completeness
c. Inventory of personnel records for individuals handling hazardous materials, particularly those requiring training.
(1) Identify personnel requiring training and certification
(2) Audit training and certification records for continuity, completeness and currency
(3) Identify personnel requiring medical monitoring and audit medical records for continuity, completeness and currency
(4) Audit personnel files for individuals working with hazardous materials for incident records and formal record keeping practices

3. Risk Assessment for Adjacent Properties
a. What is the compliance record for adjacent properties historically? (Review the data from II.)
b. If existing contamination, or portential sources of contamination exist, what is the likelihood of migration onto the subject property?
c. What is the likelihood of value damage from a neighboring release that does not contaminate the subject property?

Table 12.5 Frequently Encountered Problems at Sites

Water	
	Process discharge to the storm sewer through improper or illegal connections
	Inoperative or poorly maintained sampling equipment
	No effluent flow measurement device
	Process area diking broken or nonexistent
	Storm drains in liquid bulk transfer areas
	Oil-water separators improperly installed
	No laboratory quality-control/quality-assurance program
	Runoff from storage piles going to the storm sewer
	Improper permit reporting
	Outdated or poorly written Spill Control Plan
Air	
	Vents and exhausts not registered
	Lack of data on emission points
	"Guestimates" on performance of control equipment, lack of test results
	Improper worker protection in dusty environment
	Neighborhood complain file incomplete and/or lacking resolution of complaints
	Dust from plant operations of building roof and pavement
Solid Waste	
	Sloppy housekeeping
	Careless drum or container handling (leaks and/or dents)
	Mixing process waste with office trash
Recordkeeping	
	Unreported in-plant dump sites
	Incomplete analysis of hazardous wastes being sent off site
	Incomplete files on past disposal practices
	Unlabeled drums
	Leaking containers

REFERENCES

206. *Guidance on Landowner Liability Under Section 107(a)(1) of CERCLA, de minimus Settlements under Section 122(g)(1)(B) of CERCLA, and Settlements with Prospective Purchasers of Contaminated Property;* Memorandum issued June 6, 1989 by the Acting Administrator for Enforcement and Compliance Monitoring and the Acting Assistant Administrator for Solid Waste and Emergency Response; USEPA.
207. See the discussions in Chapter 2.
208. Author's note: There are several other professional appraisal recognition organizations that require the demonstration of a very high level of competence of their members within the context of this discussion. The author just happens to be very familiar with the requirements of the two organizations mentioned.
209. This checklist is solely the responsibility of the author and is not endorsed by or a part of any appraisal organizations standard of professional practice. Its purpose is to demonstrate that a professional, such as an experienced appraiser, can provide a signification portion of the initial data needed for a Phase I ERA.

Appraisers are generally not qualified to perform complete Phase I ERAs and should not be expected to do so.

210. The sources for this checklist are Chapters 9 and 10, *The Appraisal of Real Estate;* 9th edition; American Institute of Real Estate Appraisers; Chicago, IL; 1987; pp. 187–241; and *Introduction to Commercial Construction;* seminar, American Institute of Real Estate Appraisers; 1987; pp. 1–47. The original checklists from these sources have been extensively modified by the author.

211. From documentation sources such as building plans, blueprints, and specifications:
    a. Architectural drawings
        (1) Site plan
        (2) Elevations
        (3) Plans
        (4) Details
        (5) Sections
        (6) Schedules
    b. Mechanical drawings
    c. Electrical drawings
    d. Landscape plans
    e. Shop drawings

13 PHASE I ENVIRONMENTAL RISK AUDITS PART II — EXAMINATION OF THE LAND

INTRODUCTION

It was noted in the preceding chapter that in a recent audit of so-called Phase I inspections for a major lender, 73% of the reports failed to identify the likely presence of significant environmental risks. The purpose of this chapter is to detail the sources and application of historic information to Phase I ERAs because of this extremely poor and potentially very expensive area of weakness in the field.

HISTORICAL EXAMINATION

For those who have read the works of James Michener, the historical perspective of a Phase I ERA will be slightly less comprehensive than his works. It is not really necessary to begin the historical search during the age of the dinosaurs, although the origin of radon contamination extends even further back in time. It is necessary to identify the point at which meaningful human activities began—particularly industrial or mineral extraction—in order to properly conduct a Phase I ERA

Historical data, not only about the subject property but also surrounding properties, are often required. There are a number of sources for such data, but frequently the best sources are the knowledgeable local experts.

In Hancock County, Ohio, for example, much of the land appears to be highly productive farmland with prime development potential. A search of the Ohio EPA listings of known toxic and hazardous sites will show that the county is well below the mean of all Ohio counties for the presence of known hazardous materials sites, and relatively few problems have been reported in the media. Also frequently noted in the media is the fact that this particular county was the center of the greatest oil and gas boom of the 19th century, and that most of the well, pipelines and refinery installations were abandoned by the early 1900s. The point is that in almost any given 15-acre plot of land there may be as many as 10 or 15 abandoned wells, drilling material waste sites, or abandoned pipelines. All of these represent a minor hazardous materials problem with respect to the current use of the land (farming), but may represent major problems in terms of development for residential, commercial or industrial usage. A thorough knowledge of local history is required to even be aware of this potential set of problems.

This type of situation exists in many parts of the country from early steel mill or textile plant sites or dumps in the eastern United States to uranium or lead mines and tailing dumps in the west. Generally these sites are not formally recognized as hazardous materials sites in official listing documents at this time.

One of the major causes of mistakes found in Phase I ERAs is a natural human tendency to assume that current technological processes are also the common processes of the past. Note in the preceding paragraph the statement that mining spoils for heavy metals are to be found in the western part of the United States. This is where most individuals would expect to find them because that is where most of the heavy metals mining activities took place in the past 100 years. However, during the 1700-1880 period, most of the heavy metals used were mined from the mountain regions of the Appalachian chain, particularly in eastern Pennsylvania and western Virginia.

Another example of this problem has already been cited in the form of the materials used for the repair of body damage to automobiles and specifically, the extensive use of lead for this process prior to about 1970. The point is that a thorough Phase I ERA will require extensive knowledge of past industrial technological practices in order to determine the likelihood of the current existence of an environmental risk and without thorough examination of this aspect, a number of significant risks can be overlooked.

Older structures represent a similar set of historical investigation issues, ranging from the construction practices followed in the region in a particular time period to prior occupants' usage of the structure that may have resulted in contamination. Buildings with decorative plaster often contain a very high level of asbestos and many layers of lead paint. Underground fuel storage tanks, lead piping, asbestos covered wiring, and improperly closed septic fields represent problems often associated with older structures.

In general, historical information may be divided into two classes: (1) uncodified data representing past use of the land, and (2) codified data on known hazardous materials sites. The reason for distinguishing between the two types lies in the fact that only the latter is currently labeled as a specific source of hazardous materials information. The former is not so labeled, but may contain important information on the possible presence of hazardous materials problems. It would be fair to state that the number of hazardous materials sites recorded in the codified lists are probably less than 20% of the total number of such sites. In consequence, the currently uncodified data represents the greater source of necessary information.

DATA SOURCES FOR LAND USE STUDIES: CONTEMPORARY AND HISTORIC[212]

Land use regulations and the density of land use are an important part of the general data collection and analysis process. In this regard, land use can be defined as: The employment of a site or holding to produce revenue or other benefits; also, the designation by a governing authority of the use to which land may be put to promote the most advantageous development of the community, e.g., designation of industrial, residential, commercial, recreational, and other uses under a master plan.[213]

These and similar definitions of land use emphasize the benefit of developing or maintaining particular functions of property. However, social and economic impacts from environmental contamination of property have introduced new concerns affecting how land may be used. Because contaminations result from past behavior, historic uses of property assume potentially large—and in select instances—very great significance. This may be expressed in terms of increased liability exposure for participants in real estate transactions, decreased property values, increased health risks, and disruptions in real estate market activity.

Historic land use specifically concerns past functions of property as they affect the contemporary built environment and/or unimproved contemporary sites. Further, its investigation involves comparatively recent or remote past research, frequently requiring use of documentary resources that are less familiar or accessible.

Historic, present, and even future land use can be singular (e.g., a warehouse on a lot), combined (e.g., a gas station/grocery on one lot), or multiple (e.g., adjoining timberland parcels with ongoing mineral exploitation).[214] Historic land use also requires an identification of specific use(s) — industrial products storage, chemical manufacturing, mining, retail trade, etc. — expressed by such factors as use intensity, size of property, architecture, and occupant profile, insofar as identification is possible. The identification of specific historic use, as well as the variety of historic use, provides vital information concerning the existence and probable location of environmental hazards in or adjacent to subject properties.

HISTORIC CONTEXT OF ENVIRONMENTAL HAZARDS

Pollution of varying kinds and degrees has been evident in human cultures for millennia. In industrialized nations a major source of hazardous materials is the manufacturing sector of local, regional, and national economies. Historically, industrial development parallels the physical expansion of settlement frontiers, regional increases in population, the discovery of abundant natural resources, and an increasing emphasis on manufacturing in national affairs. Industrial hygienists, engineers, and real estate industry professionals are among those who are aware this process has, through time, produced a large number of environmentally hazardous materials, some of which have been incorporated on or into property via construction, land use activity, or transport mechanisms.

In light of this awareness, historical contexts of environmental hazards can be described in terms of rural and urban industries responsible for the hazards themselves. Rural industries such as farming and forestry demand large surfaces for production; urban industries such as chemical and textile manufacturing do not. The areal requirements of these two industry groups influence local population distributions and the degree to which manufacturing activities are concentrated in specific locales. As a result of changing conditions affecting the traditional factors of industrial location (markets, raw materials, utilities, transportation and labor), the spatial distribution of rural and urban industries varies across the American landscape over time.

For example, oil and natural gas production began in the bituminous Appalachian coal region of northwestern Pennsylvania in 1859. Oil production soon expanded into the adjacent states of Ohio and West Virginia. By 1900, increased national demand for petroleum products triggered exploration in western regions of the country. At this time, California, Texas, and Oklahoma emerged as leading producer states. In California, the largest centers of production were in the Los Angeles Basin and San Joaquin Valley.

Another example concerns the Lake Calumet region southeast of Chicago. In 1869, the U.S. Army Corps of Engineers transformed this 31-square-mile-tract from marshlands into an industrial real estate market via several Calumet River and harbor improvements. Over time, the improvements attracted manufacturers of primary metals and metal goods, industrial and agricultural chemicals, paints, and building construction products. Flour mills and meat packing plants also located here as transport routes improved. By 1897, about 34 firms were cited in the Calumet region; in 1960 this number had increased to 117.[215]

During the next hundred years an impressive list of environmentally hazardous materials derived as by-products of urban industrial activity were deposited in the Calumet region, many of which persist in local soil and groundwater today. Steel and iron manufacturers produced phenols, cyanides, naphthalene, and nitric, sulfuric, and muriatic acids. The most abundant steel manufacturing by-product was slag, which was deposited at high temperature on vacant lands adjacent to the mills. Other wastes frequently were deposited over slag to partially burn, producing other soil and groundwater contaminants.[216]

This example specifically concerns the production of hazardous wastes in a large industrial district, but in most historical contexts contaminations occur on a reduced scale, such as on or adjacent to single properties. The aggregation of individual contaminations may result in spatially defined local (i.e., block, neighborhood) public health problems.

Persons involved with historic land use research must become familiar with the development of manufacturing and other potentially harmful activities in the study area and the processes employed. It may at times be advantageous to begin an investigation on a state or regional level to determine the relative chronological appearance of the activities. Understanding the spatial distribution and chronology of environmental hazards at the level of individual sites requires a careful review of documentary resources related to a subject property and its immediate vicinity.

OVERVIEW OF DOCUMENTARY RESOURCES

Documentary sources useful for historic land use investigations fall into three general categories:

1. Governmental records
2. Corporate records
3. Personal papers

Governmental records develop from law and related manifestations of local, state, and federal authority. Abstracts of title, claims, censuses, conveyances, deeds, land grants, land patents, leases, mortgages, titles, and topographic surveys and court records (particularly nuisance claims against commercial concerns prior to 1970) are common examples. Corporate records develop in the context of commercial and industrial activities. They take numerous forms, including account books, bills of lading and sale, inventories, warehouse receipts, and waybills. Personal papers are the private documents accumulated by an individual and subject to his or her disposition. They develop in the context of an individual life and frequently contain family materials.

The risk analyst must be generally familiar with governmental records that pertain to property transactions and descriptions of property (e.g., abstracts of title, assessors' maps, deed books, land certificates, subdivision plats). These records are developed and maintained at the county level (or its equal political subdivision) through recording statutes and Torrens system provisions.

All states today have statutory provisions that designate a public official in each county to maintain public records associated with its real estate transactions. The title of this official varies by locality (*cf.* County Clerk, County Recorder, Registrar of Deeds, and in New England, Town Clerk), but data under his or her management are nearly always retrievable by parcel or transaction date.

The Torrens system developed in England in 1857 as an alternative recordkeeping process. Under this system title to property is registered instead of registering individual property transactions. Torrens statutes exist in Colorado, Georgia, Hawaii, Illinois, Massachusetts, New York, North Carolina, Ohio, Oregon, Virginia, and Washington. The system is not widely accepted in the United States, however, due to substantial costs of converting existing records and long-standing traditions of county recordkeeping. Large metropolitan counties that utilize the Torrens system are Suffolk County, New York, and Cook County, Illinois.

Property descriptions are a component of land or site descriptions. Land within a set of boundaries indicating ownership is referred to as a parcel, lot, plat, or tract. Legal descriptions of property permit the identification of individual parcels such that one cannot be confused with another. Thus every parcel of real estate is unique.

In the United States there are three common systems for legally describing property:

1. Metes and bounds
2. Rectangular survey
3. Lot and block

The oldest method for legally describing land is the metes and bounds system. Metes and bounds describe property in terms of reference points from a POB, or point of beginning. Bounds refer to the POB and all intermediate points, sometimes called monuments. The POB and intermediate points may consist of natural markers such as trees, boulders, and stream banks, or artificial markers such as mathematically calculated survey points. Metes describe the direction one moves from one bound to another and the distance between bounds.

Direction is determined by degrees, minutes and seconds of angles from north or south. Distance is measured linearly in feet or meters from the POB to each bound, until returning to the POB. The return is called closing and it determines the accuracy of the survey.

Metes and bounds were used to describe land parcels in 17th and 18th century western Europe and this method was exported to the American colonies. It is the primary descriptive system for the 13 original states and 8 others. States using metes and bounds descriptions are: Connecticut, Delaware, Georgia, Hawaii, Kentucky, Maine, Maryland, Massachusetts, New Hampshire, New Jersey, New York, North Carolina, Pennsylvania, Rhode Island, South Carolina, Tennessee, Texas, Vermont, Virginia, parts of Ohio, and parts of Oregon. The system also is used in the District of Columbia.

Early metes and bounds descriptions are difficult to verify on the present landscape due to changes in topography and original errors of measurement. Metes and bounds descriptions frequently accompany early land records (see below).

The United States Rectangular Survey, also known as the government survey system, was devised by the Continental Congress in 1785 as a more reliable and convenient alternative to metes and bounds methodology. Its development is linked with early national expansion south and west of New England and the Chesapeake states. It is the principal system of property description for land west of the Ohio and Mississippi Rivers.

The primary reference points for the rectangular survey system are true east-west and north-south lines called base lines and meridians, respectively. Other east-west lines, called township lines, are established at six-mile intervals from each base line. Other north-south lines, called range lines, are established at six-mile intervals from each principal meridian. The rectangles created where township and range lines intersect are called townships.

Thus the intersection of a base line and a principal meridian is the starting point from which range and township lines are counted to locate a specific township in a legal description. Ranges are numbered east and west from principal meridians; townships are numbered north-south from base lines. Under rectangular surveying, townships can mean either a location on a line north or south of a base line or one six-mile square of land.

In the latter case, townships are further divided into 36 sections, each a mile square and containing 640 acres. Sections, in turn, are subdivided into quarter-sections of 160 acres each, and fractions of quarter-sections, each containing less acreage than before. Townships are adjusted to the earth's curvature and some contain less than 640 acres. Irregular townships exist because of early land grant provisions and the boundaries of navigable waters. Fractional sections contain acreage shortages or overages.

The lot and block system developed from rectangular surveys in response to a need to simplify descriptions of small parcels. Its origins are traced to instances where land developers subdivided land surveyed under the rectangular system and then assigned lot numbers to identify individual sites within blocks. Blocks—usually square or rectangular areas between intersecting streets or other physical boundaries—are identified by number using ground surveys or established monuments.

Lot and block descriptions are frequently combined with rectangular survey descriptions. For example, a parcel might be legally described (in part) as Lot 6 of Block 10 of Brady Park, a Subdivision of the southeast quarter of Section 12, Township 4 North, Range 1 East.

CHECKLIST OF SELECT DOCUMENTARY SOURCES

Other kinds of governmental records, as well as varieties of corporate records and personal papers, can provide information useful to historic land use investigations. The following section presents a checklist of governmental, corporate and personal record types that potentially contain information about the presence (or likely presence) of historically deposited environmental hazards.

Group I: Governmental Records

Censuses of Manufacturers

Censuses of Manufacturers provide detailed statistics on manufacturing activities of small geographic areas, individual industries, products shipped, and materials consumed. Because they may indicate the former presence of local manufacturing operations that produced hazardous substances or waste, censuses are potentially valuable research tools. Both federal and state censuses exist.

The federal census of manufacturers was taken every ten years between 1809 and 1899 (except for the year 1829), every five years between 1899 and 1919, and every two years between 1919 and 1939. World War II disrupted the schedule until 1947, when the census resumed. Censuses are available for the years 1947,1954, 1958, 1963, 1967, 1972, 1977, 1982, and 1987. Under legislation passed in 1964, censuses of manufacturers are taken for years ending in "2" and "7". A federal census of mineral industries is compiled on the same schedule. It was first collected in 1839.[218]

State censuses of manufacturers are conducted independently of the federal census and for this reason occur on different schedules (e.g., Michigan censuses were conducted in 1837 and every year from 1854 through 1904.)[219] Where they exist, state censuses of manufacturers collect similar data as the federal census.

Locational information is not consistently reported in censuses of manufacturers. For this reason, users must carefully review the content summaries of individual census publications to determine if locational data are present and, if so, how the data are reported. For example, the 1914 federal census of manufacturers includes reports by states with statistics for principal cities and metropolitan districts, indicating the location and size of establishments. The 1958 federal census reports general statistics for the country by gross geographic divisions and states.

National and State Priorities Lists[220]

The National Priorities List (NPL) is the U.S. Environmental Protection Agency's list of uncontrolled or abandoned hazardous waste sites identified for possible long-term remedial actions under federal Superfund legislation. Select states also have developed hazardous waste site priorities lists, modeled after the Superfund program.

Sites of the NPL are ranked according to scores on the EPA's Hazard Ranking System, with the exception of top-priority sites and sites that involve the Agency for Toxic Substances and Disease Registry (ATSDR) of the U.S. Centers for Disease Control. EPA and states use HRS scores to establish priorities for response actions.

Sites of the NPL are arranged in groups of 50 with each group representing similar HRS scores. EPA considers sites with similar scores to hold approximately the same potential threat to public health or the environment. NPL sites are also numbered, with each entry containing the site name, location, and its corresponding EPA Region. Entries also include notations describing the kind(s) of remedial activities underway at the time the list was prepared. Under law the NPL is to be updated annually; however, EPA usually performs this function more frequently.

Five "response categories" describe the range of possible remedial activities:

- [R] Federal and/or state response; includes sites at which EPA or state agencies have begun or completed remedial actions.
- [F] Federal enforcement; includes sites where 1) the Department of Justice has filed a civil complaint or issued an administrative order under CERCLA or RCRA, 2) a federal court has mandated a response action following a judicial proceeding, or 3) EPA has obligated funds for a remedial study.
- [S] State enforcement; includes sites where 1) a state has filed a civil complaint or issued an administrative order under CERCLA or RCRA, 2) a state court has mandated a response action following a judicial proceeding, or 3) a state has obligated funds for a remedial study.
- [V] Voluntary or negotiated response; includes sites where responsible parties have begun or completed response action pursuant to settlement agreements or consent decrees to which EPA or a state is party.

- [D] Category to be determined; includes all sites not listed in any other category.[221]

In addition to providing response categories, the NPL also identifies "cleanup status codes" for each field activity underway or completed at a site. Sites are cleaned up in stages or *operable units*. Several operable units are often required to clean up hazardous waste sites. Three cleanup status codes appear on the NPL:

- [I] Implementation activities are taking place for one or more operable units; fieldwork is in progress at the site, but no operable units are complete.
- [O] Implementation activities are complete for one or more (but not all) operable units; some field work is complete, but additional work is necessary.
- [C] Implementation activities are complete for all operable units; all remedial actions are complete.[222]

The final set of information on the NPL is a list of all proposed and final NPL sites arranged alphabetically by state. Each entry includes the date the site was proposed or announced and, where applicable, the date on which the site appeared on the NPL. As of this writing, there are 799 NPL sites and 378 proposed sites for a total of 1,117.[223]

Property Tax Rolls

Local property tax rolls identify individuals and corporate entries that pay taxes on real property, as well as the amounts paid, at specified intervals of time. A dramatic increase in a tax obligation from one year to the next may suggest that improvements to the property have taken place, or that new kinds of activity are present on site. Although such inferences must be tested via other kinds of documentary evidence, they hold implications for land use analysis.

Federal and Local Court Records

Prior to the advent of the major environmental laws in the 1970s, contamination problems were usually addressed through nuisance actions in local, state and occasionally federal courts. Given that a proper firm name is known, likely court venue records should be considered as a potentially more useful source of data than property tax rolls.

Group 2: Corporate Records

Abstracts of Title

An abstract of title is a condensed history of title to land consisting of a synopsis or summary of material or operative portion of all the conveyances or interests associated with the subject property, including a statement of all liens, charges, or liabilities to which the property may be subject. Maps, plats, and other aids may accompany an abstract of title.

It should be noted that title insurance policies frequently exclude environmental protection from coverage, except in cases where a notice of a defect, lien, or encumbrance resulting from a violation or alleged violation affecting the land has been recorded in the public records at the date of the policy. Public records may include environmental protection liens filed with the clerk of the U.S. District Court where the land is located.[224]

Account Books

Account Books are divided into two classes: books of first entry and books into which are posted first-entry data. Journals, sometimes referred to as daybooks, wastebooks, or blotters in early contexts, are a specific kind of first book entry. Among these types of documents are cash journals (formerly called cashbooks), purchase journals (sometimes called invoice books, invoice registers, voucher books, or voucher registers), and sales journals (also known as sales books).

A ledger is a specialized book into which first entry data are posted. There are many kinds, including general ledgers, private ledgers, purchase ledgers (in modern accounting better known as payable ledgers, formerly known as bills payable books), and sales ledgers (in modern accounting better known as receivable ledgers, formerly known as bills receivable books).

Account books may contain data indicative of the transfer, sale, or purchase of hazardous materials. Unlike some other kinds of records, account books are not maintained for the consumption of posterity. Their objectivity is an important feature to business historians, but modern record retention policies may affect their availability through time.

Bills of Lading and Sale

A bill of lading is a written document that confirms receipt of goods for shipment, issued by a person engaged in the business of transporting or forwarding goods. Bills of lading are signed by the carrier or his or her agent and frequently identify the freight. The name of the consignee and terms of the contract may also be stated.

In common usage, a bill of sale confirms receipt of goods for purchase, usually issued by an agent representing a retail or wholesale establishment. The names of parties involved in the transaction may be recorded. Like bills of lading, bills of sale often identify the nature of goods associated with the transaction, possibly indicative of the transfer, sale, or purchase of hazardous materials. Bills of lading and sale are also objective business documents. They may be found in corporate archives and are an integral part of record-keeping systems in most going concerns. However, modern record retention policies may affect their availability through time.

Deeds

In simplest terms, a deed is a conveyance of realty. It is a written document signed by a grantor that transfers realty from one person or entity to another. The general requirements for a valid deed are that it must be in written form and include a legal grantor and grantee (usually specified by name), unequivocal words of conveyance, and an adequate property description. Deeds are essential documents in the chain of title searches and as such may suggest a relative chronology for the presence or suspected presence of environmental hazards.

Inventories

An inventory is a detailed list of articles of property that may include descriptions of the property, quantity of materials or goods in stock, or their value. Business inventories may reveal the presence of stored hazardous materials at particular locations. They are intended to be objective in character, but subjectivity may be introduced for a variety of reasons.

Receipts of Warehouse

This document is simply a receipt issued by a warehouseman for goods received by him for storage in that facility. Receipts of warehouse, like bills of lading and sale, may contain data indicative of transfer, sale, or purchase of hazardous materials. Their objectivity is an important feature to business historians, but modern record retention policies may affect their availability through time.

Waybills

A waybill is a written document made out by a carrier identifying a point of origin, destination, cosigner, consignee, and goods included in a shipment by railroad or motor carrier. Waybills are similar in form and content to bills of lading and sale. Modern record retention policies may also affect their availability.

Group 3: Personal Papers

Family Histories

Published family histories may contain information describing land use and/or property improvements associated with family-owned parcels. Such descriptions are often anecdotal in style, but may be highly suggestive for early periods in family affairs.

Reminiscences

These documents are akin to family histories but in published form are usually more brief and specialized in subject matter. Included in this subgroup would be transcribed or untranscribed oral histories, collected via audiotape recording by both professional and amateur historians for the purpose of preserving the recollections of individuals. Oral histories can focus on virtually any subject, but often concern workplace routines and activities.

Group 4: Cartographic Records

Cartographic records derive from a wide variety of sources, including government agencies, corporations, and individuals. At the federal level, map making has been and remains an essential function of the U.S. Geological Survey, the Bureau of Indian Affairs, and the Bureau of Land Management. Maps also play an important role in the U.S. Forest Service and the Bureau of the Census. These and other sources may contain records that indicate or suggest historic land use.

The substantive attributes of cartographic records are organizational origin, functional origin, subject, date/title/edition of publication. Organizational origin refers to the government agency or other entity responsible for the production of the record. Functional origin refers to the purpose for which the record was created. The subject of a cartographic record is the area to which it pertains.

Among these attributes, subject, date and scale are of primary importance to interpreting cartographic records. Individual maps usually have one primary subject, but they also may contain secondary subjects that indicate land use or factors affecting land use (e.g., topographic maps produced by the U.S. Geological Survey indicate cultural as well as natural features). The date of a cartographic record provides at least a relative, if not absolute, chronology for the subject(s) depicted in the record, which is important for establishing when an environmental hazard may have been present or deposited.

Archaeological Maps

Cartographic records produced by professional historic site archaeologists (also called historical archaeologists) frequently indicate the locations of late 19th and early 20th century cultural features. These include, but are not limited to, extinct residential, commercial, or industrial sites, and waste dumps. Archaeological maps pertain to individual sites, thus reflecting a precise location relative to the modern landscape or to groups of sites within a region.

Archaeological surveys and/or testing are required as part of environmental impact studies associated with federally funded construction projects. Archaeological reports containing maps are usually filed within offices of individual State Historic Preservation Officers or with cultural resource management divisions of federal and state agencies (e.g., the U.S. Department of the Interior, U.S. Army Corps of Engineers, state departments of natural resources, highways, or transportation). Reports may also be filed in university museums and libraries.

Census Bureau Maps

Two sets of cartographic records maintained by the Bureau of the Census are noteworthy for studies of historical land use. The published record set of decennial census maps for the period 1860-1945 consists of maps and atlases depicting the findings of the censuses as to population density, agriculture, irrigation, drainage, mines and quarries, manufacturers, etc. In a related manner, a small number of city maps is contained in Volumes 18 and 19 of the Tenth Census (1880) titled *Social Statistics of Cities*. These indicate municipal projects, including sewage systems.[225]

An atlas summarizing results of the 1937 census of manufacturers contains 126 maps of the United States showing the number and location of manufacturing establishments by select counties.

Geologic Maps

Cartographic records produced by the U.S. Geological Survey appear in printed reports known as Monographs, Papers, Bulletins, Geologic Folios, and Water-Supply papers. They depict areal geology, underground structure, and mineral deposits at various scales. The Geologic Folios are similar to topographic maps in appearance and were published in a "library edition" accompanied by text. However, many regions of the country were not mapped before the folios were discontinued.[226]

Locations of mineral deposits and the nature of underground geology can provide important clues to the interpretation of historic settlement patterns and ground water hydrology.

Fire Insurance Maps

These records depict the locations of potential fire hazards in residential, commercial, and industrial structures. They were developed as an aid to early insurance underwriters to distribute risk liabilities associated with the industrialization of western Europe and the United States. The earliest published fire maps in this country were issued in New York City in 1852. Fire insurance maps became increasingly important during the second half of the 19th century and are still produced today. They are a vital component in the development of the mutual insurance industry and the evolution of building construction standards.[227]

Fire insurance maps typically depict either the locations of manufacturing and industrial concerns within city limits, or potential fire hazards existing within individual structures.

A primary producer of fire insurance maps is the Sanborn Map Company. By 1875 Sanborn's cartographers produced over 600 maps of cities, towns, and individual industrial sites. By 1900, the company acquired most of its competition. Sanborn specialized in city-wide maps produced in an atlas format. Individual maps were periodically updated and shipped to subscribers, who then inserted the corrected page in the appropriate atlas. Sanborn atlases exist for all regions of the United States.[228]

Private firms that specialized in site-specific surveys of manufacturing plants were Hexamer of Philadelphia and Barlow of New York City. These maps indicate construction materials, building materials, functional areas of activity, and the locations of equipment by color codes. Map scales are sufficiently large to provide excellent detail.

Hexamer began producing fire insurance maps in the early 1850s. During the period c.1866–1896, Hexamer issued over 2,800 plates of individual sites, many in the Philadelphia and mid-Atlantic region. Hexamer city atlases were published until 1915, when Sanborn acquired the company. Barlow produced over 10,000 surveys of manufacturing sites during the same period as Hexamer, but only about 2,800 survive. These records pertain mostly to New York City and New England.[229]

As resources for identifying the relative locations of hazardous activities and/or hazardous materials in towns, cities, and individual sites, fire insurance maps are an invaluable research tool. They are relevant to historic land use analysis to the extent that locations of historic (extinct) sites can be matched to the present landscape. The greatest limitations of fire insurance maps to historic land use analysis are their sometimes restricted geographical coverage (*cf.* Hexamer and Barlow maps).

Topographic Maps

The most comprehensive series of topographic maps is produced by the U.S. Geological Survey. Begun in the 1880s, the series consists of quadrangles bounded by parallels of latitude and meridians of longitude. Two scales are represented. Quadrangles covering 7.5 minutes of latitude and longitude are published at a scale of 1:24,000 (1 inch = 2,000 feet); quadrangles covering 15 minutes are published at 1:62,500 (1 inch = approximately 1 mile). The 7.5-minute maps are generally more useful for land use study because of the greater amount of detail represented. Over 80 metropolitan areas are mapped at the 1:24,000 scale. These indicate all man-made features at the time of compilation.

Each quadrangle is identified by the name of the city, town, or prominent natural feature within it, and the names of adjoining quadrangles appear at each map margin. Contour intervals differ with the scale of the map and the relief of the geographical area covered. Topographic maps are printed in five colors. Man-made features and place names appear in black; water features appear in blue; road classifications, urban areas, and U.S. land lines appear in red; woodlands appear in green; and contour lines and their accompanying values appear in brown. On photorevised maps, areas of change are shown in purple.

Topographic maps are valuable resources for historic land use analysis because they indicate at a reasonable scale low relief areas where hazardous materials might accumulate, cultural features such as quarries and mines, and natural surface drainage patterns. The maps show the placement of buildings relative to these and other features, although individual building functions are unmarked. Of particular importance may be alterations between consecutive maps such as the filling in of quarries, ponds, lake beds or other low-lying areas. Topographic maps are also readily available from the Geological Survey and through authorized dealers in each state.[230]

Group 5: Pictorial Records

Pictorial records include photographic and drawn or painted pictures. They derive from numerous sources and are found in governmental records, corporate records, and personal papers.

Pictorial records may contain subjects useful to historic land use analysis. The presence of subjects like manufacturing facilities, drum storage areas, railroad sidings, above ground tanks, refuse dumps, and drainage ditches in a drawing or photograph may suggest the presence of potential hazards. Further, a pictorial record can be dated through analysis of its physical attributes and subject matter if an exact date does not appear on or with the document. This provides a relative chronology for land use interpretation.

Drawings, Sketches, Paintings

Drawings, sketches, and paintings depicting historic land use are occasionally found among corporate records and personal papers. Included in this group would be commercially produced lithographs of individual towns and cities known as "birdseye views" or "panoramas." These are, in visual effect, aerial views conceived by itinerant artists. Depending on the subject, drawings, sketches, and paintings may reveal relative locations of industrial and commercial sites.[231]

Historical Photographs

Photographic images may be divided into the broad categories of negatives and prints. With the exception of early kinds of photographs (i.e., daguerreotypes, ambrotypes, and tintypes), all photographic prints derive from negatives. Prints also come in all sizes and may be black and white, sepia, or any other color. Prints also appear on a wide variety of papers and cardboards.

Establishing the approximate time period when an historical photograph was made usually requires familiarity with the history of photography and knowledge of when photographic processes were in widespread use.[232] Information about photographers from city directories, tax rolls, and other public records can also be helpful in dating historical negatives and prints. However, associating individual photographers with specific historical images is in most instances, extremely difficult.

Dating historical photographs can also be accomplished by verifying known elements of an image. For example, a photograph of a vacant corner lot may reveal the names of adjacent businesses which can be dated through analysis of city directories. Under magnification, dates may also appear on the license plates of cars parked in the vicinity of the primary subject or on posters and signs. Familiarity with local landmarks can help determine the vantage point from which a photograph was taken.

Remote Sensing Products

Aerial photographs, satellite images, and other remotely sensed pictorial records can reveal historic land use information. These include, but are not limited to, the locations of former buildings, roads, railroad beds, public utilities, waste dumps, and vegetation changes due to soil and/or surface and groundwater contamination. Remotely sensed data are often digital in form and geographically referenced to the earth's surface. The U.S. Geological Survey provides air and space images information through its EROS Data Center.[233]

Group 6: Miscellaneous Documents

City and County Histories

Published city and county histories frequently contain information about the economic and commercial development of their subjects including the establishment and growth of specific industries that today are known to produce hazardous waste or other environmental contaminants. Many county histories were published between 1870 and 1920 following the American Centennial and the establishment of local historical societies. They are narrative in form and usually contain at least general locational information; however, many early county histories are anecdotal in style and lack reference notes.

City Directories and Gazetteers

City directories list the names, residence locations, and occupations of city residents. They also frequently contain similar information about commercial establishments, thus providing specific addresses of going concerns over time. Advertisements are another common feature of city directories. These advertisements sometimes include pictorial images of the advertiser's buildings, and/or lists of products. City directories were published in large numbers from the mid-19th century and are still produced today. Earlier examples exist, but are few in number.

Also appearing in large numbers from the mid-19th century through the turn of this century were commercially published gazetteers. Many were produced for individual states, but large cities are sometimes treated separately. The focus of most gazetteers is commercial activity; advertisements appear frequently. Locational data are provided, but detail varies.[234]

Local Newspapers

Advertising and business announcements in local newspapers also may reveal information pertinent to historic land use. Such information can include the availability of new products and services at particular establishments, some of which have environmental implications. Although factual errors are common in most daily newspapers, a significant advantage of information published in the newspaper is that they are date specific.[235]

AUTHENTICITY AND THE CREDIBILITY OF SOURCES

Documentary resources must meet criteria of authenticity and credibility for them to be useful in historical study. These requirements derive from the methodology of historical research and must apply to most investigations of past behavior. Two sets of criteria of authenticity exists—internal and external. External criteria refer to the questions of document chronology, authorship, and place of origin. They are as follows:

1. A manuscript or printed work should be attributed to the author whose name it bears unless there are reasons to the contrary;
2. A work should not be attributed to an author if it has already been attributed to someone else by undoubtedly trustworthy authorities;
3. When testimonies concerning the date, authorship, or localization of a document contradict one another, the degree of reliability attached to each must be accurately gauged, and the various degrees must be weighed against each other.[236]

Internal criteria of authenticity address the content or textual value of sources:

1. A work is probably to be attributed to a given author if it resembles other genuine works of his/her in internal form.
2. A work is probably genuine if its content, as regards details of time, place, conditions, etc., fits in with the environment in which it was to have been produced.
3. Discovery of a source under strange circumstances suggests spuriousness.[237]

Criteria of authenticity establish the reliability of sources from which evidence may be gleaned in support of an observation about the past. Each may involve detailed work aimed at the dating of individual documents, identification of their place of origin, or authorship.

Documents pertaining to environmental implications of historic land use also must be credible. That is, potential sources of error within each source must be understood. Error can be introduced in virtually any attempt to account for a phenomenon resulting from natural processes or occuring independently of human observation. Persons engaged in documentary research must be on guard for error introduced from mistakes of interpretation, judgment and analysis. Lies may be evident in some historical records, although documentary resources applicable to historic land use investigation (e.g., photographs, censuses of manufacturers, cartographic records) are usually objective in character.

SOURCES OF CHECKLIST DOCUMENTS

In Table 13.1, references are made to the locations of select documentary materials. This section identifies specialized sources of the checklist documents and is intended to generally acquaint the reader with the kinds of places in which documents relevant to historic land use analysis are housed. Information about several published reference works that apply to source locations are describer below.

Local sources of documentary materials include city and county governmental offices, historical societies, title companies, business and corporate offices, and municipal libraries. The size and nature of collections in local repositories and businesses varies widely. Regional sources include planning commissions, state archives, state libraries, select state agency offices (e.g., state divisions of history, natural resources, highways, etc.) and college and university libraries. At the federal level, documentary materials are housed in archives and select agency offices.

Curatorial and administrative practice is quite rigorous at most regional and federal level repositories due to the presence of professionally trained archival and library staffs. When planning to conduct research in any source of historic records, it is advisable to contact an appropriate staff member in advance and receive an orientation to the collections before work begins. At such orientations, library and/or archive users are acquainted with finding aids and reference works that can greatly aid research.

A finding aid is a descriptive research tool created by an archival agency or repository for the purpose of establishing control of a records collection. Typical finding aids include topical guides, inventories or registers, card catalogs, shelf and box lists, indexes, and calendars. Lists and calendars are more frequently published and therefore usually are more formal appearing than indexes or catalogs.

A list of record items is simply an enumeration of documents in binders, folders, or volumes, as they appear in one or more collections. A calendar contains substantially more detailed information about records as it always relates to individual documents, not document groups. In calendars, documents are listed in chronological order and each document is identified by the names of its writer and recipient. Content summaries also are included.

In addition to finding aids, published reference works available in most university reference libraries (and the reference section of many large municipal libraries) can also provide important sources for historic land use study. The following section provides brief descriptions of six standard reference sources:

- M. Balachandran, *Regional Statistics; A Guide to Information Sources* (Detroit: Gale Research Company, 1980).

  Provides bibliographic citations for U.S. regional and local data sources produced by local chambers of commerce, industrial developmental agencies, planning commissions, utility companies, market research firms, and financial institutions. Topics include general socioeconomic sources, housing and construction, demography, business directories, and financial data. Sources are indexed by author and geographical location.

- Sargent B. Child and Dorothy P. Holmes, *Check List of Historical Records Survey Publications* (Washington, D.C., 1943).

  This presents a complete listing of Historic Records Survey (HRS) publications resulting from research efforts conducted by the Federal Writers' Project staff of the Works Progress Administration. The HRS included surveys of federal, state, county, municipal, and church records. The major achievement of the HRS was its survey of county records. Field workers completed inventories of approximately 90 percent of the 3,066 counties in the United States, and approximately 20 percent of these were published. The remaining 80 percent (on mimeographed stencils) were caught in the war effort. Those which survive are housed in state archives, but not always in repositories of the states in which the documents were originally located.

  An annotated version of the *Check List* is maintained by the National Archives and is available on microfilm for purchase or loan. Field branches of the National Archives are located in Atlanta, Boston, Chicago, Denver, Fort Worth, Kansas City, Los Angeles, New York, Philadelphia, San Francisco, and Seattle. The *Check List* is also available in reference libraries of major universities.

- Brigitte T. Darnay, ed., *Directory of Special Libraries and Information Centers*, 8th ed. (Detroit: Gale Research Company, 1983).

Table 13.1 Information Types

Document	Likely Sources	Data Extracted
Aerial photograph	Air Photo Services City Engineering Department State Transp. Department City Archives State Library State Env. Agency Historical Society Utility Archives Small Conser. Services	Land Use River Forming Landfills Local Utilities Waste Storage Darkened Areas (Possibly Caused by Spills) Soil Disturbance
Land Ownership Maps	City Tax Assessor's Office City Archives Utility Archives State Archives	Land Use Land Ownership
Fire Insurance Maps (Sanborn Maps)	State Library County/Regional Library State Archives Historical Society	Plant Layout Land Use Combustion Sources Buried Utilities Water Course Locations Land Ownership
Town Atlases	City/County Library City Archives City Engineering Records	Land Use Buried Utilities Water Courses Morphology
Land Use Maps	Local/Regional Planning Agencies Local/Regional Develop-mental Agencies Historical Societies	Land Use
Utility Plans	City Engineering Dept. City Public Works Dept. Individual Utilities Information Service Archives	Buried Utilities Plant Layout
Topographic Maps	State Library Local/County Library State Geological Survey U.S. Soil Cons. Service State/Envir. Agency Historical Society U.S. Geological Survey	Land Use Water Courses Morphology Site Drainage Plant Layout Developmental Sequences
Court Records	County Courthouse State Appellate Court District Courts Lexis/Nexis	Nuisance suits relating to contamination pre-dating environmental law
Newspaper/Clippings/ Photos	City Library State Library Historial Society Newspaper Office	General Information on site incidences

Table 13-1 (cont.)

Financial & General	Dun & Bradstreet	Financial Condition Company Rating and Trends Net Assets Balance Sheet Information SEC 10K Major Financial Problems

This is a comprehensive guide to more than 16,000 information centers, archives, and special and research libraries in the United States. Libraries are grouped under five categories: 1) professional libraries in colleges and universities; 2) branches, divisions, or departments of large public library systems; 3) company libraries; 4) governmental libraries at local, state, and federal levels; and 5) libraries supported by nonprofit organizations and trade associations.

Information provided for each entry includes name of parent organization, name of library or information center, type of material in the collection (represented by key words), address, telephone, and hours of operation, staff contacts, volume of holdings, and available research services. The *Directory* consists of three volumes, the last of which is a periodic supplement describing new special libraries.

- Loretta F. Hefner, comp., *The WPA Historical Records Survey: A Guide to the Unpublished Inventories, Indexes, and Transcripts* (Chicago: The Society of American Archivists, 1980).

  This publication compliments the *Check List* (1943) by describing the locations and contents of unpublished HRS inventories, indexes and transcripts. The guide identifies repositories holding HRS materials arranged alphabetically by states or equivalent units (the District of Columbia, Northern and Southern California, and New York City were surveyed by the HRS as separate units). Entries include the name and address of each repository, descriptions of HRS materials, its inclusive dates, volume of holdings, and finding aids.

- National Historical Publications and Records Division, *Directory of Archives and Manuscript Repositories in the United States,* 2nd ed., (New York: Oryx Press, 1988).

  This work describes the locations, nature, and general contents of 4,225 repositories in the United States, District of Columbia, Virgin Islands, and Puerto Rico. Included in the *Directory* are college and university archives, local historical societies, and other institutional repositories. Individual entries contain the archive name, address, telephone number, hours of operation, volume of holdings, a brief description of collections, and available research services. The book is arranged alphabetically by state or other political unit, and by individual cities and institutions within cities. Preceding each state entry is a brief description of its program for records management. A repository index and subject index are included.

- *The County Year Book,* (Washington, D.C.: National Association of Counties and International City Management Association, annual).

  Appearing annually since 1975, this publication presents comprehensive information on most counties in the United States. Profile data and a directory of county officials (including the names of county clerks) are among its standard features. A reference section also provides sources of information on environment and energy affairs. The 1978 volume includes a comprehensive treatment of solid waste disposal practices in United States counties. Data practice, landfill operations, and hazardous waste management features.

HISTORIC LAND USE RESOURCES SUMMARY

The quantity of historical data available for a property will depend largely on the location of the property and both local and regional preservation efforts. Property impacts can result from industrial activities several hundred years ago. However, many will be the result of activities of the last hundred years.

There are two ways to view historical data: one is the source of the information; the other is the type of information. From a practitioner's standpoint, the type of information is more important. This allows the practitioner to use discretion and ingenuity in developing the actual information, adjusting for the regional differences. Table 13.1 gives a summary of the various types of information.

Establishing the history of the subject property and surrounding properties within reach of the available transportation mechanisms is the first step in a Phase I ERA. The land use history provides the investigator with the knowledge of what types of processes may have occurred. It must also give a prospective as to how development occurred and possibly identify areas or materials that need to be further studied. The data gathered through the historical review lies in the general topics of land use, plant history, industrial development, waste handling and disposal practices, existing land use, adjacent water body features and utility development. These are discussed below.

Land Use

Historical land use data provides the background of the site before, during and after the construction of the existing facilities. First, it shows the location and nature of the various structures, the changes over the years in the site topography and possibly pathways created for transport of contaminants. Second, historical land use data provides a time frame for the existence of various facilities. Third, old maps, air photos, etc., may indicate areas that were used or convenient for waste disposal. Fourth, the document review may indicate previous site landforms, such as draining ditches, ponds, or old river channels where contaminants may have collected.

The quantity of historical data available in most urban locations is staggering. The professional developing a Phase I ERA must not seek data for data's sake, but learn to filter the various informational leads as to what is important for that particular assessment. In doing so, the professional must recognize that nearly all operations have common potential contaminants such as oil, fuel, gasoline and sewage. On an individual basis, the professional must understand the specific raw materials and waste materials which were generated from each type of operation. If in legal compliance, current operations will have an inventory of this as a result of local, state and federal laws. However, at older facilities these control laws did not exist and the practices employed by individual operations varied from none to quite advanced.

Current Records

In the past decade or so, federal, state, and local regulations have resulted in the compilation of publicly available records and lists of increasingly detailed site-specific information for various manufacturers and products. Some, but not all the laws that must be dealt with are listed in Table 13.2.

One of the problems with the lists generated from these laws is that they are agency and departmental specific. There is generally no clearinghouse for all the information sources available. Thus, the professional must contact the individual agencies for available information.

Agencies which might be involved include:

1. USEPA
2. State Environmental Protection Agency
3. State Department of Health
4. State Department of Natural Resources
5. County Department of Public Health
6. County Environmental Agency
7. State Geological Survey
8. State Water Survey, etc.
9. State Department of Mines
10. Local Zoning Boards
12. State Fire Marshall (UST and LUST data)

Table 13.2 Laws Affecting Liability

Federal Water Pollution Control Act	Occupational Safety and Health Act
Solid and Hazardous Waste Amendments Act	All Similar state laws
Resource Conservation and Recovery Act	Clean Water Act
Comprehensive Environmental Response Compensation and Liability Act	Clean Water Act Amendments of 1987
Superfund Amendment and Reauthorization Act	Clean Air Act
	Safe Drinking Water Act
	Toxic Substance Control Act
	Federal Insecticide and Rodenticide Act
	Hazardous Materials Transportation Act

The most commonly available records are shown on Table 13.3.

Obviously, not every Phase I ERA will require that all of the available information be collected. However, diligence should be served in obtaining data, since a simple fact such as an underground mine and abandoned well or flood plain may cause financial harm and ultimately create a liability for the professional. The data collection process should be methodical and the information-gathering function arranged around a checklist that is generated for each project. Thus, if a data item is not applicable, then report that it is known that there is no underground mining in the area, at least the professional has addressed the potential problems.

In addition, agencies do not always have the information available. The professional needs to note this fact and the source. If agencies are contacted and the Phase I ERA report is due prior to the information being made available, this should be stated in the report with the qualification that if information has negative consequences on the site, it will be reported in an addendum to the client.

The background data will provide very useful information. For instance, a series of major construction projects, changes in topography, river channels, etc., may be identified by comparing a series of aerial photos or topographic

Table 13.3 Commmonly Available Records

Records	Source
POTW Discharge	
NPDES Discharge	USEPA and State Licensing Agency
SARA, Title II	
CERCLA	
TSCA	
Toxic Materials such as PCB's	
Underground Storage Tanks	
FIRM (Flood Insurance Maps)	Local and Regional Planning Agencies
Flood and Hazard Maps	U.S. or Regional Planning Agencies
Well Inventories	State Geological or Water Surveys
Underground Mining Activities	State Department of Mines and Minerals
Other Problems	County Health Department
Underground Storage Tanks	State Fire Marshal

maps. A series of air photographs dating back to earlier usage will show where facilities existed, how they were moved, and where potential problem materials were stored and disposed. In aerial photos, previous ponds, storage sheds, or storage areas can normally be identified. In addition, other remote sensing such as infrared pictures can show stressed vegetation when correlated to site conditions. The specific area identified may require further investigation.

The information contained in the initial data collection can result in having to alert the client that he/she may not want to continue the study. For example, if the site has been used to dispose hazardous materials or there are pending actions relating to the site.

Both on- and off-site existing wells can provide sampling points to evaluate the aquifer from which they draw. Wells can also provide conduit for contaminant flow deep beneath the ground surface into drinking water sources.

AN EXAMPLE: COMMUNITY ENVIRONMENTAL RISK INVENTORIES (CERI)

Hazardous materials present a number of problems with respect to the development of a community. In a recently completed hazardous materials risk inventory of a small city in Ohio with a development history extending back to the 1800s, it was found that there were more than 600 reasonably suspect or confirmed sites of hazardous materials ranging from old gas stations to bronze foundries, tin smith shops and industrial plating businesses.

This does not mean that each of these suspect sites contain significant health or environmental risks, simply that under the "due diligence"[238] requirements of the federal statutes these sites should be investigated prior to the transfer of ownership or as a part of prudent continuing ownership.

Most of the hazardous materials issues that will face any community today are a result of that community's economic development history. Until fairly recently, there was nothing unusual with the idea of simply replacing the leaking tank at the local gas station without removing hundreds of cubic yards of gasoline soaked soils, or of leaving the tanks in place if the station property was converted to other uses. The net result of the development process, however, is an inventory of potential problems that requires positive action in order to continue development.

A Community Environmental Risk Inventory (CERI)[239] is exactly what the title implies, an inventory of suspected or confirmed environmental risks. These risks will be initially identified through a careful search of available historic and public records. The beginning CERI would then be updated on a regular basis in the first few years with new information to provide a new CERI.

The initial CERI is a beginning inventory in the classic financial sense with two differences:

1. The inventory will be in terms of the suspected types of risks and a categorization of the likely severity of the risk a site may present, not in terms of the quantity of material present on the site.
2. The inventory will be presented in the form of two major components:
   a. A map of the suspected sites divided into four Risk Levels, the Risk Map, and
   b. A database of descriptive information on the risks; the Risk Database.

The CERI is developed with three major purposes in mind:

1. To provide an inventory of known or suspected environmental problems for area planning and management purposes.
2. To provide a uniform basis for environmental property assessments.
3. To provide realistic estimates of the actual problems after the initial CERI has been updated over a period of years with information from analytical investigations resulting from the normal property transaction processes.

The analysis of property for environmental risk is generally required for all properties without regard to the size or value of the transaction that may be involved. In the governing laws, there is little to distinguish between residential, commercial or industrial "due diligence" requirements, but from a practical point of view Phase I ERAs are usually applied to commercial and industrial transactions. This does not mean that residential or small commercial or industrial properties cannot suffer proportionate losses, rather that the cost of performing environmental audits without the existence of a CERI for the community are so high as to be unjustified in any realistic sense.

It is estimated that the public lists of environmental hazards such as the National Priorities List (NPL, the Superfund site listing) contain less than 20% of all of the existing environmental problem sites; they are lists of discovered sites. When completed according to the federal requirements, the lists of underground storage tanks will contain less than 50% of all existing underground storage tanks and very few of the sites of tanks that ceased to function prior to 1980. The sites of municipal waste disposal, industrial activities long since out of business, and others are documented in the historic records, but have not been analyzed or organized from an environmental risk point of view.

This lack of completeness of existing records has resulted in the failure of most Phase I ERAs to detect the presence of substantial risks, a lengthening of the Phase I ERA process from what should be a several weeks-long process to one extending to three or four months, and generally several thousands of dollars in additional costs per audit.

The Risk Map is the most visible part of the CERI and consists of transparent color-coded overlays to standard topographic or street maps. The overlays are divided into several Risk Levels designed to allow the user to quickly determine the presence and location of environmental risks identified from the historical data and the relative probability of an environmental problem.

The Risk Database may be examined by Risk Map key number, street address, risk type, or Risk Level assignment. The database will contain information on the available details of the risk such as the name of the organization suspected of originating the risk, the available information on the types of risk that may have resulted from the activities conducted on the site, and the types of risks that should specifically be analyzed in further environmental audits.

The updating of the CERI on a regular basis is critical to the overall value of the information. Updating will be in the form of applying newly developed information resulting from updating NPL lists or the results of specific Phase I, II, or III investigations of properties for example, in a manner similar to the development of annual running inventories.

The combination of the Risk Map and the Risk Database should supplant the need for the general area historic analysis as a part of the Phase I ERA or "due diligence" audit, allowing those audits to concentrate on the subject and immediately adjacent properties in their current condition.

INITIAL DATA DEVELOPMENT

In general, four types of information are required for the development of a CERI:

1. Governmental lists and records such as: geologic surveys; Superfund priorities lists; Department of Natural Resources mineral development data, and; local laws, rules and regulations pertaining to zoning, weights and measures, and environmental control. Included in this category are census data, aerial and geophysical satellite and other remote-sensing data, railroad maps, and industrial production data.
2. Historical records such as newspapers, local histories, local commercial listings such as the R.L. Polk series, telephone directories, and other sources.
3. Historical data on industrial processes, especially in terms of the raw and waste material streams likely to result from the processes during the time period in question.
4. Scientific data on the degradation of raw material and waste stream components under the conditions likely to have been encountered in the local environment.

The information from the first three primary sources are utilized to develop an initial CERI. At this point the CERI will contain a large number of possible risk sources that will require further evaluation.

INITIAL CERI ANALYSIS

The purpose of the analysis of the initial CERI is to develop a list of issues to be resolved prior to the assignment of Risk Levels. The CERI at this point will often contain listings for old gas station sites, for example, that may have been subsequently removed and cleaned up to where they are no longer a significant problem. Before assignment of a Risk Level to such a site, it must be examined for just such a possibility.

Additional issues to be resolved based on the initial analyses may include obtaining information on a particular industrial process at a specific period of time. A major area of concern will be the effect of time on specific substances, that is, have those substances had sufficient time to be dissolved or altered into non-hazardous substances or dispersed to an extent such that they are no longer hazardous?

The purposes of this phase of the analysis are as follows:

1. To identify all possible sites of suspected environmental risk.
2. To determine the most feasible means for resolving the suspicions.
3. To identify the waste and raw materials streams for specific industries not previously encountered and the likely pattern of distribution for any hazardous components of those streams.
4. To prepare a set of specific questions to be resolved on site.

ON-SITE EXAMINATION

Many of the issues identified above can only be answered by a physical examination of the site and possibly the construction and use records for that site. Included in this phase are interviews with current or past facilities personnel. Analytical testing of the suspect sites is not included in the development of the CERI work.

It is critical to recognize how a lack of information will be dealt with in the face of a reasonable suspicion. Actions and decisions are governed by the negative default principle.

In view of this underlying principle, an analysis strategy must be adopted for determining the likely level of risk associated with a given property that results in a conservative opinion—an opinion that favors a higher risk assignment as opposed to a lower. In the final Risk Maps and Risk Database the result will be a tendency to overstate rather than understate the likely risks, an issue that can only be resolved over time with additional information appended through the updating process.

The purposes of this phase of the analysis will be:

1. To identify changes in the physical condition of a property that may have resulted in the elimination of the risk or the reduction of the risk to a lower Risk Level.
2. To identify discrepancies in the recorded history and to resolve those discrepancies.
3. To develop a final database of historic data that is as accurate as possible given the records and the current physical condition of suspected sites.

FINAL RISK MAP DEVELOPMENT

The final Risk Map and Risk Database development is a two-step process involving the analysis of all of the remaining suspected or confirmed risk sites. The process of eliminating discrepancies is reasonably obvious, but the heart of the process, the assignment of Risk Levels, is not.

The assignment of Risk Levels to suspected sites is critical to the value of developing the inventory. Without some way to judge the level of concern, it is very difficult to determine the appropriate allocation of resources for a given transaction or planned action. One of four Risk Levels is assigned to a given situation as follows:

- *Risk Level 1:* Represents the ordinary risks associated with any given property in an area. It represents the "background," which may differ from region to region. For example, a Risk Level 1 area in the middle of the farmlands of Ohio may include the presence of low levels of pesticides, fungicides, and fertilizers as being normal and expected. In the Lake Calumet district in the Chicago-Gary, Indiana area a Risk Level 1 may indicate the presence of a fairly high level of heavy metal contamination possibly sufficient to make the area unsuitable for human habitation although not unsuitable for industrial or warehouse facilities.
- *Risk Level 2:* Represents the likelihood of levels of hazardous substances significantly above background and that a Phase I ERA of a specific property should include a more careful than normal investigation of both the subject property and adjacent properties for visible signs of problems such as underground storage tank vent piping, discolored soils or grasses, lead paint or asbestos, or filled areas that may contain wastes.
- *Risk Level 3:* Indicates an area where the suspicion of the possible presence of a problem requires at least some Phase II analytical inves-

tigation. Such investigation may take the form of soil sampling for the presence of explosive gases or heavy metals; physical examination of fills for the presence of drums; or the testing of paints or coatings for lead and asbestos.

- *Risk Level 4:* Indicates areas where the level of suspicion is so high, or confirmation by physical testing has been obtained, as to provide a virtual certainty of problems. Any Risk Level 4 area should undergo full Phase II and possibly Phase III testing and evaluation.

The assignment of a Risk Level is a complex process involving the application of a number of both mundane and sophisticated decision techniques. In all cases except Risk Level 4, primary reliance is placed on the judgment of one or more experts based on the assembled historical evidence. As with any system that relies on background data and expert judgment rather than primary analytical evidence, there is a realistic possibility that any given Risk Level assignment may be in error.

Assignment of Risk Levels

The assignment of risk levels is accomplished through a process equivalent to the ERIS development process, utilizing the following translation table of Comprehensive Risk Score (CRS) to Risk Level.

CRS Score	Risk Level
0 through 10	1
11 through 50	2
51 through 89	3
90+	4

THE COMPLETED CERI

The completed CERI will consist of a series of Risk Maps and a Risk Database. The Risk Maps will consist of color-coded transparent overlays for each of the Risk Levels identified for the particular map. The overlays will have Risk Database key numbers in the approximate location of the identified risk that will direct the user to specific entries in the Risk Database where further textual information will be located.

The Risk Database will contain the following information.

Risk Number

The Risk Number is the identification number on the Risk Map that will key the user into the specific entry in the Risk Database. If multiple risk sources exist for a specific property, all of the risks will be identified under a single Risk Number.

Risk Level

The Risk Level assignment for the specific risk. In the case of multiple risks for a given property, the Risk Level assignment on the Risk Map will be the highest Risk Level for all of the risks at that address, but each of the major risk sources at that location will be shown in the Risk Database. This should allow the user to evaluate the specific site and determine what actions may be appropriate to lower the Risk Level for that site.

Street Address

The current Street Address for the specific site of concern. The reason for specifying the current Street Address is that street addresses tend to change over time. In addition, certain types of risks such as leaking underground storage tanks may affect the risks on a number of adjacent properties that cannot be evaluated without Phase II or Phase III analytical techniques. The location of the likely source of the risk will be shown, only showing the actual risk dispersion pattern when such information is readily available in the records or as a result of updating the CERI with later Phase I, II, or III information.

Organization Names and Dates of Existence

A specific site may be occupied over time by a number of different organizations having differing risk characteristics. To the extent that the organizations and their dates of occupancy can be identified, they will be listed including the known or estimated beginning and ending dates of occupancy.

Types of Business Activity

A description based on the records of the Type of Business Activity conducted on the site that generated the suspicion of the existence of a risk.

Risk Products

Based on the type of activity in the time period(s) over which the organization functioned at a specific site, a listing of the likely risk-inducing products of the activity will be provided. This list should be utilized by any later Phase I, II or III audits to determine if the risk-inducing materials or conditions are still present and, in certain cases, if those risk-inducing materials or conditions have influenced adjacent properties.

Current Risk Status

A set of comments concerning the Current Risk Status of the site. This may include the results of the available Phase I, II, or III audits, comments on later activities on the site that may have acted to increase or decrease the risk, or the results of specific remediation efforts carried out on the site. Included in this section of the database will be indicators as to the likelihood of the materials or conditions having affected adjacent sites, the type of effect, and the level of concern over such effects.

CERI UPDATING PROCESS

As should be evident from the preceding section, the process of updating the CERI will involve the use of information from any Phase I, II, or III or "due diligence" investigations conducted in the future, as well as the results of any actions that may have tended to increase or decrease the risks associated with a particular site. Without the application of such information on a regular basis to update the CERI, the value of the entire process will be compromised over time resulting in the incorrect assessment of properties. No database of information for an entire region can be absolutely correct and comprehensive, but the accuracy of the CERI can be substantially improved, particularly now that awareness of the issues surrounding hazardous materials is increasing.

The updating process should include the agreement on the part of all the lenders and as many of the major property owners and developers in the area as possible to automatically provide a copy of any Phase I, II, or III reports that come into their possession. In addition, regular updating information from the various public sources of data such as the Department of the Environment, the USEPA, and the like will be added to the CERI. A sample risk level description for Findlay, Ohio is given in Table 13.4 with a portion of the risk database shown in Table 13.5.

Example CERI

Table 13.4 Risk Level Definition for the Findlay, Ohio Area

Radon:

No significant radon problems have yet been published.

Lead:

Numerous structures were built in the 1880–1940 period and survive to the present time. All of the structures are likely to contain lead-based paints, particularly on exterior surfaces. In the older residential sections of town, a large amount of restoration work has taken place that involved stripping paints which may have resulted in residual lead soil contamination, particularly near the foundations. In addition, lead gas pipe may be found in the walls of structures from the 1880–1920 period, and there is strong evidence of the use of lead water pipes in the municipal systems during this same time period with no corresponding evidence of removal.

Asbestos:

The extensive use of asbestos in the plaster of the more ornate residences, commercial and public buildings is a virtual certainty. The use of asbestos insulation on hot water heating systems is common, particularly if the system has not been updated since the late 1970s.

Oil and Gas Wells:

Numerous abandoned oil and gas wells may be found throughout the city and surrounding areas with abandoned pipelines lying 2′ to 4′ below the surface. In most cases these do not present an immediate environmental threat as most of the casings have filled in over the years and are now sealed shut, but some exceptions do exist and new construction can open otherwise closed casings. The closing of the casing should be examined from that point of view whenever encountered.

Oil Storage and Transport Facilities:

There are at least three major sites of crude and refined bulk storage facilities that were active to the late 1950s and have been since removed. In at least one case, a housing subdivision and country club has been constructed on the site of a previous tank farm. No information is available as to the existence of residual hydrocarbons in that area. In addition, numerous petroleum product transportation lines run through the area and little historical information with respect to leaks has been found.

Note: In the interest of clarity for this example, only selected Risk Level 3 and all Risk Level 4 have been plotted or printed out. This map and the accompanying tables are not to be used for the purposes of formulating a Phase I ERA or due diligence audit.

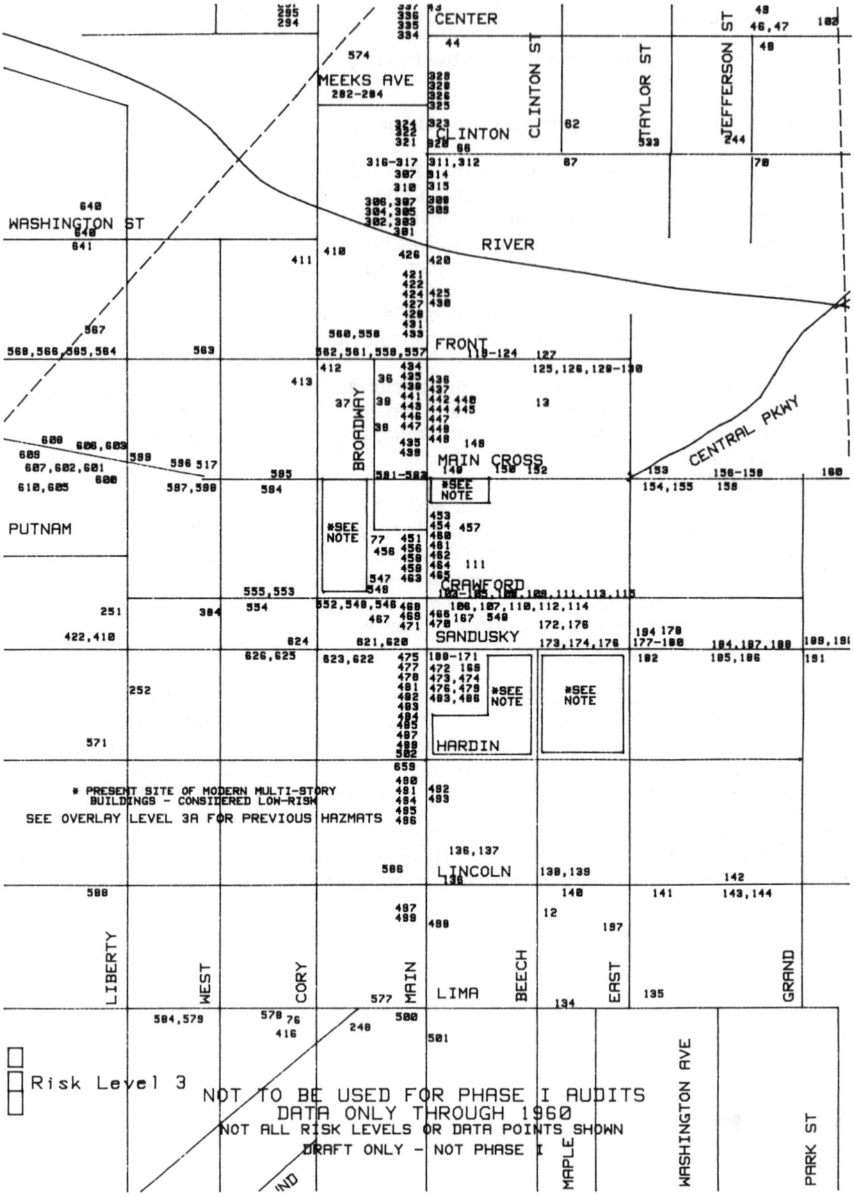

Figure 13.1 Risk Map for Downtown Findlay, Ohio.

Table 13.5 Portion of Risk Data Base for Findlay, Ohio Area

Map #: 13 Risk Level: 3 Stewart Machine Works 213 Beech Ave. 1929–approx. 1950 Machinists Comments: Heavy Metals, Solvents, PCBs
Map #: 38 Risk Level: 3 B & B Tire Service 219 Broadway 1929–approx. 1940 Gasoline and Oil Service Stations, Vulcanizers Comments: UST, Solvents, Waste Oil, Lead, Hydrocarbons
Map #: 122 Risk Level: 3 Gates, NS 118 E Front 1892–approx. 1920 Galvanized Iron Works, Roofers Comments: Heavy Metals, Solvents, Asbestos
Map #: 122 Risk Level: 3 Spotless Dry Cleaning 118 E Front 1929–1950 Cleaners and Dyers Comments: Solvents, Dye Stuffs
Map #: 157 Risk Level: 3 Union Brass Foundry 444 E Main Cross 1889–1892 Natural Gas & Water Works Supplies, Nickel and Silver Plating, Pattern Makers, Gas Mixers, Foundries Comments: Solvents, Waste Oil, Hydrocarbons, Heavy Metals
Map #: 588 Risk Level: 4 Flater H and Son 405 W Lincoln 1950–approx. 1970 Sprayers and Dusters (Confirmed DDT, Arsenate of Lead, Dry Lime, and Sulphur), Sprayers and Dusting Equipment, Insect Powders—Manufacturers of insecticides Comments: Heavy Metals, Chemicals, Insecticides, Pesticides, Rodenticides

REFERENCES

212. The major part of the balance of this chapter is based on materials supplied by Mr. W. Lee Minnerly, MA, Research Director at the American Institute of Real Estate Appraisers of the National Association of Realtors, Chicago, IL.
213. *The Dictionary of Real Estate Appraisal;* American Institute of Real Estate Appraisers, Chicago; 1984; p. 177.
214. *See* Derbes, Jr., Max J., "Highest and Best Use — What Is It?" *The Appraisal Journal;* April, 1981, pp. 166–178
215. Craig E. Colten; "Industrial Wastes in the Calumet Area, 1869–1970: An Historical Geography;" ISM Contract 175, ENR Contract HW-16; State Water Survey Division; Illinois Department of Energy and Natural Resources; September 1985; p. 6.
216. Ibid.
217. In all states except Louisiana, counties represent the largest territorial divisions of local government; in Louisiana the comparable unit is the parish. Some large cities in select states have county functions but are considered distinct political subdivisions (e.g., San Francisco, Denver, Houston, Honolulu, and St. Louis).
218. For descriptions of census data cf. *Census Catalog Guide 1988;* U.S. Department of Commerce; Washington; 1988; *Bureau of the Census Catalog of Publications 1790–1972;* U.S. Department of Commerce; Washington; 1974; and *Bureau of the Census Catalog of Publications 1946–1972;* U.S. Department of Commerce; Washington; 1974.
219. Barnett, LeRoy; "State Censuses of Michigan: A Tragedy of Lost Treasures" *Family Trails;* Vol. 6, No. 1; published by Michigan Department of Education, State Library Services; Summer-Fall 1978.
220. "National Priorities List, Supplementary Lists and Supporting Materials, June, 1988"; U.S. Environmental Protection Agency; Office of Emergency and Remedial Response; p. S-4.
221. Ibid., p. S-5.
222. Ibid., p. S-6
223. Ibid.
224. For an overview see Rifkin, Bernard M.; "Title Searchers Alert to Environmental Rules"; *Real Estate Forum;* January 1988.
225. Schmeckbier, Lawrence E., and Eastin, Roy B.; *Government Publications and Their Use;* 2nd ed.; The Brookings Institution; Washington; 1969; pp. 438–439.
226. Ibid., p. 423.
227. Wright, Helena; "Insurance Mapping and Industrial Archaeology" Paper presented at annual meeting of the Society for Industrial Archaeology, Hartford, CT, May 1981. *See also* Wright, Helena; "Insurance Mapping and Industrial Archeology," *IA: The Journal of the Society for Industrial Archeology,* Vol. 9, 1983; pp. 1–18, and Ristow, Walter W.; "United States Fire Insurance and Underwriters Maps, 1852–1968," *The Quarterly Journal of the Library of Congress,* Vol. 25, No. 3; July 1968; pp. 195–218.

228. Information on the availability of Sanborn maps is found in *Fire Insurance Maps of North American Cities and Towns Produced by the Sanborn Map Company,* Washington, DC: Library of Congress, 1981; *Union List of Sanborn Fire Insurance Maps Held by Institutions in the United States and Canada,* 2 vols.; Santa Cruz, CA: Western Association of Map Libraries, 1976, 1977; and from the Sanborn Map Company, Inc., 629 Fifth Avenue, Pelham, NY 10803.
229. Wright, ''Insurance Mapping . . .'', p. 7–8.
230. Topographic map indexes and catalogs are available from U.S. Geological Survey Map Sales, Box 25286, Federal Center, Denver, CO 80225 (303/236–7477).
231. For additional information about birdseye view and panoramic maps *see* Reps, John W.; *View and Viewmakers of Urban America: Lithographs of Towns and Cities in the United States and Canada, Notes of the Artists and Publishers, and a Union Catalog of Their Work, 1825–1925;* Columbia: University of Missouri Press; 1984; and Hebert, John R., comp., *Panoramic Maps of Cities in the United States and Canada,* 2nd ed.; Washington, DC: Library of Congress; 1984.
232. Two excellent sources on historical aspects of photography are Weinstein, Robert A. and Booth, Larry; *Collection, Use and Carae of Historical Photographs;* Nashville: American Association for State and Local History; 1977; and Pilling, Arnold; ''Dating Early Photographs by Card Mounts and Other External Evidence: Tentative Suggestions''; *Image,* Vol. 17, No. 1, March 1974.
233. The National Cartographic Information Center (NCIC) of the U.S. Geological Survey developed the Aerial Photography Summary Record Systems (APSRS) to annually catalog the extent of U.S. Aerial photography by regions. Information on APSRS publications is available from the U.S. Geological Survey, 507 National Center, Reston, VA 22902. For information about the EROS Data Center, contact the U.S. Geological Survey, Sioux Falls, SD 57198 (605/495–6151).
234. A large number of gazetteers are referenced in the *Bibliography of Place-Name Literature, United States and Canada* American Library Association, 2nd ed., Chicago; 1967.
235. *See* Anderson, Elizabeth L.; *Newspaper Libraries in the U.S. and Canada;* Special Libraries Association; New York; 1980; for locations of newspaper resources.
236. Garraghan, Gilbert J.; *A Guide to Historical Method;* Fordham University Press; New York; 1946; pp. 174–177.
237. Ibid., 177–182.
238. The phrase ''due diligence'' in the environmental context appears frequently in a number of federal, state and local laws and court cases, most notably the *Superfund Amendment and Reauthorization Act;* Public Law 99-499, 100 Statutes 1613; October 17, 1986.
239. Community Environmental Risk Inventory, Risk Mapping, Risk Map, and Risk Database(s) are registered trademarks of A.R. Wilson and Associates.

14 SUBSTANCE IDENTIFICATION: SAMPLING THEORY AND TECHNIQUES

INTRODUCTION

Before examining the Phase II and Phase III requirements, it is necessary to look at sampling techniques and the identification process for hazardous materials. Without a basic understanding of this process, Phase II and Phase III work cannot be accomplished in a reasonable manner.

GENERAL STATISTICAL SAMPLING THEORY — TWO DIMENSIONAL

Intuitively, we recognize that only the calculations made on a piece of paper are precise; the real world is never exact. Buildings are not built to precise standards, cars are not all exactly alike, materials are never absolutely uniform and, records are never filled out in exactly the same manner. From the point of view of the inspector, the problem is one of how to evaluate measurements or observations in order to reach some conclusion. Technically, drawing inferences from imprecise information is the realm of statistical sampling theory and it has three important applications in inspection.

1. The evaluation of a material to determine its characteristics.
2. The evaluation of the functioning of a positive control program in controlling the inventories of hazardous substances.
3. The evaluation of the recordkeeping process to determine its accuracy in recording the events surrounding the hazardous materials.

To deal with these issues, a common set of concepts and language must be developed and applied and without an understanding of these concepts and language the inspector is extremely vulnerable to the consequences resulting from errors, omissions and incorrect conclusions.

This manual is not intended to be a textbook on statistics or statistical quality control, a number of such texts are available[243,244,245] and should be consulted by the serious professional. The intention here is to lay out a practical and highly simplified introduction.

Randomness

The first major concept is that of randomness. If we were to measure the length of six rooms that are indicated on the blueprints as being 20 feet long, we expect the room to be approximately 20 feet in length, plus or minus some amount. Since there is no way of determining the exact length of the room prior to measuring it, we can say that the *exact* length is a random value within some limits.

This same type of randomness exists in many forms and must be dealt with regularly by the inspector in order to evaluate the information gathered from an inspection. Without statistical sampling theory, it would be necessary to examine every document, measure every room and sample 100% of all suspect materials. Using statistical sampling theory, it is possible to devise methods through which we could measure only a small number of the rooms and obtain the information we need.

Homogeneity

The second, but most important concept needed in order to measure a small number of the rooms is that the rooms are homogeneous. In this case, homogeneity lies in the fact that all of the rooms were designed to be 20 feet in length. At least with respect to this one dimension, we have reason to believe that one room is exactly like the next. That is, the rooms have a homogeneous, or common, designed length. Without being able to assume that the phenomena we want to measure is homogeneous, we cannot apply normal statistical sampling mathematics.

Homogeneity can be found in many diverse situations. For example, a pile of bills may be considered homogeneous in that they all show to whom money is owed, some description of what the money is owed for, and the amount due. An area of spray-on fire proofing that is similar in color, texture and date of application may be considered homogeneous, at least until tests are performed.

Random Selection

The next concept is the idea of random selection. Looking again at our problem of measuring only a small number of the room lengths in order to save time and obtain the needed information, the question we are going to need an answer to is, which rooms do we measure? Suppose the rooms all lie along one side of a hallway and are numbered 1 through 6. If the room lengths are assumed to be homogeneous then from a statistical sampling point of view, it makes no difference which rooms are chosen for actual measurement so long as they are not chosen according to a pattern. The choice of rooms must be by pure chance.

This suggests that one method of selecting the rooms is to throw one of a pair of dice, and the number that shows on the top is the number of the room to be measured. If we need to measure more than one room, we keep throwing until we have the required number of different room numbers, and go measure them. This is the method by which random selection takes place, whether we use dice or a table of random numbers or a so-called random number generator in a pocket calculator or personal computer.

It is certainly possible that the contractor built Room 3 four inches longer than specified and that to compensate he built Rooms 5 and 6 two inches shorter in order to come out with the right overall dimensions at the end of the hall. However, since we do not know this in advance, it will be up to the sampling process to determine this possibility.

Confidence Intervals and Confidence Levels

The next question we need to answer is: How many rooms should we measure? This question has to be answered by answering the question of: How much confidence do we need that the mean value of the measurements we have taken accurately represent the lengths of the rooms? For example, if we measured one room, and that was Room 3, how confident would we feel that the length of Room 3, 20′4″, accurately represents the length of the other five rooms? Consider Figure 14.1 that shows the exact room length of all six rooms, and Table 14.1 that mathematically summarizes the information.

The Average (or Mean) Length is simply the sum of all of the lengths divided by the number of measurements. The Difference column is the actual length minus the average length, and the Difference^2 (Difference squared) column is the difference amount squared. If we sum the square of the differences and divide the result by the number of measurements we get the Population Variance, 4 inches^2, and if we take the square root of the Population Variance, we get the Standard Deviation, 2″.

The mean value is a calculated number that tells us the value that all

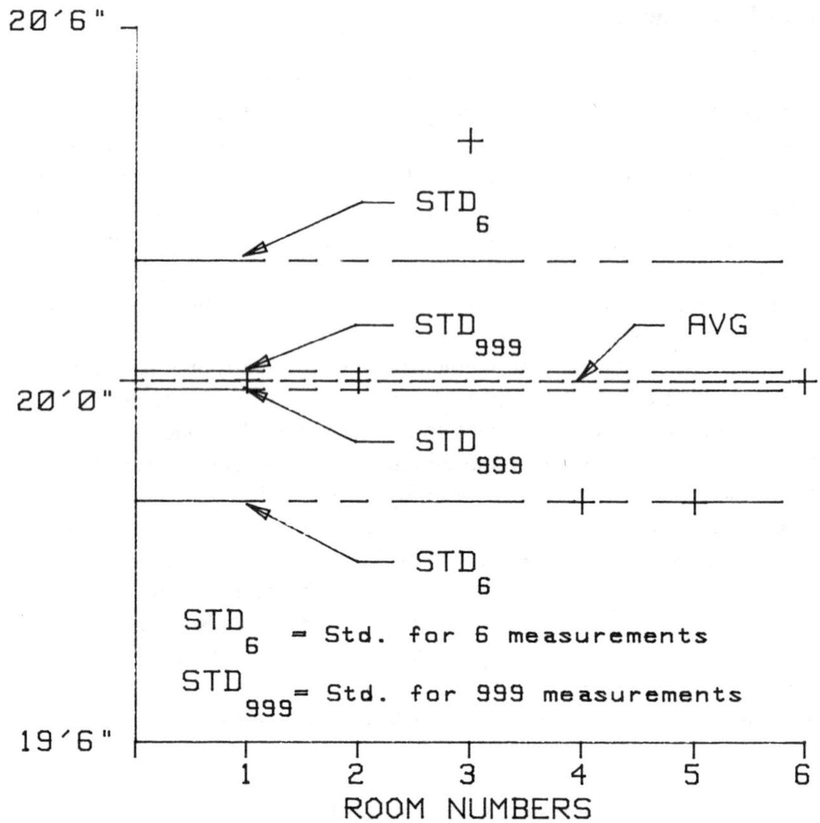

Figure 14.1 Actual Room Lengths.

of the actual measurements tend to center around. This is the number that can be said to represent the actual room lengths although a better measure of central tendency is usually the median value which is defined as the value such that one-half of the actual values are less than the median, and one-half are greater than the median value. The population variance tells us approximately how far on either side of the average value the actual measurements are likely to be. In a crude sense, but accurately enough for our purposes now, the standard deviation tells us that any randomly selected room length measurement has approximately a 66% probability of being within plus or minus 2 inches of the average room length.

It is the standard deviation that tells us how confident we can be in our use of the average length as an accurate representation of the actual room lengths. Increasing the number of measurements increases the level of confidence in the average as a truly representative and accurate measurement.

Table 14.1 Mathematical Summary of Measurements

Room Number	Actual Length	Average Length	Difference	Difference²
1	20′0″	20′0″	0	0
2	20′0″	20′0″	0	0
3	20′4″	20′0″	4″	16
4	19′10″	20′0″	−2″	4
5	19′10″	20′0″	−2″	4
6	20′0″	20′0″	0″	0
Totals	120′0″		0	24

Average Length	20′0″
Population Variance	4 inches^2
Standard Deviation	2″

To see this, let us take it to an extreme. Suppose that there were a thousand rooms and that we measured 999 of them and all except Rooms 3, 4 and 5 were 20 feet in length. What can we say about the last, unmeasured room? The average will remain relatively unchanged, the Population Variance will now be 0.02, and the Standard Deviation will be 0.14″. In other words, we can only say that the last room's length has a 66% probability of lying between 19′11.86″ and 20′0.14″. We can never say, with 100% certainty except by actual measurement, that the last room's length will be 20 feet, even though all except three of the other 999 rooms were measured to be 20'. What we can say is that within a confidence interval of 0.14″, the last room will measure 20″ with a confidence level of 66%.

The important point about these ideas is that the decrease of the size of the range within which we can expect 66% of the measurements to fall is not proportionate to the number of measurements, the size of the range decreases more slowly than the increased number of measurements. A careful study will also show that the converse is true, if we are willing to accept less accuracy, the probability that the last room is within say plus or minus 6″ of 20′, the probability of being correct may increase to 99+% (although it will never be 100%).

This is the formal basis for the Myth of Zero Existence. It is also the basis for the idea that random sampling of a homogeneous parameter will provide a meaningful result that can be used to draw conclusions.

Depending on the confidence interval we are willing to accept for the accuracy of our measurements and the confidence level we are willing to accept that our conclusion will be valid for all of the unmeasured phenomena, we can determine how many samples to take.

The design of a sampling protocol is a highly complex problem, but one that is not generally of concern to a regular inspector. The regulations, or the evidentiary requirements of the particular case, will often specify the sampling protocol and the inspector will only need to apply that protocol correctly. If there is a need to design a sampling protocol, it will be best to employ an expert to do so and the real world situations that the inspector is likely to encounter are far more complex than this simple example. Please note that the sampling protocol can be designed to provide almost any level of confidence over almost any confidence interval required, but the higher the level of confidence and the narrower the confidence interval, more samples will need to be taken and consequently the higher the cost for doing the work.

APPLICATIONS TO THE AUDITING OF RECORDS

The ideas of homogeneity, random sampling, confidence intervals and confidence levels are as applicable to the analysis of the accuracy, content and completeness of written records as to the length of rooms. With records, the concerns will be that the records are for the same purpose and are supposed to record the same information in order to satisfy our homogeneity requirement. The actual selection may be done by numbering the forms in a stack and selecting the specified form numbers from a table of random numbers.

In dealing with records sampling, the inspector may be interested in some property of the form itself; for example that the laboratory reports were all signed by the analyst performing the analysis. Generally the inspector will be required to examine the forms from a different point of view, however. For example, to determine if the sketch of the homogeneous area is an accurate description of that area, or that the recordkeeping process itself has kept accurate track of the inventory of hazardous substance and all changes in it. The ideas of random sampling still apply and in the same way, although variations in technique to improve the efficiency and/or accuracy of the inspection may be utilized.

The inspector should be aware that Certified Public Accounts have well established and court/time-tested techniques for performing audits of records and physical inventories and rather than re-invent the wheel, a CPA should be consulted in designing a records and/or physical inventory audit program.

APPLICATIONS TO THE INSPECTION OF HAZARDOUS MATERIALS

As we noted earlier, the inspector under AHERA is required to locate and identify all suspect homogeneous areas of asbestos-containing building materials. Note that the areas to be located and identified are suspect homogeneous areas. Until the materials have been tested or otherwise positively identified, they cannot be positively said to be asbestos containing or not.

There are two major objectives to sampling and testing of suspect materials:

1. To prove that the materials ARE NOT hazardous substance containing, and
2. To identify the substance if one is present.

The first step is to identify the homogeneous area. We will use the phrase ''homogeneous area'', but the concepts are equally applicable to a volume of liquid or gas, or a collection of items such as pipe covering or insulated valve covering materials. These methods are not directly applicable to volumes of solids, such as land.

> Keep carefully in mind that, unless other circumstances require sampling, sampling is not necessary at any time under the negative default principle. That is, we may always assume that a suspect material contains a hazardous substance. Only if we wish to prove that the suspect area does not contain a hazardous substance do we need a full sampling protocol and if a single sample comes back positive (i.e., the substance is present), the entire homogeneous area is confirmed as a hazardous substance-containing area without regard to the number of negative sample analysis results.

Homogeneous Areas

Homogeneity may be established through a number of mechanisms, but the decision as to the size and extent of a homogeneous area always rests with the inspector. The inspector must therefore be prepared to demonstrate that the criteria used to define the homogeneous area are valid.

There are four ''tests'' for homogeneity of an area or material grouping, as applied in AHERA.

1. That the material is uniform in color and texture throughout. Color in this instance refers to the basic material color, not the color of a surface coating such as paint. If there is reason to suspect the surface coating (say of containing lead), it must be treated as a separate homogeneous area.
2. That the material was applied or placed at approximately the same time.
3. That there is no reason to suspect the existence of ''pockets'' of differing materials within the homogeneous area, or if such exists, they are isolated from the statistical sampling scheme employed for the homogeneous area.[246]
4. The final test is that the samples taken show a uniformity of results. If this is not the case, then the area is not homogeneous and must be redefined.

STATISTICAL SAMPLING SCHEMES DEFINED BY LAW (AHERA)[247]

AHERA requires that all suspect materials be placed into one of three general groups:

1. Surfacing Materials: Materials that have been applied by spraying or toweling them on.
2. Thermal Systems Insulation (TSI): Materials used to insulate a thermal system.
3. Miscellaneous Materials: All suspect materials that are not surfacing or thermal systems materials.

SURFACING MATERIALS

Preparation of Diagrams

For each sampling area, prepare a diagram approximately to scale showing the extent of the homogeneous area. A to-scale floor plan can always be used for this purpose. An example of a diagram is shown in Figure 14.2.

The diagram should include the following information:

- An identification number;
- Brief description of the homogeneous area;
- Area dimensions and scale;
- The size of the area (square feet, cubic feet, length or number of items); and
- The date prepared.

Number of Samples

Under AHERA, nine samples per homogeneous sampling area are recommended (see Table 14.2). With nine samples, the likelihood of detecting asbestos when it is present is very high. Cost or other constraints may limit the number of samples that can be collected. If nine samples cannot be collected, use the following table to determine the minimum number as required by the AHERA Rule. This number depends on the size of the sampling area. In a very small (much less than 1,000 square feet) sampling area such as a closet, take three samples rather than the full nine.

Table 14.2 AHERA Sampling Guidelines

Size of the Sampling Area	Recommended Number of Samples to be Collected	Minimum Number of Samples to be Collected
Less than 1,000 sq.ft	9	3
Between 1,000 & 5,000 sq.ft.	9	5
Greater than 5,000 sq.ft	9	7

REPORT TITLE: ________________ REPORT PAGE NO.: ____

HOMOGENEOUS AREA SKETCH

DATE: ____ ____, ______

DOCUMENT NUMBER

Building Number and Name

Sample Area/Lot Number and Name

Drawing/Sketch Number and Name

PROVIDE THE SIZE OF THE AREA (SQ. FT., LN. FT. OR # OF ITEMS

DESCRIBE HOMOGENEOUS MATERIAL

SCALE: 1" = ____

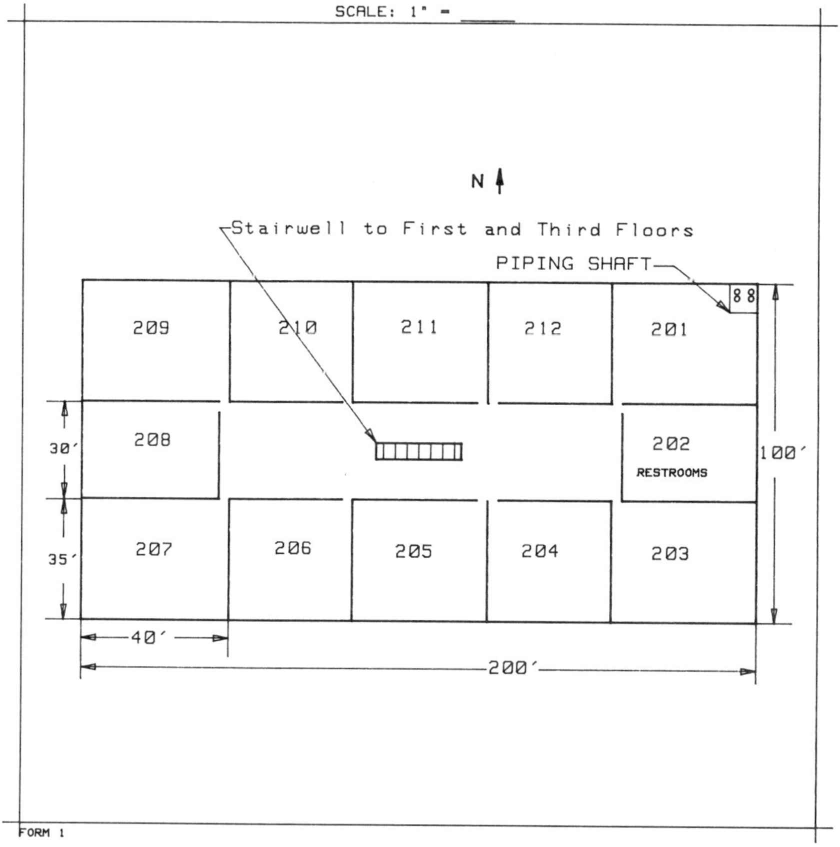

FORM 1

Figure 14.2 Example Floor Plan for Sketching Homogeneous Areas.

Selection of Sample Locations

In this sampling scheme, sample location are selected so that they are representative of the sampling area. When nine samples are collected, they are distributed evenly throughout the sampling area. If fewer than nine samples are collected, a random sampling scheme is used to determine their location. Choosing sample locations according to personal judgment produces samples which may not be representative and can lead to a wrong decision about the presence or absence of asbestos. The sampling scheme described here avoids this problem and controls the frequency of mistakes.

Divide the homogeneous area into nine equally sized subareas numbered from 1 through 9. This is done by dividing the length and breadth of the homogeneous area into three equal lengths and drawing a grid over the diagram (see Figure 14.3). This can be done carefully by eye. Exact measurements are not needed.

If the homogeneous area does not easily fit into a rectangular shape, parts of the grid might not be in the area. This is not a problem in most cases. If, however, a large part of the grid falls outside the homogeneous area (for example, if the homogeneous area is L-shaped), it is advisable to divide the homogeneous area into two or more separate sampling areas, each of which is approximately rectangular, and select sample locations by applying the sampling scheme to each area separately.

For greatest coverage, one sample from each of the nine regions should be collected. If fewer samples are to be collected, Table 14.3 may be used to obtain the number of the subarea to be sampled. The sample data should be recorded on some device such as Figure 14.4.

Table 14.3 is used as follows. Suppose that the last time you used the table you used sub-table 14.3.6. The next sub-table to use will then be Table 14.3.7 and it says that for three samples, take them from subareas 5, 8, and 1.

Take samples from approximately the center of a subarea or as close as possible to the center if accessibility, presence of light fixtures, etc., make the center location impractical. If a subarea is specified that falls entirely outside the sampling area, use the next specified subarea instead. For example, if subarea 3 falls outside the sampling area, take the third sample from subarea 4 (the next subarea number available in Table 14.3.7).

For very irregular-shaped areas, the homogeneous area may be divided into nine subareas of approximately equal size that do not necessarily form a rectangular grid. Figure 14.5 shows an example of a Y-shaped homogeneous area that was divided into nine subareas of equal size.

After using sub-table 14.3.18, start over again with sub-table 14.3.1.

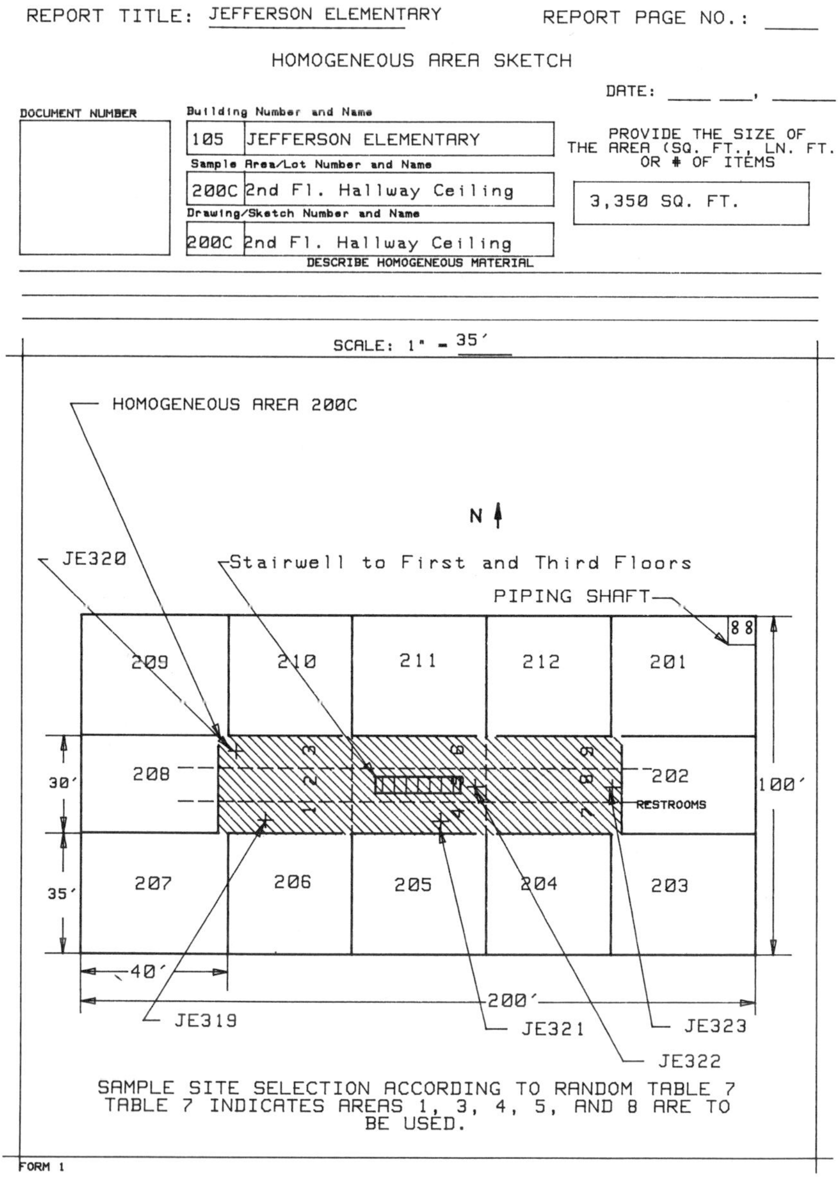

Figure 14.3 Homogeneous Area Divided Into Sampling Sub-Areas.

Table 14.3 Random Sample Site Selection Sub-Tables

RANDOM SAMPLE SITE SELECTION CHART

Choose the next Sample Table sequentially. Then choose the Sample Site Location by reading the selected chart from left to right and top to bottom.

The Sample Site Number from the Table will correspond to one of the nine sample numbers into which you have divided the Sample Area or Lot.

SAMPLE TABLE SAMPLE TABLE SAMPLE TABLE

	9	8	1
1	2	7	6
	5	3	4

	8	7	1
2	3	9	5
	4	2	6

	4	1	7
3	2	9	6
	8	5	3

	6	1	8
4	5	9	3
	2	7	4

	6	4	3
5	1	5	8
	9	2	7

	7	4	3
6	6	1	5
	2	9	8

	5	8	1
7	4	3	6
	2	7	9

	5	7	1
8	6	3	4
	2	8	9

	3	6	4
9	9	2	7
	5	8	1

	5	7	3
10	8	1	6
	2	9	4

	5	1	6
11	3	4	9
	7	8	2

	7	1	9
12	2	4	5
	6	8	3

	8	5	2
13	3	6	9
	7	1	4

	4	1	6
14	3	9	7
	8	5	2

	3	5	6
15	9	2	8
	7	4	1

	4	8	3
16	2	5	9
	7	1	6

	8	2	7
17	4	5	3
	1	9	6

	2	5	9
18	6	1	8
	4	7	3

NUMBER OF SAMPLES REQUIRED

AREAS

Size of Area	Number of Samples
LESS THAN 1,000	3
1,001 TO 5,000	5
MORE THAN 5,001	7
WHEN MORE THAN 5,000 CHECK STATE REGULATIONS.	

LOTS

Size of Lot	Number of Samples
LESS THAN 20	3
21 TO 100	5
MORE THAN 101	7
AUTHOR'S RECOMMENDATION ONLY	

FORM 3

REPORT TITLE: ______________ REPORT PAGE NO.: ____

SAMPLE LOG and SAMPLE NOTES

DOCUMENT NUMBER

Building Number and Name

Sample Area/Lot Number and Name

Drawing/Sketch Number and Name

THIS IS A CHAIN-OF-CUSTODY DOCUMENT. PLEASE FILL OUT COMPLETELY AND SIGN ACCORDINGLY

PAGE ____ OF ____

Sample Number	Receiver's Initials	Photo. Number	Description of Sampled Material	SAMPLE SITE LOCATION

NOTE: IF MAILED, HAVE CARRIER PROVIDE SIGNED PICK-UP AND DELIVERY RECEIPTS TO BE ATTACHED TO THE ORIGINAL OF THIS DOCUMENT.

Receiving Analyst's Name	Receiving Analyst's Firm	Inspector's Name	Inspector's Firm

Receiving Analyst's Signature	Date Samples Received	Inspector's Signature	Date Samples Collected

FORM 7

Figure 11.1 Sample Log.

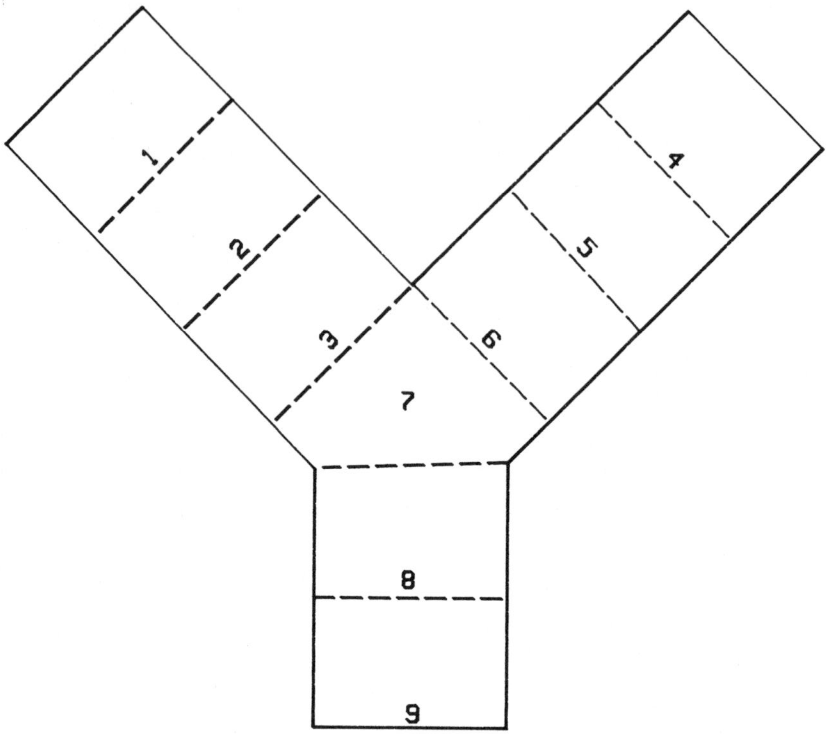

Figure 14.5. "Y" Shaped Homogeneous Area Example.

Identification of Samples

Assign a unique sample ID number to each sample location. This ID number will be on the sampling container when it goes to the laboratory for analysis. Record the ID number and the sample location on the homogeneous area diagram (see Figure 14.6) and also on a data sheet (see Sample Log, Figure 14.7). This must be done carefully so that there is no uncertainty about the location and identity of each sample. Make sure that no two samples have the same ID number. Non-sequential numbers are used to prevent the laboratories from knowing which samples come from the same sampling areas or the same buildings. The "blind" procedure helps prevent bias on the part of the analyst since there is no temptation to assume that the next sample will be similar to the previous one. On the other hand, non-sequential ID numbers make organizing the analytical results by homogeneous area much more difficult. The choice of a sample numbering scheme is up to the inspector.

REPORT TITLE: JEFFERSON ELEMENTARY REPORT PAGE NO.: ____

HOMOGENEOUS AREA SKETCH

DATE: ____ ____, ______

DOCUMENT NUMBER

Building Number and Name: 105 JEFFERSON ELEMENTARY

Sample Area/Lot Number and Name: 200C 2nd Fl. Hallway Ceiling

Drawing/Sketch Number and Name: 200C 2nd Fl. Hallway Ceiling

PROVIDE THE SIZE OF THE AREA (SQ. FT., LN. FT. OR # OF ITEMS: 3,350 SQ. FT.

DESCRIBE HOMOGENEOUS MATERIAL

SCALE: 1" = 35′

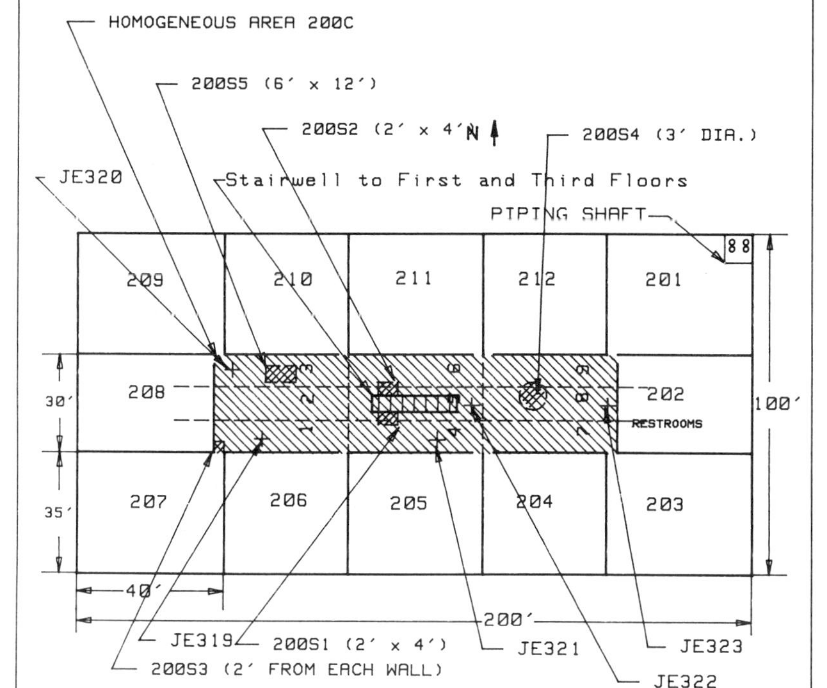

SAMPLE SITE SELECTION ACCORDING TO RANDOM TABLE 7
TABLE 7 INDICATES AREAS 1, 3, 4, 5, AND 8 ARE TO BE USED.

FORM 1

Figure 14.6 Completed Homogeneous Area Sketch.

REPORT TITLE: ______________ REPORT PAGE NO.: ____

SAMPLE LOG and SAMPLE NOTES

DOCUMENT NUMBER	Building Number and Name	
	105	Jeffeson Elementary
	Sample Area/Lot Number and Name	
	200C	2nd Fl. Hallway Ceiling
	Drawing/Sketch Number and Name	
	200C	2nd Fl. Hallway Ceiling

THIS IS A CHAIN-OF-CUSTODY DOCUMENT. PLEASE FILL OUT COMPLETELY AND SIGN ACCORDINGLY

PAGE 1 OF 1

Sample Number	Receiver's Initials	Photo. Number	Description of Sampled Material	SAMPLE SITE LOCATION
JE319	X	200 -8	Uniform sprayed-on material with painted finish. ARCHIVED SAMP	(116,28)
JE320		200 -9	SAMPLE TO BE ANALYZED This sample is similar in appearance to all others taken.	(120,0.5)
JE321	X	200 -10	ARCHIVED SAMPLE	(51,9.5)
JE322		200 -11	SAMPLE TO BE ANALYZED This sample is similar in appearance to all others taken	(47,15)
JE323	X	200 -12	ARCHIVED SAMPLE	(1,15)
			ARCHIVED SAMPLES AT BUILDING OWNER'S OFFICES	
			SAMPLE SITE GRID REFERENCES ARE IN FEET FROM THE CORNER OF THE HALLWAY AT ROOMS 208, 209	

NOTE: IF MAILED, HAVE CARRIER PROVIDE SIGNED PICK-UP AND DELIVERY RECEIPTS TO BE ATTACHED TO THE ORIGINAL OF THIS DOCUMENT.

Receiving Analyst's Name	Receiving Analyst's Firm	Inspector's Name	Inspector's Firm
		James Jackson	Jackson & Assoc
Receiving Analyst's Signature	**Date Samples Received**	**Inspector's Signature**	**Date Samples Collected**
			6/10/1987

FORM 7

Figure 14.7 Completed Sample Log.

Example: The sampling procedure is illustrated by this example. A school was visually inspected for ACBMs. The Second Floor Hallway was found to contain suspect surfacing materials. All materials were believed to be the same, and thus comprise one sampling (homogeneous) area.

Approximate room dimensions were obtained from the building floor plans, and diagramed as shown in Figure 14.6.

There were not enough funds for nine samples to be collected in every sampling area. Therefore, the minimum number of samples required, based on area, was determined according to Table 14.2.

The total area of friable materials is 10,080 ft^2 (square feet) as calculated by:

Area = (30' x 120')
LESS: (10' x 25') for stairwell
= 3,350 ft^2

Since this area is greater than 1,000 square feet, 5 samples should be collected.

The sampling area diagram was divided into nine subareas. Assuming this is the seventh sampling area to be sampled, sub-table 14.3.7 is used. Therefore subareas 1, 3, 4, 5 and 8 were selected, the samples taken, the location from which the samples were taken recorded on the homogeneous area sketch, and the information recorded on the Sample Log.

THERMAL SYSTEM INSULATION

The concept of homogeneous areas applies equally well to thermal insulation. The major difference is that insulation on thermal systems is likely to be much more varied than materials on surfaces. A "typical" building may contain multiple insulated pipe runs from any combination of the following major categories:

- Hot water supply and/or return
- Cold water supply
- Chilled water supply
- Steam supply and/or return (watch for different pressure steam lines)
- Roof or system drain
- Chemical or waste transport

Each of these systems may have been installed at different times and insulated with different materials. Therefore, it is best to first identify the building system in question and use this information in conjunction with the physical appearance of the insulation to delineate homogeneous sampling areas.

Each "system" may be composed of a variety of materials. For example, the following list contains nine different types of thermal insulation:

- Corrugated cardboard-type pipe wrap
- White chalky pipe wrap
- Fibrous glass insulation covering a pipe wrap of unknown characteristics
- Cementatious "mud" around pipe fittings
- Hard, canvas-wrapped insulation on pipe elbows
- Block insulation on boilers
- White batt insulation on boiler breeching
- Black batt insulation inside ducts
- Rope around pipe sleeves in ceiling and floor slabs

Each of these insulation types should be considered a separate component of the system and a separate homogeneous area for sampling purposes. Fibrous glass, foam glass, rubber, and styrofoam are not suspect materials. Note, however, that they may cover up ACBMs or be wrapped in asbestos-containing canvas wrap.

The number of samples and the sample locations will depend on local circumstances. Try to take at least three samples in each sampling area. For long pipe runs or risers, more samples should be taken, especially if the piping extends to more than one functional area. Pay special attention to any change in the appearance of the insulation on long pipe runs. This would indicate a possible change in insulation type and the need to delineate a new sampling area. The AHERA rules require at least three random samples for thermal insulation. Exceptions are: (1) small (less than 6 linear or square feet) areas of patched insulation (at least 1 sample), and (2) areas of insulating cement (the number of samples to be determined by the Building Inspector).

Random sampling may be accomplished by dividing the length of the pipe covering, or the number of fittings, into nine groups by length or by sequential fitting number, and then applying the next available sub-table from Table 14.3 to pick the sub-group to be sampled. Any point within the sub-group chosen may be used to obtain the sample.

Normally, samples should be collected at locations where minimal damage will be inflicted on the insulation. Choose exposed ends, damaged areas, or areas where the protective covering or jacket is missing. This is called "convenience sampling." The AHERA Rule, however, requires random sampling. Thus, samples will have to be taken from intact insulation. Of course, the building owner always has the option of assuming the insulation contains asbestos instead of sampling and analyzing for it.

MISCELLANEOUS MATERIALS

Miscellaneous suspect materials are, for the most part, non-friable (ceiling tiles are an exception). As such, sampling is more difficult and destructive. EPA does not recommend sampling these materials. Instead, they should be identified as suspect and documented as such in permanent records.

Some building owners wish to have miscellaneous materials sampled and analyzed anyway. Ceiling and floor tiles are probably the most frequently sampled of materials in the miscellaneous category. If sampling is desired, try to identify separate homogeneous areas just as you would for surfacing material and thermal insulation. (You will probably find that many different types, colors, and vintages of both floor and ceiling tile in a building.) Take a convenience sample in an inconspicuous location. Very hard materials like Transite wallboard should not be sampled. Not only is sampling needlessly destructive and hazardous, Transite (or Transite-type) wallboard can easily be identified.

STATISTICAL SAMPLING SCHEMES DEFINED BY EVIDENTIARY REQUIREMENTS

For evidentiary purposes the statistical techniques are the same, but the number of samples to be taken from a given homogeneous area may be considerably greater in order to increase the level of confidence or narrow the confidence interval. In this case, the technique will be to divide the homogeneous area into subareas of uniform size and the required number, and take a sample from each of the subareas. The use of the random tables will not apply simply because the required number of samples will have been specified and the division of an area into uniform sized subareas is believed to serve the random selection purpose, as it is in AHERA when nine samples are to be taken. Note that the handling and analytical procedures will be defined by the court having jurisdiction or the legal counsel directing the work.

SAMPLING TECHNIQUES AND PRECAUTIONS[248]

Personal Protection Equipment

Since inhalation of asbestos fibers during inspection and sampling may pose a serious health hazard, the use of personal protective equipment by building inspectors is crucial during the sampling process. As a minimum level of protection, inspectors should wear a respirator, either full or half-face mask with high efficiency (HEPA) disposable filter cartridges. Full face masks will also prevent eye irritation from dust, fibers, and debris released during the sampling operation. Disposable clothing should be worn during sampling if the sampling operation is likely to dislodge pieces of suspect material or if the environment is extremely dusty (e.g., crawl space, dirty mechanical room). Inspectors should have plastic bags, twisters, and labels with them to handle the disposal of cartridges, protective clothing, wet cloths, and debris. These waste materials should be stored pending survey results. If laboratory reports establish the presence of asbestos-containing materials, these waste materials should be disposed of as asbestos-containing wastes.

Sampling Equipment

Inspectors will need various tools and aids to accomplish sampling tasks:

- A ladder and flashlight are needed to access and aid visibility;
- Airtight, rigid sampling containers (e.g., film canisters);
- A plastic spray mister bottle with water to spray the area to be sampled;
- Plastic drop cloths to spread beneath the area to be sampled;
- A knife, linoleum cutter, cork borer, or other tool appropriate for extracting samples;
- A caulking gun and compound for filling holes once a sample has been extracted;
- Spray acrylic or adhesive to encapsulate sample extractions;
- Duct tape for repairing thermal system insulation jackets;
- Cloths (pre-moistened) for cleaning up debris and tools;
- A vacuum cleaner equipped with high efficiency particulate air (HEPA) filters, if available;
- Indelible ink pen for labeling sample containers; and
- Camera for photographic documentation.

Administrative Supplies

In addition to homogeneous area diagrams, data forms for bulk samples will be needed in the field. An example data form was shown in Figure 14.4. This form is intended as a guide; you will no doubt want to tailor the form to your needs and working style. Identification labels for sample containers, packing enclosure warnings and forms, plastic bags, sturdy cartons, sealing tape, and writing materials (pens, pencils, clipboard) are also needed.

Given the amount of equipment needed for bulk sampling, the inspector will likely need assistance. A push cart or table on wheels is worth considering to mobilize the inspection and sampling operation.

SAMPLING PROCEDURES

If possible, collect samples after hours or when the building is not in use. Steps for sampling surfacing material, thermal insulation, and miscellaneous materials are set forth below.

Surfacing Materials

1. Spread the plastic drop cloth and set up other equipment, e.g., ladder.
2. Put on protective equipment (respirator at all times, protective clothing if needed).
3. Label container with its identification number and record number, sample location, and type of material sampled on a sampling data form. Always place the label on the container itself, not on the lid, as lids can be inadvertently switched by a laboratory when handling numerous sample containers.
4. Mark the location of the sample on the sampling diagram and record the sample identification number on the plan diagram as well.
5. Moisten area where sample is to be extracted (spray the immediate area with water).
6. Extract sample using a clean knife to cut out or scrape off a small piece of the material. Be sure to penetrate all layers of material. Be careful not to disturb adjacent material.
7. Place sample in a container and tightly seal it.
8. Wipe the exterior of the container with a wet wipe to remove any material which may have adhered to it during sampling.
9. Clean your tools with wet wipes and wet mop or vacuum area with a HEPA vacuum to clean all debris.
10. Fill hole with caulking compound on highly friable material and/or spray with an encapsulant (to minimize subsequent fiber release) or for appearance.
11. Repeat the above steps at each sample location. Place sample containers in plastic bags.
12. Discard protective clothing, wet wipes and rags, cartridge filters, and drop cloth in a labeled plastic bag. Seal and retain until lab results are received, at which time dispose of the bag as asbestos-contaminated waste if tests were positive for asbestos. (Labels should read: *"Danger — Contains Asbestos Fibers — Avoid Creating Dust — Cancer and Lung Disease Hazard".)* Disposal should be made in a state-approved landfill. Note: Unless every sample tests negative for asbestos, discard waste as asbestos-containing material.

Thermal System Insulation

Sampling thermal insulation materials follows the same procedural sequence as laid out above. Obtain samples from exposed/damaged areas if possible. However, random sampling will require sampling of some intact material. Sampling holes can be patched with plastic spackling, caulk, or fibrous glass plus wetable fibrous glass cloth.

SAMPLING THEORY AND TECHNIQUES FOR LAND, AIR AND WATER

There are fundamental differences between two-dimensional materials or materials that may be considered to be two-dimensional, and three-dimensional materials such as air, water and land. The latter materials tend to be both non-homogeneous (i.e., heterogeneous) and anisotropic. The non-homogeneous characteristic shows up as thermal inversions in air and water, and as layers of materials in land (topsoil, hardpan and bedrock for example). Anisotropy means that the exact amount of a contaminant, for example, is dependent on the distance and/or direction from the source.

Because land, air and water exhibit both heterogeneity and anisotropy, very different sampling and statistical techniques are frequently required. This is only an introduction to the subject and additional sources[249,250,251,252,253] of more detailed information should be sought by the serious practitioner. Of even greater importance than additional study will be the advice and counsel of an experienced professional or professionals in the field. By some stretch of the imagination, an amateur may be able to undertake a building inspection and come off reasonably well, but experience has shown beyond doubt that this possibility is so remote as to be non-existent when it comes to detailed air, water or land inspections.

Because of the very complex nature of site inspection involving three-dimensional subjects, a set of guidelines for site investigation must be set up and utilized that are agreed upon with the client, but subject to change and adjustment as data from the early stages of investigation become available. There are several excellent sets of guidelines available which should be consulted in addition to this brief discussion.[254,255,256]

"Homogeneous" Areas

The equivalent of a two-dimensional homogeneous area when dealing with three dimensions is a cube of uniform materials. This concept is only rarely used but can provide a useful explanatory starting point.

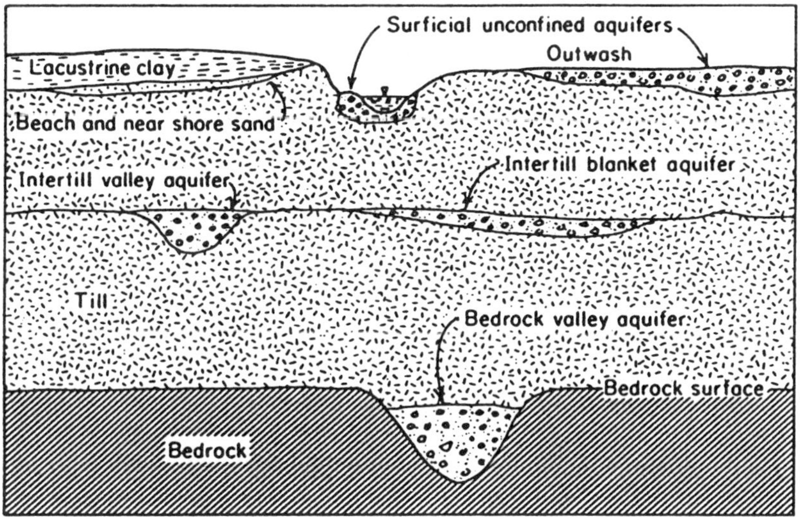

Figure 14.8 "Typical" Subsurface Characteristics.

Imagine that the land (or air or water) to be examined has no distinguishing features; it is flat and uniform on the surface with no wells, streams, low-lying areas or other features. As far as we know, the ground underneath the surface is uniform also. In this case an initial search pattern, utilizing core borings, could be set up exactly as it was in sampling a homogeneous area of suspected ACBM and with the same techniques including random selection of the sub-areas.

At this point, if the sampling were to simply determine the "baseline" condition of the land, or to demonstrate to some acceptable level of confidence that the land was uncontaminated, then the same random sampling schemes as are used for two-dimensional materials, with the addition of requiring some depth parameter to the core borings, would be all that is required. Unfortunately, this will only rarely be the case.

Heterogenous Areas

The heterogeneous nature may show up in a number of ways. For example, on the surface it may show up in the form of filled areas, areas of discolored foliage or distressed plant life. Under the surface, it may show up in contained pockets of gravel in a bedrock "bowl" or "trough" (very frequently found in heavily glaciated areas such as the northern third of the United States). It may also appear in the form of contaminated areas lying adjacent to uncontaminated areas. The contaminated areas are generally known as plumes, as in plumes from smokestacks because they are often directional in character. Heterogeneity is both a disadvantage and an advantage. It is disadvantage in that it causes deviation from a purely random sampling pattern to obtain maximum sampling information, but an advantage in the sense that if detected and properly analyzed by the inspector, the sampling process can be altered to provide much more efficient use of the sampling and analysis budget.

Anisotropic Areas

Anisotropy is essentially the plume effect—the substance of interest has a directional characteristic. Plumes can be readily seen in the form of smokestack effects or the effect of a river entering a lake as seen from an airplane. Given the differences between a foreign substance in a medium such as air or water and the resistance to movement and the often confused geology of a site, the plume effect can be very apparent in land as well.

In general there are two basic types of plumes, denoted by the physical phase of the contaminant—gas or liquid (or solids carried in liquid). Both will follow the path of least resistance in forming a plume, but for a gas this may mean flowing "uphill," if that is the direction of least resistance, while liquids will seek to flow "downhill." Both liquids and gases will flow towards areas of lower pressure, whether that area of lower pressure is generated by an open basement or an active well, or simply an area of discharge to a stream, lake or pond. The tendency of fluids to move in the direction of least resistance and toward areas of lower pressure is what causes the plume effect.

Homogeneity, heterogeneity, and anisotropy all strongly influence both the sampling strategies to be used and the sampling statistical analysis techniques.

SAMPLING STRATEGIES

The EPA recognizes five basic sampling strategies.

Random

The taking of samples from randomly selected locations throughout the site. The locations are selected in much the same manner as we have already described for one- and two-dimensional sampling—by random chance. Random sampling is usually not effective in controlling sampling error unless the population being sampled is large and large numbers of samples can be taken.

Systematic

Systematic random sampling is the taking of samples in regular subdivisions or at specific intervals throughout a site. This system is frequently used on sites where there is no obvious contamination, but some sampling is warranted.

Stratified

Stratified sampling involves the division of sample populations into groups with similar characteristics resulting in subdivisions that are more homogeneous than the sampling population as a whole. This is one method for dealing with the heterogeneous character of most sites.

Judgment

Judgment sampling is the taking of samples and the placement of sampling locations based on specific site information and the knowledge and experience of the investigator. Judgment will enter into every sampling strategy but should not be the sole criteria, because if used exclusively as a sampling strategy, judgment sampling the biases of the investigator will eventually be imposed on the results and lead to prejudiced conclusions.

Judgment sampling can be significantly improved if the Phase I ERA has carefully defined the areas of highest suspect risk, as call the stratified and hybrid sampling techniques. This combination of judgment based on sound historic data and statistical sampling techniques applied to specific portions of the total area can provide significant information at relatively low cost.

Hybrid Sampling Strategy

The placement of sampling locations will often involve a combination of the preceding four techniques. For example, the AHERA asbestos-sampling strategy is a hybrid of judgment sampling (the limitation of sampling to only those materials suspected of being asbestos containing) and systematic random sampling.

STATISTICAL SAMPLING CONCEPTS

Representative Samples

A sample is supposed to be representative of the whole. In sampling materials for environmental analysis, the sample should be a representative part of the suspect materials. Samples can be discrete—taken from a single location or point in time, or composite—a blending of samples from a general area or over a period of time. Discrete sampling gives a clearer pictures of the concentration and extent of contamination, but several samples must be taken from the area and analyzed. A composite sample is useful in determining the average concentration of a contaminant in a specific area but generally cannot pinpoint the extent or variation of the contaminant over the area.

If a material is homogeneous, a single sample may be representative of the whole. If the material is heterogeneous, several samples from different locations may be required to assure representation.

To deal with the problems of analysis associated with complex heterogeneous and anisotropic situations a whole specialized field, called "geostatistics" has evolved. The literature of this field has its own language and specialized mathematics that are far too complex to deal with here, but a few of the terms and concepts will be introduced here to acquaint the inspector with the ideas and results that might be obtained.

Variograms

A plot of the variance (one-half the mean squared difference) of paired sample measurements as a function of the distance (and optionally of the direction) between samples is called a variogram, or more accurately, a semivariogram. Typically, all possible sample pairs are examined and grouped into classes (lags) of approximately equal distance and direction. Variograms provide a means of quantifying the commonly observed relationship that samples close together will tend to have more similar values than samples far apart. An example of a variogram is shown in Figure 14.9.

Kriging

A weighted-moving-average interpolation method where the set of weights assigned to samples minimizes the estimation variance, which is computed as a function of the variogram model and locations of the samples relative to each other, and to the point or block being estimated.

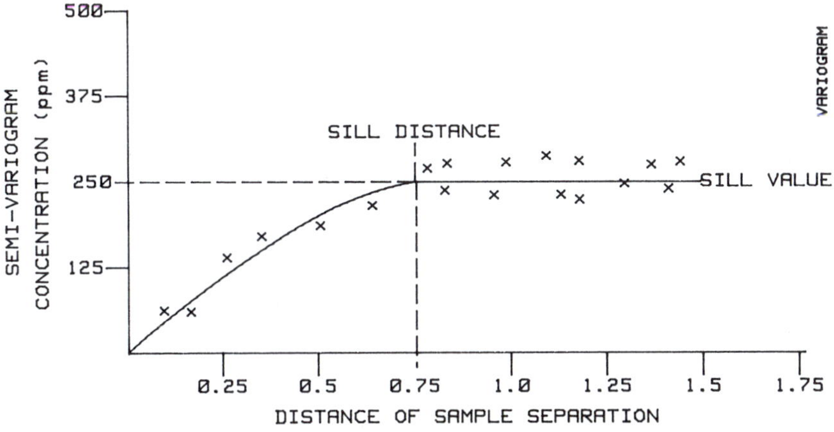

Figure 14.9 Variogram Example.

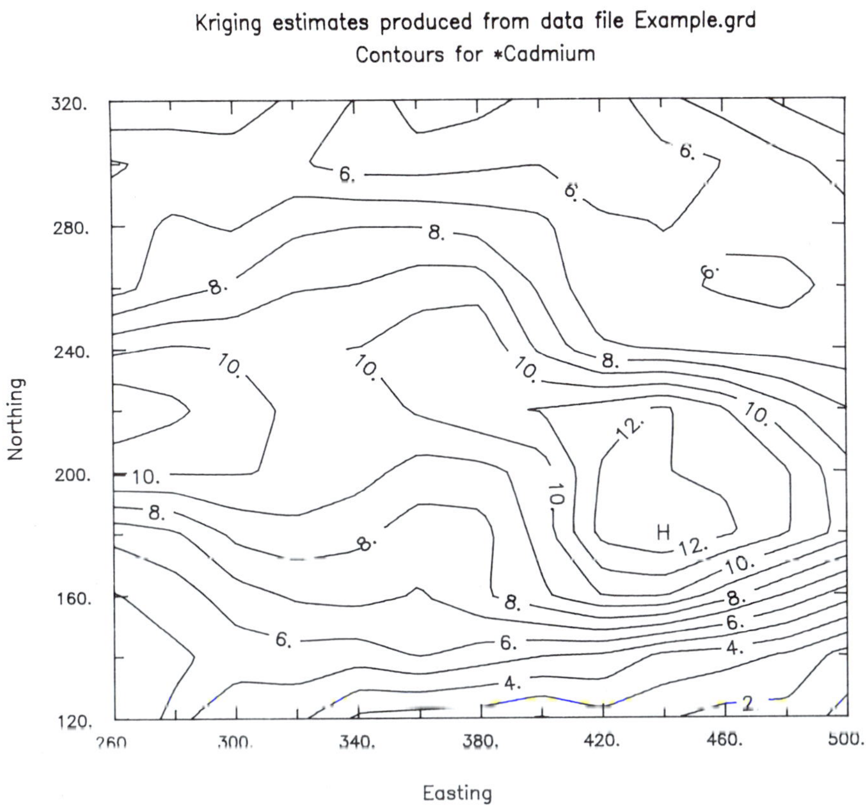

Figure 14.10 Kriged Plot of Contaminant Concentrations.

Kriging Standard Deviation

The standard error of estimation computed for a kriged estimate. By definition, kriging is the weighted linear estimate with the particular set of weights which minimizes the computed estimation variance (standard error squared). The relationship of the kriging standard deviation to the actual error of estimation is very dependent on the variogram model used and the validity of the underlying assumptions; therefore kriging standard deviations should be interpreted with caution.

The application of the techniques of variograms and kriging can result in a contamination plot that looks very much like a topographic map (see Figure 14.10) which, in a sense, it is. Instead of plotting the heights of land above sea level, the kriged map plots contaminant concentration profiles from sample data.

TOPOGRAPHY, GEOLOGY AND HYDROLOGY

Topography

Topography is the study of the surface features of the land and includes drainage patterns, land relief (or natural features) and man-made features such as buildings, parking lots, drainage ditches and man-made ponds. All of these along with slope gradients and soil geology/morphology can influence the direction and rate of movement of contaminant plumes. Together topology and geology can effect ground stability, permeability, site and groundwater contamination.

Surface Geology and Topology

Drainage patterns and gradients on the subject property, onto the subject property from adjacent property, and from the subject property to adjacent property are important with respect to the movement of contamination. The geology of the surface soils can provide data with respect to the rate of runoff, the existence of flood plains, and the existence of dumps.

Wetlands

Wetlands include coastal and inland marshes, wet meadows, mudflats, ponds, bogs, bottomland hardwood forest, wooded swamps, and fens.

Wetlands introduce special considerations with respect to a site. They are considered one of the most important parts of the ecosystem and as of 1984, it was estimated that approximately 50% of the wetlands in the United States had been destroyed. The importance of preserving the remaining wetlands has resulted in a variety of state, local and federal regulations.

The U.S. Fish and Wildlife Service defines wetlands as:

> Wetlands are lands transitional between terrestrial and aquatic systems where the water table is usually at or near the surface or the land is covered by shallow water . . . Wetlands must have one or more of the following three attributes: (1) at least periodically, the land supports predominantly hydrophytes, (2) the substrate is predominantly undrained hydric soil, and (3) the substrate is nonsoil and is saturated with water or covered by shallow water at some time during the growing season of each year.

The U.S. Army Corps of Engineers, which regulates development of wetlands under the *Clean Water Act,* gives a simpler definition as:

> . . . those areas that are inundated or saturated by surface or groundwater at a frequency and duration sufficient to support vegetation adapted for life in saturated soil conditions. Wetlands include such areas as swamps, marshes, bogs, estuaries, certain unique pond systems, and inland and coastal shallows.

Many additional agencies are concerned with the regulation of wetlands, including water resource agencies, the EPA, the Bureau of Land Management, Office of Coastal Zone Management, Federal Emergency Management Agency, National Park Service, and others.

A permit is required from the U.S. Army Corps of Engineers before any dredging or filling takes place. In addition, other laws may effect the development of wetlands, streams, ponds, lakes or estuaries, such as:

- *The National Environmental Policy Act*
- *The Coastal Zone Management Act*
- *The National Historic Preservation Act*
- *The Deepwater Port Act*
- *The Federal Power Act*
- *The Fish and Wildlife Coordination Act*
- *The Endangered Species Act*
- *The Marine Mammal Protection Act*
- *The Wild and Scenic Rivers Act*
- *The National Fishing Enhancement Act of 1984*

Threatened and Endangered Species

Over 280 species have been placed on the U.S. List of Endangered and Threatened Wildlife and Plants under the *Endangered Species Act of 1973* through the direction of the Secretary of the Interior (U.S. Fish and Wildlife Service) and the Secretary of Commerce (National Marine Fisheries Service). The presence of an endangered or threatened species on a site will almost certainly prevent development of a subject property.

Geology

Geology is the study of the structure of the earth's crust, rocks and sediments. Surface geology will determine the initial rate of absorption of a contaminant into the ground and consequently the rate of runoff of a spill.

Subsurface geology will govern the depth of contamination and the rate, direction and concentrations of contaminant spread (or pluming). Subsurface problems may be indicated by:

1. Prior property usage and materials handling.
2. Analysis of residues and/or vapors.
3. Reports of health problems for individuals frequently in the area.
4. Stressed vegetation.
5. Downstream aquatic life damage (fish kills, for example).
6. Downstream water discoloration or contamination.
7. Areas of unusual soils (filled areas), unnatural subsidence or mounding.

Most water migration is through soils with widely varying rates of migration. In general, water in soils does not flow in the form of underground streams, but rather is filtered through loose grains of soil or fractures in the rock.

The site geology will determine the depth, location and extent of the water-bearing saturated zone and underlying aquifers. Contamination from a single source can be found on a site in the form of widely spaced pockets of contamination joined only by a thin strip of contaminant which has followed a fault or fracture.

Generally, the ground of a site may be classified into unconsolidated deposits and consolidated rocks. Unconsolidated deposits consist of mineral grains that have disintegrated from consolidated rocks or have been redeposited from other unconsolidated deposits. Unconsolidated deposits are collections of grains with pore spaces between the grains and are further classified as aquifers or confining beds. Aquifers are unconsolidated deposits for fractured rocks that transmit water and pollutants. A confining bed, or aquiclude, stores water but does not transmit significant quantities.

Three important terms that relate to the water storing and transmitting capacity of unconsolidated deposits are porosity, specific yield and hydraulic conductivity.

1. *Porosity* is the measure of total voids in the ground and is measured as the percentage of void space in a volume of rock. Unconsolidated deposit porosity can range from 20% in course gravel to 90% in mud and dry organic material.[257]
2. *Specific yield* is a measure of the usable stored water that will drain under the influence of gravity from the pore spaces. Some water and pollutants will be retained on the surface of the grains through the action of surface tension; this is called the specific retention.
3. The quantitative capacity of a unit area to transmit a fluid is called the *hydraulic conductivity,* and varies for different fluids such as water, oil, gasoline, dry cleaning fluid, etc. The hydraulic conductivity varies widely from rock type to rock type and from place to place within an unconsolidated deposit and directly effects the rate of migration of contaminants.

Consolidated rocks, often referred to as bedrock, consist of bonded minerals that have formed under pressure, heat and/or chemical processes. Bedrock is usually considered to be an aquifuge—a rock that does not store or transmit water.

Hydrology

Hydrology involves all aspects of water and its recycling. During real estate transactions the hydrologic cycle with respect to streams, lakes, ponds, seas and groundwater must be addressed and evaluated.

Surface waters generally consist of rivers, streams, lakes, ponds and seas. Contamination of surface waters can lead to destruction of aquatic habitat and to the contamination of groundwater and water supplies and can fall under the *Clean Water Act,* the *Water Quality Act* and the *Safe Drinking Water Act.*

The basic parameters for freshwater streams are elevation, stream flow rate, and physical and chemical characteristics of the water, and are measured regularly by the U.S. Geologic Survey (USGS) at established gauging stations. The basic chemical and physical parameters of frequent concern are outlined in Table 14.4.

Permits for the discharge of pollutants from point sources into navigable waters are required by the *Clean Water Act* under the National Pollution Discharge Elimination System (NPDES)

Groundwater is water below the water table. Basic data on groundwater flow direction and quality is most important since groundwater quality may effect drinking water supplies and is generally discharged at some point into surface water systems.

Groundwater may move in any direction in an aquifer, the primary controlling factors being areas of recharge and discharge. A gradient, or pressure slope, is established from areas of recharge to areas of discharge. If

Table 14.4

Parameter	Unit of Measure	Remarks
Solids	mg/L	Pollutants suspended in water, organic and inorganic
Conductivity	Micro ohms/cm	Dependent on dissolved solid in water
Turbidity	NTU's	Cloudiness due to fine suspended particles
Temperature	F or C	
Color, Taste, Odor	Subjective	Aesthetics
pH	Scale: 0–14	Acid or alkaline, measures the concentration of hydrogen ions
Dissolved Oxygen (DO)	mg/L	Can indicate pollution, important for aquatic life
Biochemical Oxygen Diminution (BOD)	mg/L oxygen	Measures oxygen uptake by bacteria
Chemical Oxygen Diminution (COD)	mg/L oxygen	An indirect measure of BOD
Nitrogen	mg/L	Measurement of pollution from ammonia, nitrates and nitrides (primarily from fertilizers)
Chlorides	mg/L	Measurement of salts (NaCL and CaCL) generally indications of raw sewage
Bacteria	MPN/100 ml	Measurement of sewage pollution

the aquifer geometry and a hydraulic gradient are established, a water level contour (piezometric) map may be generated. If such a map can be constructed, then the influence of the introduction of a contaminant into the groundwater system may be evaluated with considerable accuracy.

When water-soluble contaminants are suspected, it may be sufficient to obtain a single sample from contained water. With streams or aquifers, however, both upstream and downstream sampling should occur. When nonsoluble contaminants are suspected, either they will float on the water or will sink. If they float, surface samples from the stream or top of the aquifer will be required. If the contaminant will sink, samples from the bottom of the stream or aquifer will be required.

CONTAMINANTS

Each contaminant will have specific chemical and physical properties that will determine its migration in water and through soils. The chemical properties may cause the contaminant to form compounds with naturally occurring chemicals at the site, resulting in a retarded or advanced rate of migration and possibly an alteration in the chemical form of the contaminant. Other chemical properties may allow the contaminant to easily pass through or dissolve underlying geologic strata and change the rate of migration. Chemical and physical properties of the contaminant will also effect their movement through surface and subsurface waters, affecting both migration and deposition on the bottom or top of streams and aquifers.

Contaminant Migration

The potential migration of contaminants on the surface, through subsurface soils and in the groundwater must be considered in order to increase the probability that the placement of sampling locations will define the presence and extent of contamination.

Contaminant migration is affected by a number of factors in addition to porosity, hydraulic gradient, hydraulic conductivity and specific yield. For example, if the density of the material is less than that of water (gasoline for example), it will float on top of the water table and move through the capillary fringe area. If the density is greater than that of water, it will tend to sink to the bottom of the aquifer, to just above the bedrock, and move with the water in the aquifer. The rate of migration may be influenced by the seasons (rainy versus dry), and the concentration with distance from the site may be strongly influenced by the specific retention.

The term "fluid" was deliberately chosen for this discussion because gases are also fluids and the underground migration of gases is of increasing concern, particularly gases from improperly sited or operated landfills, or gases escaping from natural gas or oil fields, and radon gas. All of these gases tend to migrate rather quickly through soils and enter buildings through cracks or drains in floors directly connected to the ground.

SAMPLING QUALITY ASSURANCE/QUALITY CONTROL (QA/QC)

Field Quality Control

Field quality control concerns are directed toward sample collection procedures with the goal of collecting representative samples. This will include decontamination procedures for sampling equipment. Decontamination procedures should be designed to prevent cross-contamination between samples and between sample corings.

Sample Handling Procedures

Sample handling procedures will include specifications for the types of sample containers which must be compatible with the suspected contaminants, sample preservation techniques such as reduction to and storage at low temperatures, the number of samples to be taken, the labeling system, and chain-of-custody procedures.

Document Control

Document control will include the recording of observations in a field log, photograph logs, and sample data sheet maintenance.

Site Safety Guidelines[258,259]

Hazardous materials assessment involves the handling and transportation of hazardous materials. Site safety guidelines should include individual personnel protection, emergency procedures, and procedures to prevent or minimize the emissions of hazardous substances that might endanger human health and/or the environment.

> Safety consideration: Check for the location of underground structures, utilities, electrical vaults, pipes, tunnels, and sewers. Sending a hand-powered auger into an underground high-tension line can ruin your whole day!

SAMPLING TECHNIQUE

Sample Integrity

Making sure the sample has not changed since collection is just as important as sample representation for the drawing of valid conclusions. Sample integrity is a function of a number of factors such as:

1. **Handling techniques and equipment**
   The collection of relatively undisturbed samples can be important in maintaining sample integrity, especially if the suspected contaminants are volatile, such as solvents or fuels. Generally the less a sample is handled the better, especially if it contains organic matter.
   Some sample container and gathering tool materials can interfere with sample integrity. For example, if toxic metals or corrosive materials are suspected, stainless steel sampling tools and containers should be used to prevent the contaminants from attacking the container or tool and causing additional contamination or a change in the character of the contaminant.

2. **Proper sample containers**
   Plastic containers may contain volatile solvents and other organics; glass containers will dissolve in the presence of hydrofluoric acid, and some contaminants can break down when exposed to light and must be held in containers that will not transmit light to the contents.
   In general, glass or plastic containers can be used when only metal contamination is suspected, glass vials with "Teflon" seals can be used when volatile organics are suspected. Glass jars with "Teflon" lined lids can be used where semi-volatile and non-volatile organics are suspected; plastic jars where hydrofluoric or phosphoric acids or strong bases are suspected; glass jars where strong acids are suspected. In the collection of solid samples, especially from deep soils, sampling tools with sampling sleeves, that is removable sample containers inside the sampling tool, will result in the minimum disturbance of the sample and should be made of stainless steel or brass. Such devices serve as both sampling tool and sample container.
3. **Methods of preserving the sample**
   Samples may not be stable and will start to degrade or stratify once sampled and placed in a sample container. Depending on the type of contaminant suspected, it may be necessary to add preservatives (which should be clearly identified on the sample label, sample log, and chain-of-custody document), or to cool the sample. In general, if cooling is necessary, most samples should be held at approximately 4° C.
   In general, there should be as little time as feasible between sampling and laboratory analysis. The laboratory protocols will specify maximum holding times for various tests, and these holding times can greatly affect the sampling procedures and the analysis costs, particularly at remote sites.
4. **Decontaminating sampling equipment between samples**
   Good field decontamination of equipment can be achieved as follows:
   Small tools — The four-bucket system generally works well, where the tool is first washed in a bucket of detergent-containing water, then rinsed in two separate buckets of tap water with a final rinse in distilled water. The cleanliness of the tools will then be primarily dependent on the frequency with which the bucket contents are changed with a thorough washout and rinsing between changes.
   Large Equipment and Tools — A portable steam cleaning system is a virtual necessity.
5. **Labeling and chain-of-custody**
   The documentation of the sampling and sample analysis of land, air and open water areas will require both greater detail and greater inspector/client interaction than for buildings. The initial sampling strategy, whether based on a featureless cube such as was discussed above, or on target areas considered suspicious, should be carefully planned in such a way that the early sampling results will provide direction to later sampling. Because of the high cost of such sampling after the initial samples have been taken, virtually all pretense toward overall random sampling of the complete site is surrendered. Only if no contamination is found during the sampling process does the application of random sampling strategies continue in a effort to demonstrate that the area is not contaminated. This means that the sampling

strategy is very time dependent, earlier results cause modification to later sampling, and the reasons for deviating from a pure random sampling scheme should be carefully documented.

Without regard to how well designed a sampling strategy or how well executed the sampling technique may be, failure to properly document the samples can invalidate all of the results. Sample labeling is the front-end of the process and sample labels should be printed and filled out in indelible ink and attached to sample containers with a permanent fixative resistant to water and chemicals. The labels should show:

a. Boring number
b. Sample location
c. Sample depth
d. Collection date and time
e. Collector's name
f. Unique sample number

As soon as a sample is collected, it should be entered on a chain-of-custody form that will remain with the samples at all times and document all handling and changes of possession. In addition, the sample information, along with pertinent comments of collection such as circumstances, general observations on site and conditions, gross physical appearance and the like should be entered on the sample log and possibly in the field log.

Surface Sample Collection

There are several tools commonly used to obtain surface samples.

Stainless steel scoops or shovels, with the sample being transferred to an appropriate container for transportation to a laboratory may be used to scrape a shallow surface sample.

Stainless steel ladles (much like soup ladles) may be used to obtain a surface sample from a liquid.

A six-inch sampling tube with a removable sampling sleeve may be used to obtain a shallow core sample, the removable tube being capped with Teflon-lined caps to form a relatively undisturbed sample in a transport container.

To obtain composite samples of surface soils over a large area, a device similar to a mechanical broom can be used, the sweepings being transferred to an appropriate container. Please note that decontamination between each use is extremely important and somewhat difficult with this device.

Composite samples of surface soils can also be obtained by using the scoop or shovel in several locations and then blending the materials. Neither the mechanical broom nor the mixing of scoops of materials will provide undisturbed samples.

Shallow Subsurface Collection

Various augers may be used to obtain soil samples to a depth of several feet; the hand auger being particularly recommended when ignitable vapors are a concern. These devices will generally not work at a depth much beyond 10–15 feet, and frequently they are uneconomical for a large number of samples at depths exceeding 5 feet.

A device that may be driven into the ground and having a removable sample sleeve may prove particularly useful for a large number of shallow samples, say of less than 5 feet.

Deep Subsurface Collection

Truck-mounted rotary coring machines are the norm when a large number of samples, or deep samples, are required. In general, these coring machines will have a hollow drilling tool through which the actual sampling device is inserted. The general procedure is to drill to the top of the desired sample depth, and then insert a sampling device with a removable sample sleeve into the hollow tool; the sampling device being driven by hammer blows into the undisturbed soil beyond the end of the coring tool. The coring tool is then driven past the sampling tool, thereby loosening it and allow extraction of a relatively undisturbed sample. This procedure may be repeated at a number of depths within a single boring.

Sampling of liquids requires specialized tools designed to trap a sample from the required depth through means of ball cocks, remotely operable valve systems, or one-way valve systems that allow the fluid to pass through the device freely as the sampler moves down, but closes off the flow as the sampler is withdrawn. The latter type of device may present some contamination issues from the fluid at lesser depths flowing through the device.

Groundwater sampling is generally accomplished through the use of monitoring wells, which once completed and developed (meaning that the well is emptied and allowed to refill several times prior to sampling) may then have samples extracted through pumping or the use of the aforementioned water sampling devices. Depending on the contaminants of concern, the well casing materials may need to be plastic or stainless steel to prevent contamination of the samples from the casing. If undisturbed samples are required, they cannot be obtained by pumping.

The analysis techniques used, both to analyze samples in the laboratory and to statistically analyze the laboratory results to form a picture of the contamination and the conditions discovered relating to the availability of transport mechanisms should be most carefully documented. In particular, the raw data obtained, along with the laboratory analysis of both the contaminants and the soil samples for type and permeability of those soils, and ideally the archiving of important representative cores, should be of major concern.

GENERAL SAMPLING CONSIDERATIONS

The first characteristic to be looked for is contamination of the soil or surface water. Such contamination may come in a number of forms and may be transported through the soil in a number of ways, but all of the transport mechanisms depend on the soil permeability to both liquids and gases.

Almost all urban top soils (i.e., the top layer of dirt, not necessarily topsoil, which is a specific material) have been contaminated with hydrocarbons to some degree. There may also be heavy metal contamination resulting from the spreading of used motor oil (now considered by the courts to be a hazardous material), or dumping of heavy metal containing wastes. Farmland may be heavily contaminated with fertilizers and insecticides.

Some of the top layer of contamination may have been carried down into the next layer through the action of rainwater seeping down or it may have entered the next layer from a specific contamination site. Among other substances that may have entered from a specific site are methane gas from an improperly designed garbage dump, benzene from leaking underground pipelines or storage tanks, and other contaminants from leaking dumps.

The material from surface contamination of the inspection area will tend to be relatively uniform in concentration from sample site to sample site, given the site topography. Material leaking from a specific site will tend to show relatively uniform changes in concentration from sample site to sample site along the major axis of the plume (which is very often *not* a straight line), and tapering off away from the major axis.

Any areas of relatively lower pressure in the neighborhood of a contaminant will tend to set up a flow in the liquids and gases contained in the soil and the contaminant will be carried along with that flow. It is this phenomena that causes the ''plume'' effect for contaminants, whether from artificially induced pressure gradients or simply the natural effects of hot air or water rising or groundwater flowing downhill.

ANALYTICAL SUBSTANCE IDENTIFICATION

Laboratory Selection

The first item in the analytical identification of substances is the selection of a laboratory. A basic checklist of items for such selection would be as follows:

- Does the laboratory have proper certification?
- Is an acceptable quality control program in place?
- Is the laboratory keeping proper records?
- Does the laboratory use accepted handling and analytical methods?
- Is the laboratory competent?
- Can the laboratory meet the schedule?
- Can the laboratory defend its results?
- Is the laboratory independent?

Laboratory Certification

There are two national certification programs for laboratories. The American Industrial Hygiene Association (AIHA) provides a very highly respected certificate primarily designed for industrial hygiene applications. However, if a laboratory has an AIHA certification, this is strong evidence of a very high-quality standard of work which probably carries over into all other areas of that laboratory's functions. In addition, the National Institute of Occupational Safety and Health (NIOSH) manages a Proficiency Analytical Testing (PAT) program that commercial laboratories may participate in. The PAT system requires a series of quality control checks to be performed on the work of one laboratory by other laboratories and if a laboratory is successful in this peer review process, it may be classified as "proficient" in the testing procedure by NIOSH although this is not strictly a certification of the laboratory.

The EPA does not certify laboratories, but generally refuses to contract with laboratories that cannot perform the work to their quality control standards and evidence that a laboratory is qualified to perform work for the EPA may be taken as a reasonable indicator of proficiency. Laboratories to be used for asbestos bulk sampling and TEM work should be chosen from the list of laboratories participating in EPA's quality assurance (round robin) program. The list is frequently updated; to obtain the most recent list call: 800/334-8571. EPA's listing of accredited laboratories should be used until the National Bureau of Standard's Laboratory Accreditation Program is in effect.

Various state agencies also run laboratory certification or proficiency programs, but the exact requirements of such programs should be verified for the particular testing desired as the state programs contain a high level of qualification and quality control variance.

Quality Assurance

The single most important issue in laboratory competence is its Quality Assurance program. Such a QA/QC program may be evaluated in terms of several components.

Organization

If the laboratory has a clearly defined organization with well-defined job descriptions and assignments and the internal discipline to enforce those definitions and assignments, a major component of the QA/QC program is in place. There should be a specified QA/QC officer responsible for the management of the program and insuring that the quality and accuracy standards are maintained. This QA/QC manager should not report to the laboratory manager. The laboratory manager is responsible for actual work assignments and should insure that unqualified analysts are not used to perform the work. In addition, there should be a laboratory sample custodian responsible for receiving, logging in, and logging out the samples.

QA/QC Measurement

Laboratory personnel should be fully aware of the detection limits of their equipment and the methods for calibration of such equipment and standard laboratory procedures for calibration should be evident.

Methods The methods to be used by the laboratory in the performance of an analysis should be written in the form of a checklist to be used by the analyst with each sample in performing a particular analysis. The existence and use of a checklist for each sample is particularly critical when unusual testing is performed or extremely repetitive testing is performed to prevent errors in or omissions of critical steps.

Standards Reference samples of known purity are used to develop calibration curves and are used to identify and/or quantify an unknown material. Evidence of proper use and control of the standards should be available.

Blanks Blanks are used by the laboratory to test the purity of the laboratory reagents or the media used to collect the samples. In some cases, blanks will need to be supplied by the field inspection team in order to calibrate background curves for very frequently occurring materials such as asbestos.

Blind Samples Blind samples contain a known amount (but not to the analyst) of material and are used to test the accuracy of both the analyst and the precision of measurement.

Duplicates Duplicates are generally created by dividing a sample into two parts and having both tested, then comparing the results to determine if any errors or inaccuracies are entering into the analysis. This is particularly valid if the duplicates are sent to another, unrelated, laboratory. Duplicates should be randomly selected from the set of all samples destined to undergo the same analysis procedure.

Inter- and Intra-laboratory Sampling Analytical uncertainty resulting from the variability of results can be evaluated by exchange programs between laboratories (inter-laboratory sampling) and between analysts with the same laboratory (intra-laboratory sampling). Knowledge of the amount of variability is essential to the proper interpretation of laboratory results. This is the purpose of sending duplicate samples to different laboratories and the NIOSH-PAT and EPA "round robin" programs.

Laboratory Recordkeeping

Laboratory records of analyses should also provide at least the following information:

- A description of or reference to the method used.
- Deviations or special circumstances encountered during the analysis.
- Estimates of the limits of detection and quantification.
- The results of the analysis.
- The report should be signed by *the analyst and the laboratory manager*. The signature of the analyst is strictly required under AHERA and some other laws.
- The date of the analysis.
- The sample identification.

In addition, the laboratory should generally be keeping the following records.

- Laboratory notebooks, similar in form and content to the inspector's field log. The laboratory notebooks should be maintained by each analyst and should have sufficient information on the work performed and the special circumstances encountered to allow any other analyst to perform the work and arrive at the same conclusions.
- Instrumentation maintenance records showing all work performed to maintain and calibrate the instruments.
- Sample tracking records to clearly show where each sample was at each point in time while it was in the laboratory. This is the laboratories internal chain-of-custody record.

- QA/QC records showing all of the QA/QC pertinent events such as the results of certification testing or inspections, proficiency testing of individuals, calibration of equipment, etc.

Use of Accepted Handling and Analytical Methods

Both NIOSH[260] and EPA[261] provide guidance documents for handling and analytical methods. Whenever an evaluation involves a hazardous waste under RCRA, these procedures should be used. Some states have developed their own standards and test methods that may also need to be followed. It must be kept in mind that each state may set rules more stringent than the federal standards, and several have done so.

Laboratory Competency

In general, the laboratory director should have a high-level academic degree, preferably a doctorate in chemistry, bio-chemistry, physics or related discipline. The laboratory manager should have a qualified academic background; the analysts should have qualified backgrounds including specific instrumentation qualifications, and frequently qualifications in the particular testing to be performed.

Laboratory reference checks should allow you to answer the following questions.

- Who performed the tests?
- Were variances checked?
- Were duplicate samples sent to another laboratory and with what result?
- Did the laboratory have an explicit commitment to excellence?
- Were the services satisfactory?
- Were deadlines met?
- Do you use the laboratory often?

Meeting Deadlines

Laboratories are frequently research organizations, and "rush" jobs tend to both confuse and upset the laboratory's functioning. While research qualifications and the implicit insistence on excellence of results are admirable, if a "rush" job is to be undertaken, the ability of the laboratory to perform the work to a high-quality level should be examined. Some laboratories can deal with this type of situation successfully and others cannot, without any reflection on the competence of the laboratory.

Defense of Results

Litigation is a fact of life in the hazardous materials investigation business and one of the first points of attack will be the competence and performance record of the laboratory. If all of the preceding considerations have been satisfactorily dealt with, defense of results should be straight-forward. However, if litigation is considered particularly likely, an additional qualification of experience in successfully defending the laboratory's results may be appropriate.

Laboratory Independence

While many consultants may differ with this author, the independence of the laboratory from all concerns relating to the outcome of the analysis, except those relating to the quality and accuracy of those results, is paramount. If there is a financial connection between any of the parties in an inspection and the laboratory, there is likely to be a presumption of collusion that is undesirable, and in many cases is totally unfair, but is certainly unnecessary and consequently to be avoided.

ANALYTICAL TESTING METHODS

Instrumentation

Spectrophotometers

When elements or compounds are exposed to electromagnetic radiation, the amount of radiation at a particular part of the spectrum that is reflected can be detected and measured by spectrophotometers. Frequently the precision and accuracy of the measurement of such reflected (or refracted) radiation can identify both the compound and the quantity of the compound. There are five commonly used types of spectrophotometers:

1. *UV/Vis Spectrophotometers* provide qualitative and quantitative analysis for both organic (high carbon content) and inorganic (low or not carbon content) compounds in solution. UV/Vis is most often used as a screening instrument for identifying the categories of hazardous materials present for additional and more specific testing and analysis.
2. *Infrared (IR) spectrophotometers* provide qualitative and quantitative results for both organic and inorganic compounds that are liquid or solid. Again, IR is most frequently used as a screening device.

3. *Mass Spectrophotometers* (MS) analyze for organic compounds and are most frequently used in conjunction with gas chromatography (GC), sometimes in a combined instrument known as a GC/MS. MS is particularly useful in the identification of uncommon or difficult to analyze substances.
4. *Atomic Absorption Spectrophotometers* (AA) are used to identify heavy metals in liquid and solid materials, particularly in the parts per million (ppm) range.
5. *Inductively Coupled Plasma* (ICP) *Spectrophotometers* also analyze for the presence of metals and are replacing MS devices because they can analyze for several metals simultaneously and over a wide sensitivity range. Because of their efficiency, they are the instrument of choice for fast turnaround dependent analyses.

Chromatographs

Chromatography is based on the fact that different materials will pass through a column of material (gas, liquid or solid) at different rates and may thereby be separated and detected independently at the end of the column. This allows for both the identification of the material and its quantification. The type of chromatography depends on the media in the column with *gas chromatography* being used to detect organic compounds in the gaseous state. Gas chromatography in conjunction with a mass spectrophotometer is the most commonly used and definitive technique for identifying and quantifying organic materials. Ion chromatography (IC) is used to identify and quantify organic and inorganic substances. Because of the fact that IC uses a liquid medium, it is frequently used for the analysis of water samples, but there are few EPA-approved uses of the technique at present.

Microscopes

Two common techniques are in use:

1. *Polarized Light Microscopy* (PLM) is used in conjunction with staining techniques applied to the sample that cause the material to appear in unique colors under the microscope when polarized light is used. Phase Contrast Microscopy (PCM) is used to analyze airborne and biologic materials.
2. *Electron microscopy* is used to analyze very small particles of materials. Whereas optical light microscopy may have an effective upper magnification limit of several thousand (2,000x), electron microscopy techniques can exceed one hundred thousand (100,000x). This greater resolution is particularly applicable to very tiny airborne fibers such as asbestos.

EPA Protocols

As the primary agency concerned with environmental contaminants, the EPA establishes and recommends standard analytical procedures. Some of the EPA methods more frequently used on hazardous waste sites are:

EPA Method	Analysis For:
413.2	Total hydrocarbons
608/8080	Chlorinated pesticides (PCBs)
614/8140	Organophosphate pesticides
624/8240	Volatile organic compounds
8020	benzene, toluene, xylene
9040	pH
RCRA	Arsenic, barium, cadmium, chromium, lead, mercury, selenium, silver

Whenever possible, EPA-approved methods should be used, but if there are no approved methods, the method having the highest reliability should be employed.

AHERA-SPECIFIC PRACTICES AND TECHNIQUES

EPA recommends that for every 20th bulk sample that is collected, a QA sample be taken immediately adjacent to the 20th sample. Thus the 20th and 21st samples of every group of 20 are side-by-side samples. Many building inspectors will collect a QA sample for every 10 samples collected for a greater degree of assurance. Laboratory analyses of these 2 samples are expected to closely agree. Each sample is labeled independently so that the identity of QA samples cannot be determined except by reference to records kept by the building inspector.

QA samples can be handled in one of two ways. They can be sent together with all the samples to a single laboratory for analysis. Alternatively, they can be sent to a second laboratory and analyzed independently. The first checks on analytical variability within the same laboratory. The second checks on variability between laboratories. Using a second laboratory is appropriate for large studies.

Laboratory results on the QA samples should not disagree on the presence or absence of asbestos (i.e., 1% or less vs. more than 1% asbestos). If significant disagreement occurs, additional samples should be taken and analyzed.

There may be discrepancies in estimating the exact percentage of asbestos in side-by-side samples. These discrepancies are not as serious as the presence/absence result since any sample of friable material which contains more than 1% asbestos is designated as ACM. However, the comparison of the asbestos percentage estimated by the testing laboratory(ies) can provide useful information on the reliability of the analysis. Discrepancies may occur as a result of sample contamination, inconsistent procedures, differences in technique, or mistakes (e.g., mislabeling of samples). Of course, some variability in the "true" asbestos content of ACM would be expected from one location to another. Ordinarily the percentage for each QA sample as compared with the percentage for its corresponding regular sample should not exceed 10%.

Any disagreement about the type of asbestos mineral (chrysotile, amosite, crocidolite) present should be resolved by additional analyses. Information on mineral type may be important when evaluating alternative methods of managing ACM, especially if removal of the ACM is under consideration.

Procedures to assure the integrity of the samples are also a component of the QA program. Strict chain-of-custody procedures should be followed. See Section K for a description of chain-of-custody procedures and data forms.

Analytical Techniques for Bulk Samples

Polarized light microscopy (PLM) is the EPA-approved method for analyzing bulk materials for asbestos. This method of analysis is relatively inexpensive. PLM utilizes a light microscope equipped with polarizing filters. The identification of asbestos fiber bundles is determined by the visual properties displayed when the sample is treated with various dispersion staining liquids. Identification is substantiated by the actual structure of the fiber and the effect of polarized light on the fiber, all of which is viewed by the trained technician. The limit of detection of asbestos by PLM is about 1% by area. Samples containing lower levels of asbestos are not reliably detected by this technique.

X-ray diffraction (XRD) is another method used for analyzing bulk materials for asbestos. It is sometimes utilized to confirm the presence of asbestos in a sample already analyzed by PLM if the identity of the fibers remains ambiguous. XRD is not used routinely since it is not as sensitive as PLM in detecting asbestos; its limit of sensitivity is approximately 3%.

Other analytical methods employed to detect the presence of asbestos in bulk samples include electron microscopy and infrared spectroscopy. *Electron microscopy* should be used to detect smaller fibers of asbestos, such as those found in dusts. Both techniques are sometimes used to confirm the presence of asbestos in a sample analyzed by PLM, rather than as routine analytical methods.

BULK SAMPLE COMPARATIVE ANALYSIS REPORT FORM
For ACM Bulk Sample Analyses

REPORT PFGE NO.: ____

REPORT TITLE: ______________

DATE: ____ ____, ______

DOCUMENT NUMBER

Building Number and Name

Sample Area/Lot Number and Name

Drawing/Sketch Number and Name

SAMPLE NUMBER	LAB. ID NUMBER	1 Chrsotile	2 Amosite	3 Crocidolite	4 Cellulose	5 Fiberglass	6 Min. Wool	7 Perlite	8 Vermiculite	9 Limestone	10 Quartz	11 Binders	12 Other (Specify)	13 Anthophylite	14 Actinolite	15 Tremolite	16 Mica	17 Opaques	18 Synthetic Fibers	19 Gypsum	20 Diatoms	21 Pumice	22 Paint	23 Cement	24 Other (Specify)
COUNT OF POSITIVES																									
AVERAGE PERCENTAGES																									

FORM 8

Figure 14.11 Bulk Sample Analysis Report Form.

Laboratory Reporting

A competent, analytical laboratory should provide a detailed bulk sample analysis report that includes the following information, at a minimum (see Figure 14.11):

- Client sample identification number.
- Laboratory sample identification number.
- Analytical technique used.
- Laboratory quality control procedure.
- Physical description of sample, as received.
- Type(s) and estimated percentage of asbestos.
- Type(s) and estimated percentage of non-asbestos fibers.
- Type(s) (if known) and percentage of other components.
- Date of analysis.
- Analyst's signature.

This information, along with data generated in the field (e.g., location of sample, type of material, photo references, etc.), should be maintained as part of an overall building inspection, recordkeeping program.

REFERENCES

243. Grant, Eugene L., and Leavenworth, Richard S., *Statistical Quality Control,* Third Edition; McGraw-Hill Book Company; New York; 1972.
244. Mendenhall, William, and Reinmuth, James E., *Statistics for Management and Economics,* Fourth Edition; Duxbury Press; Boston, MA; 1978.
245. *Statistical Quality Control Handbook;* Western Electric Company; Indianapolis, IN; 1956. (This is probably the finest applied process quality control handbook available and written for the non-statistician.)
246. This is one of the major applications of the "salient" concept.
247. This section is extensively adapted from *Section I, Model EPA Curriculum for Training Building Inspectors;* ibid.
248. *Section I;* ibid.
249. Journel, A. G., and C. H. Huijbregts, *Mining Geostatistics;* Academic Press; London; 1978.
250. David, M., *Geostatistical Ore Reserve Estimation:* Elsevier, Amsterdam; 1984.
251. Clark, I., *Practical Geostatistics;* Applied Science Publishers; London; 1979.
252. Rendu, J. M., *An Introduction to Geostatistical Methods of Mineral Evaluation;* South African Institute of Mining and Metalurgy; Johannesburg; 1981.
253. Strivastava, R. M., "A Non-ergodic Framework for Variograms and Covariance Functions"; *SIMS Technical Report No. 114;* Department of Applied Earth Sciences; Stanford University; 1988.

254. *Characterization of Hazardous Waste Sites — A Methods Manual, Volume I, Site Investigations;* EPA Document 600/4–84/075; April 1985.
255. *Guidance for Conducting Remedial Investigations and Feasibility Studies Under CERCLA;* OSWER Directive 9335.3-01; March 1988.
256. *Test Methods for Evaluating Solid Waste;* EPA Manual SW 846.
257. *Hydrogeology;* S. N. Davis and R. J. M. DeWest.
258. *Standard Operating Safety Guidelines;* EPA; July 1988.
259. *Occupational Safety and Health Guidance Manual for Hazardous Waste Site Activities;* US Department of Health and Human Services; October, 1985.
260. *Manual of Analytical Methods;* NNIOSH Publication No. 84-100; 3rd Edition.
261. *Test Methods for Evaluating Solid Waste, Physical/Chemical Methods;* EPA Document SW-846.

15 PHASE II AND PHASE III ENVIRONMENTAL RISK AUDITS

DEFINITION OF PHASE II AND PHASE III

Much earlier definitions of Phase II and Phase III ERAs were offered along the following lines.

Phase II ERA:

To confirm or deny, by scientifically and legally accepted methods, the existence of a suspected environmental risk.

Phase III ERA:

Given the presence of an environmental risk, to determine the extent of the risk and develop an appropriate plan of action for reduction of the risk.

The obvious difference between Phase II and Phase III ERAs lies in the amount of quantitative data required to achieve the Phase III objective. Phase II concerns are automatically included in Phase III work and for this reason both will be discussed as a single subject. This presents a problem in the sense that the two phases should not be confused. Phase II work is considerably less extensive than Phase III and far less expensive to perform. Essentially, Phase II work stops at the risk source identification stage.

In some situations the client will not require the quantitative information supplied by a Phase III ERA, the confirmation or denial of the existence of a suspected risk being sufficient for the client's purposes. On the other hand, Phase II information is of very little use if there is a need to develop valuation or cost impact information or a risk management program (ERMP). Phase II data alone will not supply the necessary details on the quantity of the risk source, condition of the primary control mechanisms, availability of transport mechanisms, or target data necessary to develop the alternatives for dealing with the risk.

As the marketplace is coming to realize, there are many alternative remediation methods applicable to any given situation and the highest cost/highest risk alternative is often removal of the risk source. It has been shown[262] that other alternatives for dealing with asbestos in a commercial building may be as much as 50% less expensive, and, considering the risks, insistence on removal may be inappropriate and counterproductive. Phase III ERAs are required to develop the necessary information to define the appropriate strategy for dealing with such situations.

USE OF PHASE I RESULTS TO DEVELOP PHASE II/III RISK SOURCE SAMPLING STRATEGIES

The Phase I ERA will have developed a list of suspected environmental risks, the likely location of those risks, and the general characteristics of the risk source. In the form of a plot plan, the Phase I report may look like Figure 15.1, where the suspect locations are shown, and Table 15.1 providing narrative detail on the suspected risks.

Based on the Phase I information, a strategy for the confirmation or denial of the presence of an environmental risk can be developed that is both efficient and legally and scientifically sound.

Proof of the Non-Existence of a Risk Source: Risk Source Identification

The negative default principle stated earlier must be kept clearly in mind during the development of a sampling plan because it imposes both positive and negative limitations on the sampling strategy. Fundamentally, the negative default principle requires that because the presence of a risk is suspected one of three conclusions with regard to the risk must be reached:

1. The existence of the risk source will be assumed and that all actions appropriate to the actual existence of the risk source will be taken.
2. The existence of the risk source will be confirmed by a single positive analytical finding within a given homogeneous area.
3. The existence of the risk source will be denied only by a negative finding on all appropriate analytical tests of samples taken based on a statistically meaningful sampling scheme.

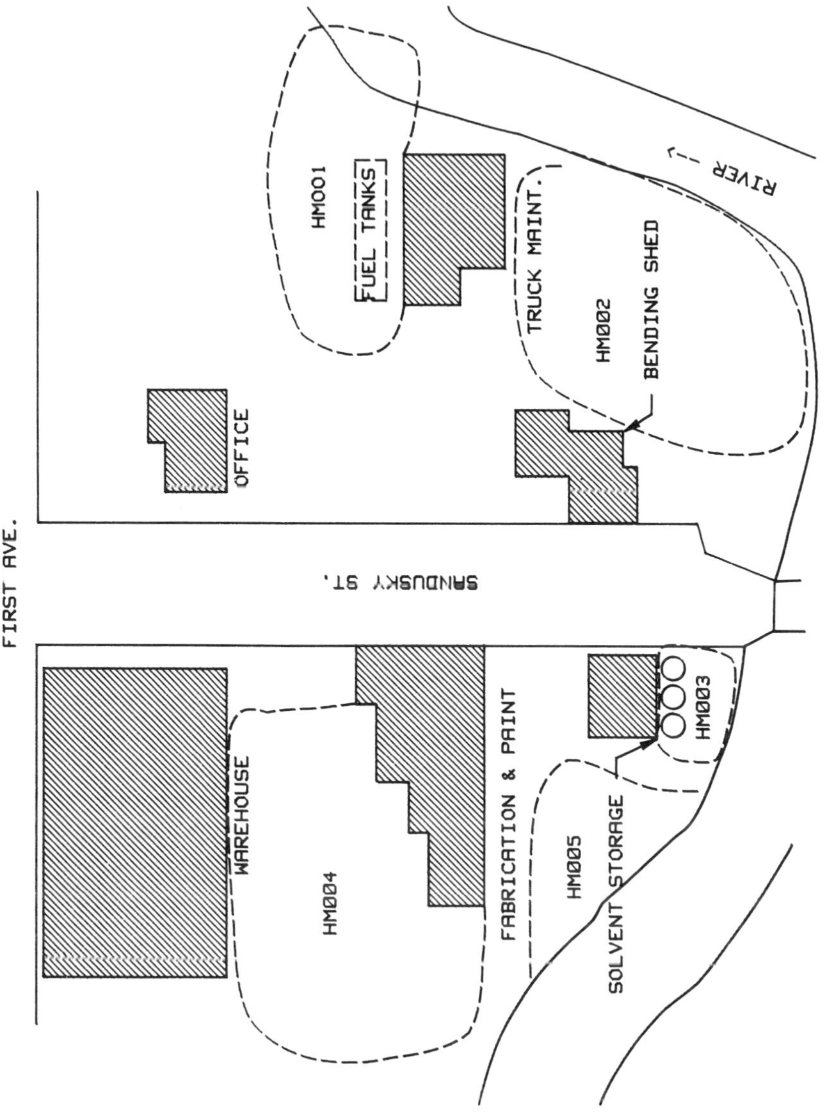

Figure 15.1 Phase I Suspected Environmental Risk Plot Plan.

Table 15.1 Description of Suspected Environmental Risks

Homogenous Area	Risk Source	Description
HM001	Leaking UST	Petroleum fractions, evidence of spills around tank islands, dead grasses on river bank. Tanks first placed in 1930, replaced 1950 and 1970. Currently four 10,000-gallon tanks, three for diesel fuel and one for gasoline.
HM002	Creosote	Apparent site of old creosote soaking pits and drying yards. In operation from 1880–1950 approx.
HM003	Solvents and paint primers	Three 10,000-gallon storage tanks currently used for toluene, sulfuric acid, and lead primer. In place since 1973, but current tanks replaced older tanks with solvents and cleaning compounds. Inventory records for the period are missing.
HM004	Waste Paints	Evidence of the dumping of waste paints and solvents. Primarily lead-based paints used until 1975 when replaced with non-lead paints.
HM005	Filled area	Interviews with retired employees including plant superintendent indicates that this area was filled with construction wastes, refuse drums, solvent and paint drums, lubricating grease and oil drums, and other refuse.

The negative default principle identifies the basic philosophy of sampling: Sampling is required to prove that an environmental risk does *not* exist or to quantify the extent of the risk.

Sampling, particularly sampling of soils, underground aquifers, air, and open bodies of surface water, can be an extremely expensive proposition with costs exceeding $500,000 for a forty-acre plot occurring with some frequency. In addition, the sampling process itself can be both physically dangerous to the investigators and may trigger the uncontrolled release of the risk source. Sampling should not be undertaken unless there is a strong reason to do so, and such reasons will primarily derive from both the operational characteristics of the facility and the objectives of the investigation.

Evidentiary Sampling

Sampling of a risk source for operational purposes is very different from sampling for purposes of developing litigation data. Litigation normally will involve an attempt to establish the source of the contaminate material or a very precise definition of the extent of contamination, or both. The requirements of the specific case will usually determine the sampling protocols in virtually all respects and involve a larger number of samples, extreme care in sample handling procedures, and often a completely separate set of samples for each party to the litigation.

Investigation Objectives

If the objective of the investigation is to establish the quantity of contaminated materials for the development of remediation costs (i.e., a Phase III ERA), the sampling must generally be undertaken to determine the boundaries of the contaminate plume. Note that in this case, the purpose is not to sample the contaminate but to sample until uncontaminated materials are identified, and the sampling scheme must be adjusted to this purpose.

In the case of homogeneous area HM001 in Figure 15.1, where contamination from an underground storage tank leak is present (or assumed to be present), sampling in the immediate vicinity of the tank itself is not particularly necessary except to formally establish the identity and initial concentration of the contaminate. Sampling should be designed to establish two key factors:

1. The "background" level of the contaminate naturally occurring in the soils, if the material is naturally occurring. Note that naturally occurring does not necessarily mean that the substance derived solely from nature. In most urban areas surface deposition of hydrocarbons and lead resulting from the burning of fossil fuels and internal combustion engine by-products have resulted in elevated levels of these substances close to the surface over very wide areas, and these may influence strongly the definition of "background" levels.
2. The boundaries of the contaminated zone, where the zone is defined as having a level of contaminate present that is elevated above "background" by a statistically significant amount.

The topography, geology and hydrology of the site, the quantity of contaminate, the length of time over which the contaminate may have migrated, and the presence and strength of the natural driving forces for migration (rainfall, air movement, evaporation, etc,) will all influence the shape and size of the contaminate plume. The sampling scheme should be designed with the influence of these factors in mind. To aid in the design of the sampling strategy, computer dispersion modeling techniques may be employed in an attempt to predict the likely location of the boundaries.

As an alternative to the employment of dispersion modeling (a subject still in its infancy) an initial regular grid pattern of samples may be used, the results of the analytical tests from the samples being plotted using geostatistics (kriging) to define the rough boundaries. A second round of sampling directed by the results of the first may then be employed to further define the boundaries.

Although not particularly elegant in the scientific sense, the boundaries can, and in the end will be defined by the "dig until no contamination is present" sampling scheme, and this approach may be the most reasonable when the contaminate is known to be present and is believed to be of relatively limited extent. A word of caution: This is a blind system of approach and any estimates of the costs or time to remediate must be considered to be very approximate.

Facility Operation Considerations

The time at which a sampling scheme should be employed, or even if one should be employed at all, may be determined by the facility's operating characteristics. The primary purpose of sampling is to provide the information needed to reduce risk. If sampling will not serve this purpose, it may not be a useful employment of scarce resources.

Consider a building built in the middle 1960's with a spray-on fireproofing that was specified to be asbestos containing. Suppose further that no major renovation or demolition or other activity is planned or considered likely in the near future that will disturb the fireproofing. Under these circumstances, other than possibly taking a single sample simply to remove any remaining doubt and to establish the mineral type of asbestos used, no meaningful operational purpose will be served by executing an extensive sampling program. The implementation of a sound Operations and Maintenance (O&M) program will significantly reduce risk based on the assumption that all of the material contains asbestos,.

Another scenario is presented by the case of a facility that must have an Operations and Maintenance program because of the presence of some risk source (for example a chemical refinery). The assumption that the piping is covered with asbestos, for example, will not significantly increase operations and maintenance costs over the base already required by the O&M program implemented for other reasons.

In general, if the assumption that a risk source is present, or if the confirmation of the presence of the risk source will not significantly alter the operation of the facility, extensive sampling is not generally necessary. Note however that it is the sampling of the risk source that has been the topic of discussion here, not the sampling of the primary control, transport, or secondary control mechanisms.

PHASE III ENVIRONMENTAL RISK ASSESSMENTS

The Phase III ERA assumes that the presence of an environmental risk is known, either as a result of records uncovered during the Phase I ERA, as a result of assuming the presence of a risk at the end of a Phase I ERA, or as the result of a confirmation of the presence of the risk through Phase II ERA techniques. The objective then becomes to quantify the environmental risk as quantification has been discussed in the development of the ERIS. The basic Phase III ERA work will develop a large body of detailed data that is summarized in the ERIS. The following discussion will apply to each risk source identified.

Integral to the concept of the development of an ERIS Detailed Inventory is another concept borrowed from the financial case; the worksheet. A worksheet is the detailed background information collected as a part of the analysis process and a detailing of the methods used to analyze the data to form the final conclusions represented by the ERIS Detailed Inventory. The worksheets are an integral part of the ERIS and are different from the accountant's worksheets in that they must be readily available to the risk manager. In effect, the worksheets form the operating basis for the ERMP, and the ERIS is the current condition statement. All worksheets should be keyed to the ERIS Detailed Inventory statements for ready reference. Examples of worksheets that may also be used in conjunction with the approach of Chapters 5 through 9 are given in Figures 15.2 through 15.5.

Risk Source Quantification (Figure 15.2)

There is an important distinction to be made concerning the quantity of a risk source that must be documented as part of the ERIS work. RCRA and other rules and regulations speak in terms of a *reportable quantity* (RQ) of a risk source, meaning the actual quantity of the risk source itself, not the total quantity of the risk source and the medium containing the risk source. For cost and control purposes, the reportable quantity is important but is not generally the governing quantity. If the risk source cannot be separated from the medium containing it, the total quantity of risk source and medium will be of primary management concern. If the risk source can be readily separated from the medium, or if it is already separate from a medium, the reportable quantity will be appropriate. The analyst should clearly document which concept is being used at any particular point during the construction of the ERIS. In general, both types of quantities will be included in the ERIS.

The risk source should be quantified in "natural" units of measure on worksheets subsidiary to the ERIS and then converted into ERIS units just prior to recording the information on the ERIS Detailed Inventory Worksheet. This may at first appear confusing but the "natural" units (e.g., gallons, barrels, cubic feet or yards, tons, etc.) will be the units recognized by the management control process and the individuals who will be working with the risk source on a daily basis. The ERIS units are used for reporting and environmental risk reduction strategy development purposes where a common unit of measure is far easier to work with to perform the necessary calculations.

The recording of the information and calculations used to determine the quantity are very important, especially the often implicit assumptions used in the calculations. As time progresses and the ERIS develops, the quantity estimation information should become significantly refined and more precise. This may result in an increase or decrease in the recorded inventory due to better information, but the change in an inventory once an ERIS is in place must be based on formally documented data that substantiates the changes. This can only be accomplished if the initial quantity derivation data, assumptions, and calculations have been clearly specified.

A significant portion of the quantity estimation will be based on the validity of the sampling plan, the documentation of the area (volume) assumed to contain the risk source, and the assumed density or density gradient of the risk source within the volume. This information should be clearly outlined on the worksheets.

Primary Control Mechanisms (Figure 15.4)

Physical Control Mechanism Component

The development and documentation of the physical control mechanism condition may range from the relatively simple use of the AHERA type of decision tree to full fault tree analysis utilizing quantitatively developed mean-time-between-failure (MTBF) data. Whatever the method, the method and the steps taken to achieve a conclusion must be carefully documented in the worksheets. If a Delphi approach is utilized, the credentials of the individual contributors to the group conclusion should also be documented in the worksheets.

Each of the elements going into the ERIS score for current condition should be carefully outlined in the worksheets and the basis for the final condition assessment and probable future conditions should be carefully discussed. The major factors important to the future condition assessment should be identified and the reasons supporting the conclusion should be discussed in order that the risk manager will be able to evaluate options for controlling the future condition.

Because many factors, some of them highly technical in nature, may influence the present and future condition of the control mechanisms, each of those factors analyzed should be carefully identified for two reasons. First, it is entirely possible that some other critical factor has been overlooked and there should be no reason whatever for the manager to have to guess whether or not that factor was taken into account in the analysis. To miss a critical factor through ignorance or lack of information is one thing, to allow the missing of that factor to mislead users of the analysis is another. Second, identification of the factors allows the ERMP to concentrate on those factors in controlling the risk in the future, improving the likelihood of success of the ERMP. To simply say that a piping system has a low probability of failure is insufficient; the physical and operating conditions that lead to that conclusion pinpoint the areas of concern if those conditions change.

Human Control Mechanism Component

Analysis of the human component of the primary control mechanism is one of the most difficult parts of ERIS development and must be based on a number of quantitative and non-quantitative sources of data. Personnel medical and injury records; training programs including the competency of the trainers and training materials and conditions; retraining protocols; communications efficiency; site admittance procedures; short-term worker competency analysis procedures; adequacy of emergency response equipment and drills; and other areas must all be evaluated. The record of disciplinary actions taken for violation of critical work procedures and the attitude of workers toward the use of work practices must also be assessed. Both the quantitative and non-quantitative data used to develop the ERIS Human Component scores must be documented for the same reasons as the physical component analysis was documented.

Transport and Secondary Control Mechanism (Figure 15.5)

It is at this point that the full value of the careful mapping of risk sources will be evident. Without mapping, the importance of physical location cannot be evaluated with respect to transport and secondary control mechanism analysis or the proximity to targets appreciated. Carefully denoted on the maps and the accompanying documentation should be such items as the slope of the land, permeability of the soils, depths to aquifers, locations and flow rates of surface waters, types and descriptions of secondary containment devices, and the basis for the conclusions concerning each relative to the ERIS scores. As with all other factors that result in an ERIS score, each element of the individual score should be carefully identified and documented to support the conclusion provided to management.

If recommendations are to be a part of the Phase III report, the basis for the recommendations should be founded on the documentation of the transport and secondary control mechanisms. If a recommendation is made to increase the height of a containment wall around an above-ground storage tank, the basis should be in terms of the existing situation and the reasons for that situation leading to the specific ERIS score for containment. The

Figure 15.2 ERIS Risk Source Worksheet.

Report: ____________________ **Page** ____ **of** ____

Date: ____/____/____ **Homogeneous Area No.** ____________

Analyst: ____________________ **Reference Drawing No.** ____________

Risk Source Description: ____________________________________

Laboratory Report Identification: ____________

Substance Concentration: ____________ (ppm, ppb, other ____________)

Toxicity Rating: ____________

Persistence Rating: ____________

ERIS Toxicity/Persistence Score: ____________

Risk Source Quantity: ____________

"Natural" Units Calculations: Reportable Quantity Calculation? Y N (Circle one)

Basis for Calculations: (Describe reference resources, e.g., aerial photographs, inventory, scaled drawing of site, etc.) ____________________________________

"Natural" Units of Measure: (e.g., cubic feet, gallons, tons, barrels, etc.) ____________

Calculations:

"Natural" Quantity: ____________ in ____________ (units)

ERIS Quantity: ____________

catastrophic failure of an above-ground storage tank is a relatively rare event, and the justification for a higher containment wall will most likely be based on the transport mechanism analysis and availability of targets rather than the probability of occurrence of the event itself. A rare event times a very high target damage potential can lead to a relatively high expected value for the event and therefore a high level of justification for a higher containment wall.

Diagrams and calculations that do not leave out steps are critical to the understanding of the conclusions and to the management of the risks. The Phase III ERA report must be written for an individual who will read and use it 20 years in the future and completely unknown to the author.

Figure 15.3 ERIS Target Worksheet.

Report: ____________ **Page** ____ **of** ____
Date: ____/____/____ **Homogeneous Area No.** ____________
Analyst: ____________ **Reference Drawing No.** ____________
Risk Source/Primary Control Mechanism Identification: ____________

Human Target Analysis

Surface Water Route:

If route not applicable, state reason: ____________

Surface Water Use Score: ________
Population/Distance to Surface Water Score: ________
Raw Score: ________
ERIS Score (= Raw Score × 2.04) ________

		Number		Weight	Total
Current Population (1,000's):	0–2,000′	____	×	____	____
Est. Date: ____	2,001′–1 mile	____	×	____	____
____	1–2 miles	____	×	____	____
____	2–3 miles	____	×	____	____
____	3+ miles	____	×	____	____
				TOTAL	____
t_1 Population (1,000's):	0–2,000′	____	×	____	____
Est. Date: ____	2,001′–1 mile	____	×	____	____
Est. Basis: ____	1–2 miles	____	×	____	____
____	2–3 miles	____	×	____	____
____	3+ miles	____	×	____	____
				TOTAL	____
t_2 Population (1,000's):	0–2,000′	____	×	____	____
Est. Date: ____	2,001′–1 mile	____	×	____	____
Est. Basis: ____	1–2 miles	____	×	____	____
____	2–3 miles	____	×	____	____
____	3+ miles	____	×	____	____
				TOTAL	____
t_3 Population (1,000's):	0–2,000′	____	×	____	____
Est. Date: ____	2,001′–1 mile	____	×	____	____
Est. Basis: ____	1–2 miles	____	×	____	____
____	2–3 miles		×	____	____
____	3+ miles	____	×	____	____
				TOTAL	____

Figure 15.3a ERIS Target Worksheet.

Report: ______________________ **Page** _____ **of** _____

Date: _____/_____/_____ **Homogeneous Area No.** ______________

Analyst: ______________________ **Reference Drawing No.** ______________

Risk Source/Primary Control Mechanism Identification: ______________

Ground Water Route:

If route not applicable, state reason: ______________________________

Ground Water Use Score: ______________

Population/Distance to Surface Water Score: ______________

Raw Score: ______________

ERIS Score (=Raw Score × 2.04) ______________

		Number		Weight	Total
Current Population (1,000's):	0–2,000′	_____	×	_____	_____
Est. Date: ________	2,001′–1 mile	_____	×	_____	_____
Est. Basis: ____________	1–2 miles	_____	×	_____	_____
______________________	2–3 miles	_____	×	_____	_____
______________________	3+ miles	_____	×	_____	_____
				TOTAL	_____
t_1 Population (1,000's):	0–2,000′	_____	×	_____	_____
Est. Date: ________	2,001′–1 mile	_____	×	_____	_____
Est. Basis: ____________	1–2 miles	_____	×	_____	_____
______________________	2–3 miles	_____	×	_____	_____
______________________	3+ miles	_____	×	_____	_____
				TOTAL	_____
t_2 Population (1,000's):	0–2,000′	_____	×	_____	_____
Est. Date: ________	2,001′–1 mile	_____	×	_____	_____
Est. Basis: ____________	1–2 miles	_____	×	_____	_____
______________________	2–3 miles	_____	×	_____	_____
______________________	3+ miles	_____	×	_____	_____
				TOTAL	_____
t_3 Population (1,000's):	0–2,000′	_____	×	_____	_____
Est. Date: ________	2,001′–1 mile	_____	×	_____	_____
Est. Basis: ____________	1–2 miles	_____	×	_____	_____
______________________	2–3 miles	_____	×	_____	_____
______________________	3+ miles	_____	×	_____	_____
				TOTAL	_____

Target Analysis (Figure 15.3)

Again the importance of mapping is emphasized. The number and location of targets both present and in the future will depend on the density of targets within reach of the transport mechanisms. The number of individuals within a specified distance must be carefully documented to provide a basis

Figure 15.3b ERIS Target Worksheet.

Report: ______________________ **Page** ____ **of** ____
Date: ____/____/____ **Homogeneous Area No.** ____________
Analyst: ______________________ **Reference Drawing No.** ____________
Risk Source/Primary Control Mechanism Identification: ____________

Air Route:
If route not applicable, state reason: ____________

Air Route Use Score:	________
Population/Distance Score:	________
Land Use Score:	________
Raw Score:	________
ERIS Score (= Raw Score × 3.03)	________

		Number		Weight	Total
Current Population (1,000's):	0–1/4 mile	____	×	____	____
Est. Date: ______	0–1/2 mile	____	×	____	____
Est. Basis: ______	0–1 mile	____	×	____	____
______	0–4 miles	____	×	____	____

(Highest Total Score is the score to be used on the ERIS Detailed Inventory.)

t_1 Population (1,000's):	0–1/4 mile	____	×	____	____
Est. Date: ______	0–1/2 mile	____	×	____	____
Est. Basis: ______	0–1 mile	____	×	____	____
______	0–4 miles	____	×	____	____

(Highest Total Score is the score to be used on the ERIS Detailed Inventory.)

t_2 Population (1,000's):	0–1/4 mile	____	×	____	____
Est. Date: ______	0–1/2 mile	____	×	____	____
Est. Basis: ______	0–1 mile	____	×	____	____
______	0–4 miles	____	×	____	____

(Highest Total Score is the score to be used on the ERIS Detailed Inventory.)

t_3 Population (1,000's):	0–1/4 mile	____	×	____	____
Est. Date: ______	0–1/2 mile	____	×	____	____
Est. Basis: ______	0–1 mile	____	×	____	____
______	0–4 miles	____	×	____	____

(Highest Total Score is the score to be used on the ERIS Detailed Inventory.)

for decision making. The elements going into the ERIS score must likewise be carefully documented. If aerial photographs or census data is used, they must be referenced and the original data preserved if it is not likely to be readily available in the future.

The following example worksheets should provide an idea as to the type of data to be gathered and the information required to support each item in the ERIS Detailed Inventory Worksheet.

Figure 15.3c ERIS Target Worksheet.

Report: ____________ **Page** ____ **of** ____
Date: ____/____/____ **Homogeneous Area No.** ____________
Analyst: ____________ **Reference Drawing No.** ____________
Risk Source/Primary Control Mechanism Identification: ____________

Direct Contact:
If route not applicable, state reason: ____________

Direct Contact Score: ____________
Population Within One Mile Score: ____________
Raw Score: ____________
ERIS Score (= Raw Score × 20.00) ____________

	Number		Weight	Total
Current Population (1,000's):	______	×	______	______
Est. Date: ______ Est. Basis: ______				
t_1 Population (1,000's):	______	×	______	______
Est. Date: ______ Est. Basis: ______				
t_2 Population (1,000's):	______	×	______	______
Est. Date: ______ Est. Basis: ______				
t_3 Population (1,000's):	______	×	______	______
Est. Date: ______ Est. Basis: ______				

Fire/Explosion:
If route not applicable, state reason: ____________

Fire/Explosion Score: ____________
Population Within Two Miles Score: ____________
Land Use Score: ____________
TOTAL RAW SCORE: ____________
ERIS Score (= Total Raw × 12.50) ____________

	Number		Weight	Total
Current Population (1,000's):	______	×	______	______
Est. Date: ______ Est. Basis: ______				
t_1 Population (1,000's):	______	×	______	______
Est. Date: ______ Est. Basis: ______				
t_2 Population (1,000's):	______	×	______	______
Est. Date: ______ Est. Basis: ______				
t_3 Population (1,000's):	______	×	______	______
Est. Date: ______ Est. Basis: ______				

Figure 15.3d ERIS Target Worksheet.

Report: ______________________ **Page** _____ **of** _____
Date: _____/_____/_____ **Homogeneous Area No.** ______________
Analyst: ______________________ **Reference Drawing No.** ______________
Risk Source/Primary Control Mechanism Identification: ______________

Sensitive Environment Target Analysis

Surface Water Route:
If route not applicable, state reason: ______________________________

Surface Water Score: ______________
ERIS Score (=SW Score × 33.33) ______________

Ground Water Route: (This route not applicable for sensitive environments)

Air Route:
If route not applicable, state reason: ______________________________

Air Route (Land Use) Score: ______________
ERIS Score (=AR Score × 33.33) ______________

Direct Contact:
If route not applicable, state reason: ______________________________

Distance to Critical Habitat: ______________
ERIS Score (=DC Score × 33.33) ______________

Fire/Explosion:
If route not applicable, state reason: ______________________________

Sensitive Environments: ______________
Number of Buildings: ______________
Distance to Nearest Building: ______________
Total Raw Score: ______________
ERIS Score (=Raw Score × 9.09) ______________

Figure 15.4 ERIS Primary Control Mechanism Worksheet.

Report: ______________________ **Page** _____ **of** _____

Date: _____/_____/_____ **Homogeneous Area No.** ______________

Analyst: ______________________ **Reference Drawing No.** ______________

Risk Source/Primary Control Mechanism Identification: ______________

Primary Physical Control Mechanism Condition

Primary Control Mechanism Description: ______________________________

____

Condition Analysis Method: ______________________________________

____

Condition Analysis Reference Document Identification: ______________

Observed Release: ______________

Surface Water Route: ______________

Ground Water Route: ______________

Direct Contact: ______________

Fire/Explosion: ______________

ERIS Score: ______________

Primary Control Mechanism Probability of Failure Analysis:

Current Condition:

Description of Failure Factors: (1) ______________________

(Rank Most to Least Likely) (2) ______________________

(3) ______________________

(4) ______________________

(5) ______________________

Current Condition is Rated Good/Fair/Poor: ________ (G/F/P)

Reasons for Assignment: ________________________________

____

____

____

t_1 *Condition Assessment:*

Est. Date: ________ (1) ______________________

Description of Failure Factors: (2) ______________________

(Rank Most to Least Likely) (3) ______________________

(4) ______________________

(5) ______________________

Probability of Failure is Rated High/Moderate/Low: ________ (H/M/L)

Reasons for Assignment: ________________________________

____

____

____

Figure 15.4a ERIS Primary Control Mechanism Worksheet.

Report: ______________ **Page** ____ **of** ____
Date: ____/____/____ **Homogeneous Area No.** ______________
Analyst: ______________ **Reference Drawing No.** ______________
Risk Source/Primary Control Mechanism Identification: ______________

t_2 *Condition Assessment:*

Est. Date: ________
Description of Failure Factors: (1) ______________
(Rank Most to Least Likely) (2) ______________
(3) ______________
(4) ______________
(5) ______________
Probability of Failure is Rated High/Moderate/Low: ________ (H/M/L)
Reasons for Assignment: ______________

t_3 *Condition Assessment:*

Est. Date: ________
Description of Failure Factors: (1) ______________
(Rank Most to Least Likely) (2) ______________
(3) ______________
(4) ______________
(5) ______________
Probability of Failure is Rated High/Moderate/Low: ________ (H/M/L)
Reasons for Assignment: ______________

Human Component Analysis:

Awareness:
Condition Analysis Method: ______________
Condition Analysis Reference Document Identification: ______________
Awareness Rating (Good/Fair/Poor): ________
Reasons for Rating: ______________

Training:
Condition Analysis Method: ______________
Condition Analysis Reference Document Identification: ______________
Training Rating (Good/Fair/Poor): ________
Reasons for Rating: ______________

Figure 15.4b ERIS Primary Control Mechanism Worksheet.

Report: ____________________ **Page** _____ **of** _____
Date: _____/_____/_____ **Homogeneous Area No.** ____________
Analyst: ____________________ **Reference Drawing No.** ____________
Risk Source/Primary Control Mechanism Identification: ____________

Equipment:
Condition Analysis Method: ____

____

Condition Analysis Reference Document Identification: ____________
Equipment Rating (Good/Fair/Poor): ________
Reasons for Rating: ____

____

Discipline:
Condition Analysis Method: ____

____

Condition Analysis Reference Document Identification: ____________
Discipline Rating (Good/Fair/Poor): ________
Reasons for Rating: ____

____

Comments: ____

____

____

____

____

Figure 15.5 ERIS Transport/Secondary Control Mechanism Worksheet.

Report: ____________________ **Page** _____ **of** _____

Date: _____/_____/_____ **Homogeneous Area No.** ____________

Analyst: ____________________ **Reference Drawing No.** ____________

Risk Source/Primary Control Mechanism Identification: ____________

Surface Water:

1-yr, 24-hour Rainfall	____________	(Note 1)
Facility Slope and Terrain	____________	(Note 2)
Distance to Surface Water	____________	(Note 3)
Physical State of Soils	____________	(Note 4)
Raw Score	____________	
ERIS Score (= Raw Score × 1.33)	____________	

Probability of Use (High/Moderate/Low/0): ____________

Reasons for assignment of probability of use: ____________

NOTE 1: (Describe Source of Data) ____________

NOTE 2: (Describe and Reference Data Sources) ____________

NOTE 3: (Describe and Reference Data Sources, show distance measurement)

Drawing Number: ____________

NOTE 4: (Describe and Reference Data Sources)

Drawing Number: ____________

Laboratory Report Reference: ____________

Figure 15.5a ERIS Transport/Secondary Control Mechanism Worksheet.

Report: ____________ **Page** ____ **of** ____
Date: ____/____/____ **Homogeneous Area No.** ____________
Analyst: ____________ **Reference Drawing No.** ____________
Risk Source/Primary Control Mechanism Identification: ____________

Ground Water:

Net Precipitation	____________	(Note 1)
Depth to Aquifer	____________	(Note 2)
Permeability of Soil	____________	(Note 3)
Physical State of Soils	____________	(Note 4)
Dist. to Nearest Well	____________	(Note 5)
Raw Score	____________	
ERIS Score (= Raw Score × 0.91)		____________

Probability of Use (High/Moderate/Low/0): ____________
Reasons for assignment of probability of use: ____________

NOTE 1: (Describe Source of Data) ____________

NOTE 2: (Describe and Reference Data Sources, include drawing references)
Drawing Number: ____________
Well Log References: ____________

NOTE 3: (Describe and Reference Data Sources, show distance measurement)
Drawing Number: ____________

NOTE 4: (Describe and Reference Data Sources)
Drawing Number: ____________
Laboratory Report Reference: ____________

NOTE 5: (Describe and Reference Data Sources)
Drawing Number: ____________
Measured Distance from plume to nearest well ____________ (ft./mi.)

Figure 15.5b ERIS Transport/Secondary Control Mechanism Worksheet.

Report: ______________________ **Page** _____ **of** _____

Date: _____/_____/_____ **Homogeneous Area No.** ______________

Analyst: ______________________ **Reference Drawing No.** ______________

Risk Source/Primary Control Mechanism Identification: ______________

Air Route:

Reactivity and Incompatibility ______________

ERIS Score (=R & I × 6.67) ______________

Probability of Use (High/Moderate/Low/0): ______________

Reasons for assignment of probability of use:

____

____

____

NFPA Reactivity Rating: ______________

Incompatible Materials Groups: ______________

Comments: ____

____

____

____

Direct Contact:

Accessibility Rating ______________

ERIS Score (Acces. × 6.67) ______________

Probability of Use (High/Moderate/Low/0): ______________

Reasons for assignment of probability of use:

____

____

____

Comments: ____

____

____

____

Figure 15.5c ERIS Transport/Secondary Control Mechanism Worksheet.

Report: ____________________ **Page** _____ **of** _____

Date: _____/_____/_____ **Homogeneous Area No.** ______________

Analyst: ____________________ **Reference Drawing No.** ______________

Risk Source/Primary Control Mechanism Identification: ______________

Fire/Explosion:

Ignitability Rating ______________

Reactivity Rating ______________

Incompatibility Rating ______________

Raw Score ______________

ERIS Score (= Raw Score × 2.22) ______________

Probability of Use (High/Moderate/Low/0) ______________

Reasons for assignment of probability of use:

____

____

____

Comments: ____

____

____

____

General Comments on Transportation/Secondary Control Mechanisms:

____

____

____

____

____

____

____

____

____

____

RISK REDUCTION BENEFIT AND COST ESTIMATION REPORTING

The reporting of estimates for the various applicable risk reduction alternatives requires that the benefits be derived from a re-analysis of the factors involved in the ERIS Detailed Inventory for each alternative. This will provide a new Human or Sensitive Environment Risk Score for the alternative as if the alternative were the current condition. The analysis of the costs to implement the alternative require a careful examination of both the direct and indirect costs and some dollar-denominated benefits.

The recording of the information surrounding the costs for various alternatives must take place with great care to avoid several pitfalls common to such analyses:

1. Care should be taken to be certain that all of the risk source/primary control mechanism combinations that will be affected by a given alternative have been properly accounted for and identified. If an alternative involves more than one risk source/primary control mechanism, the benefits analysis should also reflect that fact.
2. Care should be taken to be certain that fixed, variable and particularly incremental (to existing budgets) costs are identified and properly analyzed. It is very easy to assume that other costs will not be influenced by the taking of a particular action, but this may not be true and could result in a considerable source of analysis error.
3. Mistakes in the calculation of net present values for various alternatives are a common source of error that should be guarded against. It is the date on which an expenditure is assumed to occur that will govern the present value, and especially with respect to programs having a long execution time (such as an O&M program), the expenditures may occur at many different times. Likewise benefits may accrue at different times in the future as various actions are taken. In general, the expenditures should be tied to the benefits to be derived from a completed project. This may entail the use of several different cost/benefit analyses to arrive at a conclusion.

Of particular importance in developing alternative cost data is a recording of the assumptions involved and the source of the data used to make the estimation. The cost model proposed in an earlier chapter should be consistently used. If the data source is a published document (such as construction costs published by Means Data Services or the like), or a specific contractor with experience in the field, the source should be quoted as a reference. The calculations involved in making the estimation should be carefully outlined so that someone else can duplicate the work if necessary. Any drawings, plans, specifications, and work schedules used to develop estimates should be carefully referenced.

The combination of the current risk status as provided by the ERIS Detailed Inventory and the alternative risk reduction analysis, result in a proposed plan of action for implementation by the client. If the proposed plan or a variant is adopted, the information provided must be sufficient to allow the implementation and control of the ERMP.

AUTHOR'S NOTE

This book deals with a subject in its infancy and presents a different point of view for environmental risk analysis and management. This presentation is necessarily incomplete, if for no other reason than that it proposes the development of the equivalent of financial statements for the management of environmental risks. The development of our modern system of accounting has taken hundreds of years, so it should not be surprising that this is just a beginning, albeit hopefully a sound beginning.

During the process of teaching a number of seminars on this subject, and assisting in the development of this concept in several corporations, I have found that I can learn more from the audience that I have to teach. You are therefore invited and encouraged to provide comments directly to me concerning the Environmental Risk Inventory System and Environmental Risk Management Program concepts presented here. In addition, if the ideas have merit in your situation, I would be very happy to explore your application with you either informally or formally.

I may be contacted at:

A. R. Wilson & Associates
1550 Old Henderson Rd., Suite N-232
Columbus, Ohio 43220

or through the

Hazardous Materials Institute, Inc.
614/459-1105

Al Wilson
May, 1990

REFERENCES

262. Wilson, A. R., "The Appraisal Journal"; *Probable Financial Effect of Asbestos Removal on Real Estate;* The American Institute of Real Estate Appraisers; Chicago, IL. July, 1989.

INDEX

Elements of Overseas Trade Series

Elements of International Trade and Payments

Other books in the
'Elements of Overseas Trade Series'

Elements of Cargo Insurance, Dennis Badger and Geoffrey Whitehead
Elements of Export Law, Geoffrey Whitehead
Elements of Transportation and Documentation, Ralph Bugg and Geoffrey Whitehead
Elements of Export Marketing, John Stapleton